CONTROL AND ANALYSIS OF NOISY PROCESSES

David M. Koenig

Corning, Inc.

Prentice Hall
Englewood Cliffs, New Jersey 07632

Library of Congress Cataloging-in-Publication Data

Koenig, David M.
    Control and analysis of noisy processes / David M. Koenig.
        p.   cm.
    Includes bibliographical references and index.
    ISBN 0-13-033366-2
    1. Process control--Statistical methods.   2. Process control-
-Mathematical models.   I. Title.
TS156.8.K64 1991
670.42--dc20   90-39137
                                                                   CIP

Editorial/production supervision
  and interior design: Harriet Tellem
Cover design: Ben Santora
Manufacturing buyers: Kelly Behr/Susan Brunke

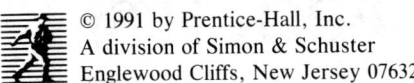 © 1991 by Prentice-Hall, Inc.
A division of Simon & Schuster
Englewood Cliffs, New Jersey 07632

The publisher offers discounts on this book when ordered
in bulk quantities. For more information, write:
    Special Sales/College Marketing
    Prentice-Hall, Inc.
    College Technical and Reference Division
    Englewood Cliffs, NJ 07632

Printed in the United States of America
10 9 8 7 6 5 4 3 2 1

ISBN 0-13-033366-2

Prentice-Hall International (UK) Limited, *London*
Prentice-Hall of Australia Pty. Limited, *Sydney*
Prentice-Hall Canada Inc., *Toronto*
Prentice-Hall Hispanoamericana, S.A., *Mexico*
Prentice-Hall of India Private Limited, *New Delhi*
Prentice-Hall of Japan, Inc., *Tokyo*
Simon & Schuster Asia Pte. Ltd., *Singapore*
Editora Prentice-Hall do Brasil, Ltda., *Rio de Janeiro*

To Ryan, Jennifer, Julie, Bertha, and Rudy
and in memory of Wilda

CONTENTS

CHAPTER 5   THEORETICAL SPECTRAL ANALYSIS OF NOISY PROCESSES   263

PREFACE

For decades industrial process control was the province of control engineers. However, in the last few years, because the process industries have turned to statistical process control (SPC) for help in tightening production variances, statisticians have begun to take a larger role in process control. At the same time, control engineers have begun to become more aware of the stochastic or non-deterministic nature of the process variables they are attempting to control. This book addresses process control from the viewpoint of someone at the interface between the statistician and the control engineer.

The first chapter on process dynamics is written with a minimum use of mathematics and with an emphasis on discussion rather than derivation. The benchmark for characterizing processes is the first order with deadtime (FOWDT) model, which is an excellent reference point even when the process deviates from FOWDT behavior. The benchmark for characterizing noise is the Box and Jenkins ARMA model, which accurately describes much of the noise contaminating industrial processes. The first tool for analyzing noise is the autocorrelation. Since it is assumed that digital computers will be used to collect the data, analyze the data, and control the process, all the development is done in the discrete time domain. This is realistic for other reasons as well. The mathematics of describing discrete time-domain processes is simpler. The concepts of the integral and derivative are replaced with the sum and the difference. The main tool for solving discrete-time problems, the Z-transform, is, in the opinion of many analysts, easier to grasp than the Laplace transform. Finally, in the continuous time domain, the concept of white noise is an idealistic abstract concept, while in the discrete time domain it is a useful practical quantity.

The second chapter introduces control algorithms. Because of their widespread use and success over the years, the PI and PID control algorithms are dealt with extensively, although some time is spent discussing the Smith predictor and Box–Jenkins algorithms. The emphasis in Chapter Two is on "regulation," where a control algorithm is tasked with keeping the controlled variable satisfactorily near a constant target in the face of autocorrelated stochastic disturbances. This is in contrast to many control theory texts, where the control algorithm is designed to keep the controlled variable near a moving target and where disturbances are considered relatively insignificant. In keeping with the approach of Chapter One, the control algorithms are presented in the discrete time domain. Time-domain simulation coupled with the autocorrelation is used to illustrate their features. Because of the influence of SPC, a section in Chapter Two is devoted to demonstating the ability of PID to control processes subject to white noise. In the same section a comparison of PI and control chart based strategies is made.

The third chapter is devoted to demonstrating the power of the Fourier line spectrum as a tool for analyzing noisy processes. In effect, this chapter transforms the discrete time-domain data into the discrete frequency domain, where each frequency is a harmonic of the fundamental frequency. In this new domain, some of the problems uncovered in Chapter Two are analyzed in an attempt to develop further insight. This chapter should show that the line spectrum, the cumulative line spectrum, and the autocorrelation are useful as tools for finding problems hidden in noise and for evaluating processes under feedback control.

The fourth chapter increases the mathematical level of the book one notch by introducing the Z-transform and demonstrating its application to a variety of topics related to control and process analysis. Unlike most control theory texts, the traditional preliminary discussion of the Laplace transform is bypassed (with exception of a short section that compares the Lapace and Z-transforms) since none of the analysis is done in the continuous time domain. As a sidelight, it is shown how the Z-transform can be used to gain insight into the continuous frequency-domain performance of filters, smoothers, and differencers.

Chapter Four is designed to lay the groundwork for Chapter Five, which introduces the power spectral density, defined on the continuous frequency domain, as a tool for studying controlled processes and as a means of integrating much of what has been presented in the first three chapters. With the new tools of the Z-transform and power spectral density in the reader's grasp, second looks are given the PID, the Smith predictor and the Box–Jenkins control algorithms and a new approach, Internal Model Control (IMC), is introduced and examined.

Therefore, in the five chapters of this book there is a transition from the discrete time domain to the discrete frequency domain and finally to the continuous frequency domain.

Throughout the book, the control and analysis techniques are applied to 13 different process models. The Z-transform, which is not presented until Chapter Four, is required for the understanding of the mathematical construction of these models. Therefore, the details of their construction are presented in an appendix so that the reader who is familiar with the Z-transform can refer to the appendix while reading the earlier chapters.

This book is designed to be read by novices in the field of control engineering, as well as practicing control engineers and statisticians. With the exception of a few sections, the first two chapters are designed to emphasize the phenomenological aspects of process dynamics and control. The idea is to get the reader interested and reasonably comfortable with some basic concepts before bringing on the heavy artillery. Chapter Three is more mathematical than Chapters One and Two and should give the reader not only a working knowledge of the line spectrum, but should also demonstrate how useful it can be in an industrial environment. The final two chapters use more mathematics than the first three, but all new mathematical tools are introduced at an elementary level so that the reader need not look elsewhere to grasp the essentials.

CHAPTER 1

FUNDAMENTALS
OF PROCESS DYNAMICS

A sequence of concepts in process dynamics will be introduced by using a corresponding sequence of simple examples, each slightly different from its predecessor. This introduction to the fundamentals of process dynamics will supply the tools necessary to assess the difficulty of a pending control problem, as well as to evaluate the quality of an installed control system.

With the exception of the sections on stochastic disturbances, the FOWDT model, and the determination of the dynamic parameters from experimental data, great care has been taken to avoid most of the mathematics usually associated with a presentation of process dynamics. Although the material in these sections is referred to in later chapters, depending on the mathematical skill of the reader, it might be best skipped during a first reading.

1.1 THE PROCESS GAIN AND TIME CONSTANT

The first example process, shown in Figure 1.1, consists of a tank of water and serves to introduce some nomenclature and two basic parameters in process dynamics.

The level in the tank is the *controlled variable*, the process variable that is to be kept on target. It is represented by the symbol C (for *Controlled variable*). The valve on the pipe, through which the inlet liquid flows, will be adjusted in

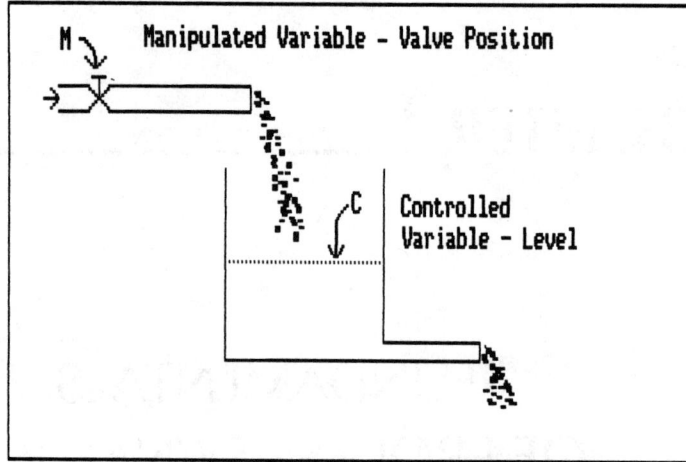

Figure 1.1   Example process 1.

order to keep the controlled variable on target. The valve position is the *manipulated variable* and is represented by the symbol M (for *Manipulated* variable). (Many textbooks on control use the symbol Y in place of C and X or U in place of M; however, this book will adopt the nomenclature used by one of the classics in the field of process control: Eckman, 1962.)

Figure 1.2 shows the *step change response* of the tank in Figure 1.1. The manipulated variable is given a step change in the amount of ΔM, where the Δ symbol indicates a change in the quantity that follows it, in this case M, the manipulated variable.

The transient response of the tank level, which is the controlled variable, can be characterized by two parameters. The first is the *process gain*, G, which is the ratio of the ultimate change in the controlled variable, ΔC_{ul}, to the step change in the manipulated variable, ΔM:

$$G = \frac{\Delta C_{ul}}{\Delta M}$$

The second characteristic parameter is the *process time constant*, T, which is defined as the time required for the controlled variable to reach 63% of its final value in response to a step change in the manipulated variable. The time constant could have been defined as the time required for the controlled variable to reach 50% of its final value, that is, a process half-time, in deference to the nuclear physicists, or as the time required for the controlled variable to reach, say, 95% of its final value. The mathematics of process dynamics (as will be seen in Section 1.6) is simpler if the first definition is used.

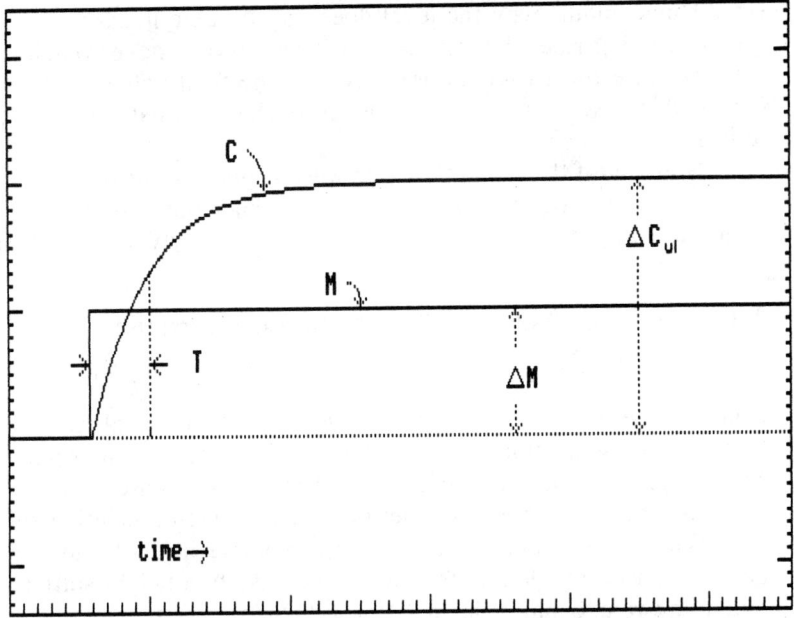

Figure 1.2 Step change response.

1.1.1 Some Qualitative Aspects of the Tank's Dynamics

Consider a couple of commonsense questions about the behavior of the example tank.

First, why, when a step change is applied to the inlet valve position, does the tank level move from one steady state to another, instead of continuing to change indefinitely? Second, why does the level reach the new steady state asymptotically rather than immediately or, say, linearly?

The answers to these questions can be determined by the following chain of logic. In the initial steady state, the inlet flow rate equals the outlet flow rate and the tank level is constant (except for small disturbances, which will be considered in the next section). After the inlet flow rate is increased, the mismatch between the inlet and outlet flow rates causes the level to increase at a *rate of change* directly proportional to this mismatch in flow rates and inversely proportional to the planar surface area of the tank. This explains why the level does not change to a new constant value immediately; that is, the step in the inlet flow rate causes a step in the rate of change of the level but not in the level itself.

The higher tank level causes the hydrostatic head at the outlet to increase, which in turn causes the outlet flow rate to increase. As the tank level and the outlet flow rate increase, the difference between the inlet and outlet flow rates decreases. Consequently, the tank level continues to increase but at a decreasing rate. This explains why the level does not increase linearly. In fact, because of this decreasing rate, the tank level will, in theory, never reach the final steady state because the closer the level gets to the final value the further the rate of change decreases. Therefore, the level reaches its final steady-state value asymptotically.

Note that if the outlet flow rate were kept constant, say by means of a pump inserted into the outlet pipeline, and if the inlet flow rate were greater than the outlet flow rate, then the level would increase linearly until the tank overflowed.

1.1.2 Effect of Physical Characteristics on Dynamic Parameters

Consider the two tanks shown in Figure 1.3. The top tank has the larger volume (or planar cross-sectional area) and a smaller-diameter outlet pipe. How does the process gain and time constant of the first tank compare to those of the second? Consider the case where the inlet flow rate is given a positive step change.

Will the first tank with the larger volume have the longer time constant because it will take longer for the level to rise to a height sufficient to cause the outlet flow rate to match the inlet flow rate? If the two tanks had the same-diameter outlet pipe, this would be true. Because of the first tank's greater flow resistance in the outlet pipe, a greater hydrostatic head and hence a higher level will be required in order for the outlet flow to match the inlet flow. The extra time required for the level to reach this higher value means an even larger time constant for the first tank than if it had the same outlet restriction as the second tank. Therefore, the time constant depends on the volume (or cross-sectional area) as well as the outlet flow resistance, and the second tank will have a smaller time constant.

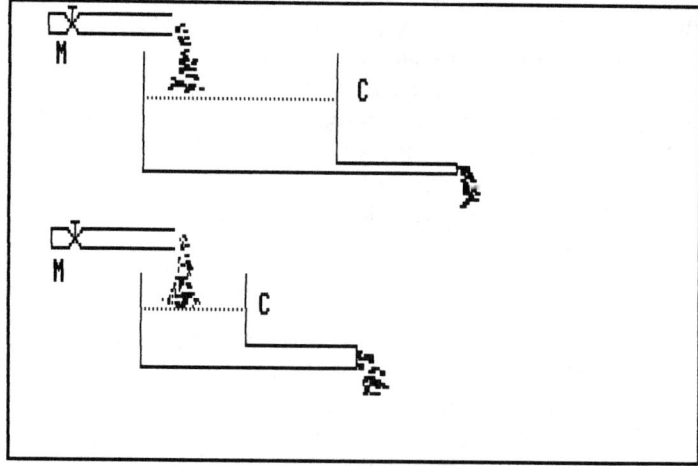

Figure 1.3 Two tanks with different characteristics.

Now, how about the process gain? The discussion in the previous paragraph has already let the cat out of the bag. Since the first tank has a smaller outlet pipe diameter, in order for the outlet flow rate to match the inlet flow rate after a step change in the latter, the level will have to rise to a higher value. Therefore, the process gain of the first tank is higher. In Section 1.6, these same concepts will be reviewed from a mathematical viewpoint.

1.1.3 Intermediate Summary I

The first goal of Section 1.1 was to introduce some nomenclature: C for the controlled variable and M for the manipulated variable. The second goal was to introduce the concepts of a process gain G and a process time constant T using a simple example process consisting of a tank of water. It was shown that there is a commonsense basis for a dynamical concept like the process time constant and that both the process gain and the process time constant can depend in a logical way on the physical parameters of the process. Finally, attention was given to a characteristic common to many industrial processes: after a step change in a manipulated variable, the process usually moves to a new steady state.

1.2 PROCESS DISTURBANCES

The controlled variable can be considered to consist of two components: one due to the effect of the manipulated variable and one due to disturbances that will be classified as either stochastic or deterministic. This is shown in Figure 1.4, which illustrates a fundamental assumption that the process and the disturbance are completely separable; that is, any erratic behavior is not related to the process but has an external source. This is a realistic and useful assumption and it will be used throughout the text, but the reader should be aware that the recently developed science of chaos (Gleich, 1987) questions this fundamental assumption and may lead to a completely different approach to the origin of disturbances.

Figure 1.4 defines the relationship between the manipulated variable, the controlled variable, and the disturbance. Implicit in this structure is the assump-

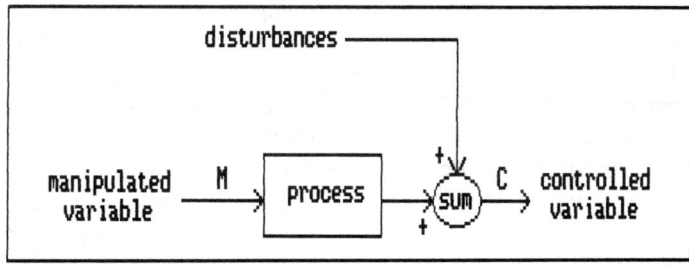

Figure 1.4 Effect of disturbances.

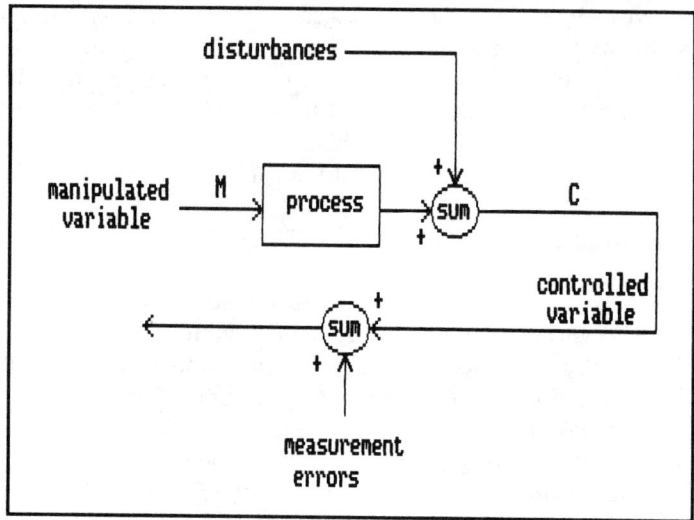

Figure 1.5   Relationship between disturbances and measurement errors.

tion that either the measurement of the controlled variable is accurate, at least relative to the magnitude of the disturbances, or that the measurement error is included in the disturbance. Throughout this book this assumption will be made. However, many authors (for example, Maciejowski, 1989) separate the measurement error from the disturbance as shown in Figure 1.5.

The difference between the two quantities can be illustrated by the following example. Consider a process with a zero time constant and a positive process gain of G. First, assume that the disturbance consists of a positive offset, δ, and that there is no measurement error. To compensate for this disturbance, one might change the manipulated variable by an amount equal to $-\delta/G$, which would drive the controlled variable back to the desired value. Next, assume that there is no disturbance but that there is a constant measurement error equal to δ. Unlike the previous case, decreasing the manipulated variable by an amount equal to $-\delta/G$ in order to compensate for this offset would be the wrong move.

These two extremely simple cases suggest that compensating for disturbances is inherently different from compensating for measurement errors. They also suggest that before proceeding with a process analysis and a control algorithm design, the control engineer or statistician should correct as many problems with the measurements of the process variables as possible. In the remainder of the book, process disturbances, measurement errors, and noise (whether process or sensor based) will be dealt with under the general heading of disturbances.

Consider Figure 1.6, which shows the second example process, another tank of water. This time, assume it has the task of supplying water to the rooms of a large hotel on whose roof it is located.

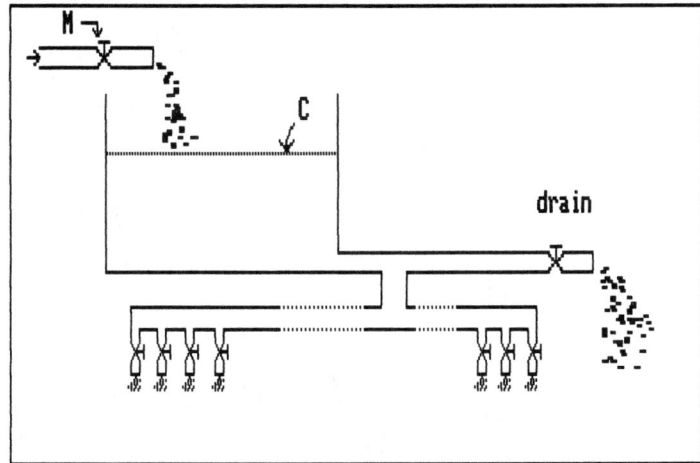

Figure 1.6 Example process 2.

1.2.1 Stochastic Disturbances

The level in this tank is subject to *stochastic* disturbances caused by the unpredictable operation of shower heads, faucets, and toilets in the rooms of the hotel. The adjective stochastic (defined by one edition of Webster's dictionary as "conjectural") is used because the exact cause of the disturbances in the tank level is not determinable; one could only "conjecture" or guess as to their origin.

The first part of Figure 1.7 shows how the level varies when the stochastic disturbances are *unautocorrelated* in time. At any instant in time the deviation in the level from some nominal value is independent of any past level variation. This type of stochastic disturbance is also called *white noise*. White noise will be

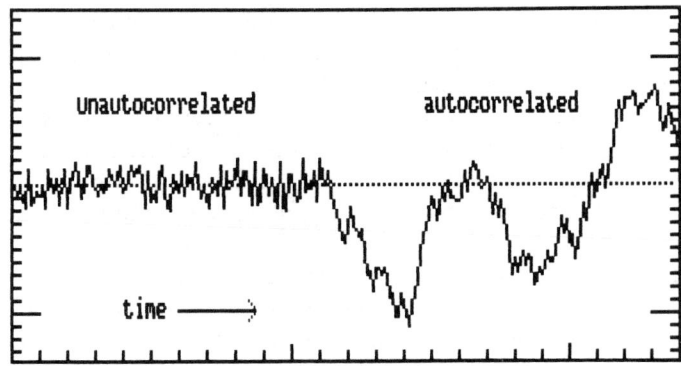

Figure 1.7 Autocorrelated and unautocorrelated stochastic disturbances.

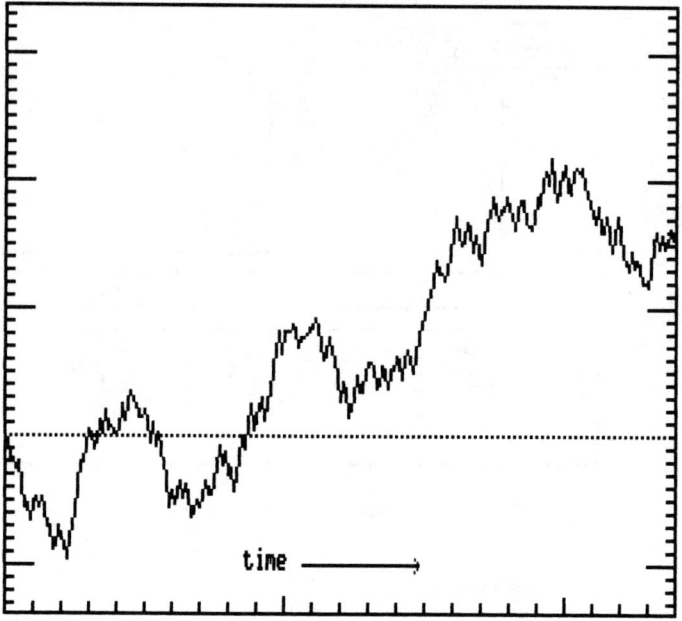

time ———→

Figure 1.8 Nonstationary disturbance.

considered in Section 1.3.1 and again in Chapter Three where the basis of the word white will be discussed.

The second half of Figure 1.7 shows how the level varies when the stochastic disturbances are *autocorrelated* in time. Here the reader should perceive that at any instant in time the deviation of the level from some nominal value appears to be dependent on past deviations. Drift in a sensor reading is a common example of an autocorrelated stochastic disturbance.

Although the concepts of autocorrelated and unautocorrelated disturbances are by far the most difficult to grasp, the issue is further complicated by considering autocorrelated stochastic disturbances that are *stationary* or *nonstationary*. If over a long period of time the disturbances in the tank level appear to vary about some nominal value, then the disturbances are said to be stationary. Conversely, if over a long period of time the tank level were to drift away from some initial value and not appear to vary about any nominal value for any length of time, then the disturbances would be said to be nonstationary. Figure 1.8 shows an example of a nonstationary stochastic disturbance.

Using the mathematical tools of autocorrelation and line spectrum, it is possible to quantitatively characterize stochastic disturbances, and although such a description of stochastic disturbances would definitely add insight, it will be deferred until later in the book.

1.2.2 Deterministic Disturbances

The reader should return to Figure 1.6, which schematically shows the outlet pipe of the hotel water supply tank attached to many small valves symbolizing the faucets, shower heads, and toilets in the hotel rooms. Look now at the large valve (labeled "drain") at the far right of the outlet pipe. Imagine that for some reason the water in the tank has become discolored and that the hotel manager wishes to flush out the discolored water while maintaining service to the hotel. He or she might do this by going up to the roof and opening the drain valve until the remaining water in the tank appeared clear enough. In this case the level in the tank would undergo a *deterministic* disturbance.

Figure 1.9 shows how the level responds to step changes in the position of this drain valve. Note that the response of the level is similar to that when the inlet flow valve position was changed in Figure 1.2.

1.2.3 Intermediate Summary II

There are but two reasons why the controlled variable might not be at its target value: The target may have been recently changed or the controlled variable may be subject to disturbances. There are two kinds of disturbances: stochastic and

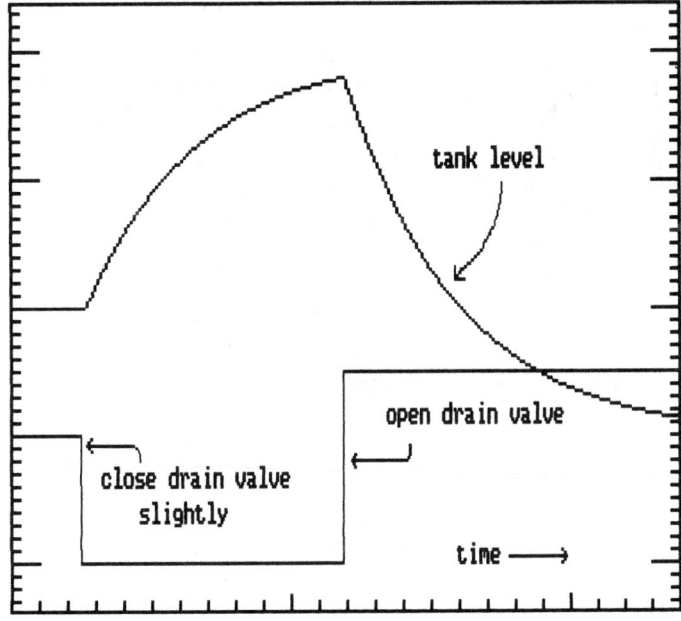

Figure 1.9   Deterministic disturbances.

deterministic. The exact origin of stochastic disturbances is, by definition, unknown. In the latter case, they can be stationary or nonstationary.

Deterministic disturbances have a known origin and, in the example of the hotel water supply, they were step changes in the outlet flow rate; however, in general they can take on any form as long as they can be predicted with certainty.

As will be seen later on, each of these disturbances presents a slightly different control challenge.

1.3 CHARACTERIZING STOCHASTIC DISTURBANCES

Having described stochastic disturbances qualitatively in the previous section, an attempt will now be made to put them on a firmer mathematical foundation. The readers may wish to pass through this section quickly if they are reading this text for the first time or if their mathematical footing is not strong. However, this section should be read in detail before tackling Chapters Three, Four, and Five.

1.3.1 White Noise and the Autocorrelation

First, the concept of an unautocorrelated stochastic disturbance, also often called *white noise*, has to be redefined. Such a sequence of unautocorrelated disturbances is denoted by $w(t_k)$, where t_k represents time such that

$$t_k = t_{k-1} + h$$

where $k = 1, 2, \ldots$, and where h is the constant sampling interval. It is assumed that the sequence is infinite in extent and that the first N values have been sampled for analysis. To make the mathematical manipulations less cumbersome, w_k will be used in place of $w(t_k)$ from now on.

By definition, the average, w, of the infinitely long w_k sequence is zero; that is, for large N, the average, $\overline{w}$ (or the sample estimate of the mean), will be effectively zero:

$$\overline{w} = \frac{1}{N} \sum_{k=1}^{N} w_k \approx 0$$

The sample estimate of the variance about the mean, s_w^2, of the w_k sequence will have a finite value given by

$$s_w^2 = \frac{1}{N} \sum_{k=1}^{N} [w_k - \overline{w}]^2 \tag{1.1}$$

For small N, a better estimate of the variance about the mean is

$$s_w^2 = \frac{1}{N-1} \sum_{k=1}^{N} [w_k - \overline{w}]^2 \tag{1.2}$$

However, in this book, Equation (1.1) will be used for the sample estimate of the variance. Note that, since the white noise sequence is assumed to have zero mean, the value of $\overline{w}$ in Equation (1.1) would be effectively zero.

The idea of an unautocorrelated sequence means that every w_i is independent from every other w_k as long as i does not equal k. Alternatively, consider two samples w_k and w_{k+i} separated from each other by i samples. If, for every separation i and every base index k, these two quantities are independent, then the w_k sequence can be said to be unautocorrelated. Another way of stating this independence is to say that, for large N, the sample estimate of the autocorrelation, which will be defined immediately, is effectively zero for every nonzero separation. Frequently, the separation i is referred to as the lag; that is, w_k is said to lag w_{k+i} by i samples. As promised, the *sample estimate of the autocorrelation*, or just the autocorrelation, for lag i is defined as

$$r_w(i) = \frac{1}{N} \sum_{k=1}^{N-i} \frac{[w_k - \overline{w}][w_{k+i} - \overline{w}]}{s_w^2} \tag{1.3}$$

This quantity can be considered as the average of the products $[w_k - \overline{w}][w_{k+i} - \overline{w}]$ normalized by s_w^2. By definition, $r_w(0)$ will equal unity. More comments will be made about this definition in Section 1.3.8 when the concept of the expected value is discussed.

If the w_k sequence is unautocorrelated, then for large N the product $[w_k - \overline{w}][w_{k+i} - \overline{w}]$, on the average, will be nearly zero. On the other hand, if the sequence is autocorrelated, then, for some of the i's, w_{k+i} would depend on w_k for every k, and the average of the products would be significantly different from zero.

Figure 1.10 shows the autocorrelation of an unautocorrelated sequence. Note, first, that the autocorrelation for zero lag is unity (by definition) and that, second, the autocorrelation is noisy just like the sequence from which it was generated, but there appears to be no pattern. Third, note that all the autocorrelations with lags greater than zero are effectively zero; that is, they fall within

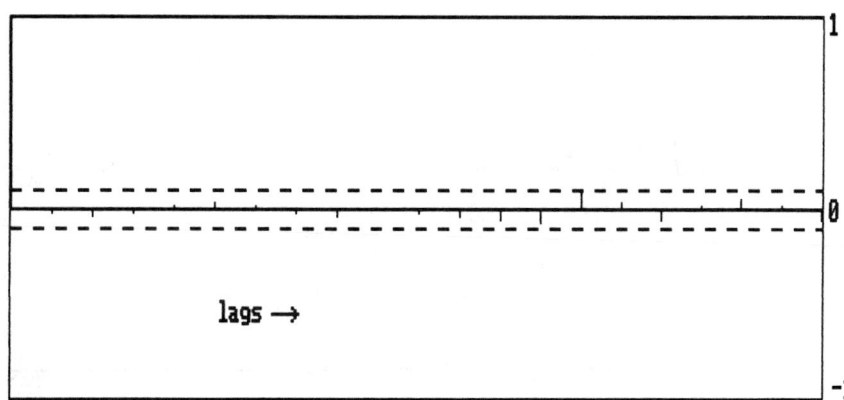

Figure 1.10   Autocorrelation of white noise.

the dashed lines that represent autocorrelation values of $\pm \overline{3}/\sqrt{(N)}$. These two lines define a band that has a width of six times the standard deviation of auto-correlations for every nonzero lag under the assumption that the sequence in question is completely unautocorrelated. In other words, if the sequence is white, then there is a 99% probability that the autocorrelations for all lags will lie within the band defined by $\pm 3/\sqrt{(N)}$. A more thorough development can be found in Section 2.1.6 in the text by Box and Jenkins (1970).

In passing, note that if there are two stochastic sequences x_k and y_k, for k = 1, 2, . . . , N, then the *sample estimate of the cross-correlation* of x and y for lag i would be given by

$$r_{xy}(i) = \frac{1}{N} \sum_{k=1}^{N-i} \frac{[x_k - \bar{x}][y_{k+i} - \bar{y}]}{s_x s_y} \qquad (1.4)$$

where s_x and s_y are the sample estimates of the standard deviations of x and y. It follows that if x_k and y_k are independent of each other then the cross-correlation of x and y would be effectively zero for all lags.

1.3.2 Autoregressive Stochastic Disturbances

Generate a new sequence c_k from the white noise sequence w_k by the following equation:

$$c_k = \alpha c_{k-1} + w_k \qquad (1.5)$$

where the coefficient α is the autoregressive parameter. (Note that since the mean of w_k is zero so is the mean of c_k.) This autocorrelated stochastic sequence is often called *autoregressive* since it depends on its previous value as well as on white noise. Figure 1.11 shows such a sequence where $\alpha = 0.9$. Figure 1.12 shows the autocorrelation of this sequence. Note that the autocorrelation drops off slowly

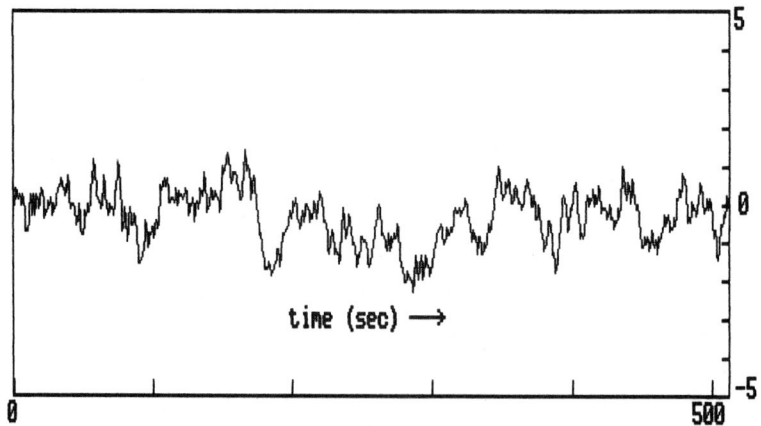

Figure 1.11   Autoregressive stochastic sequence (coefficient = 0.9).

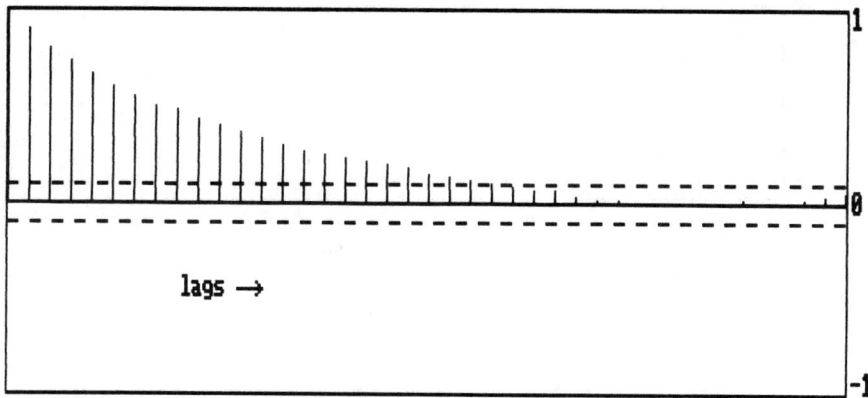

Figure 1.12   Autocorrelation of an autoregressive stochastic sequence (coefficient = 0.9).

with the lag. In fact, the autocorrelation at each lag is approximately 90% of the autocorrelation at the previous lag. This is an example of a positively autocorrelated sequence.

Figure 1.13 shows another autoregressive sequence where $\alpha = -0.9$. Note how each value of the sequence tends to have a sign opposite that of its predecessor. Figure 1.14 shows the autocorrelation of this negatively autocorrelated sequence. The value of the autocorrelation for lag 1 is negative, while that for lag 2 is positive. This alternation in sign of the autocorrelations, along with the way they drop off slowly, is a characteristic of a negatively autocorrelated autogressive stochastic sequence.

If $\alpha = 1$, the c_k sequence simply becomes the sum of the unautocorrelated sequence w_k, that is, the sum of white noise. This type of sequence is illustrated

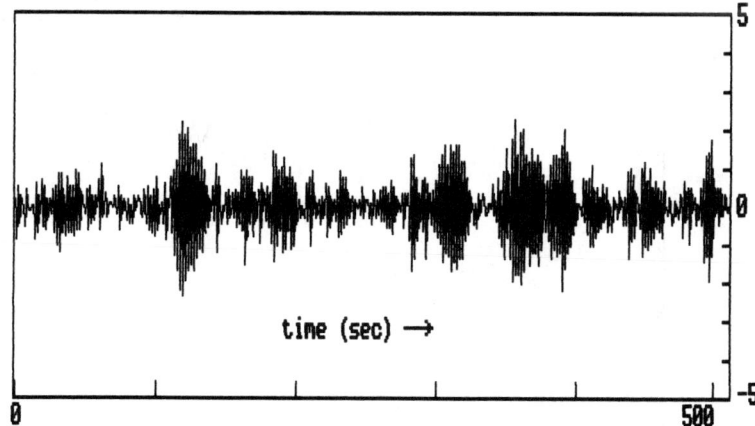

Figure 1.13   Negatively autocorrelated stochastic sequence (coefficient = -0.9).

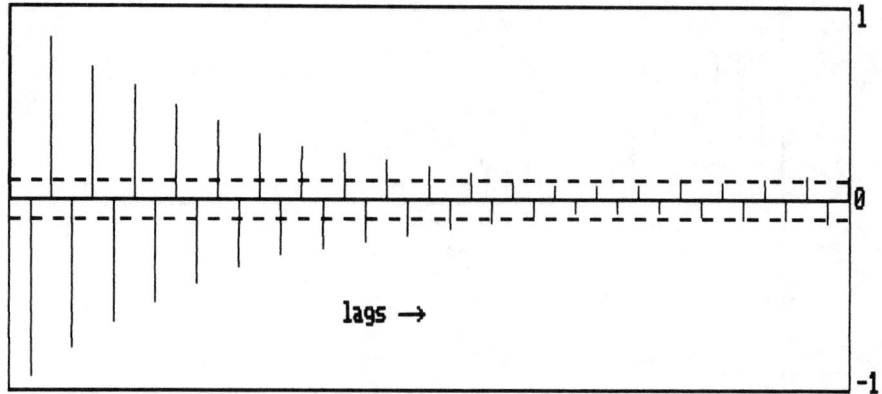

Figure 1.14 Autocorrelation of a negatively autocorrelated sequence (coefficient = −0.9).

in Figure 1.15. Note that, unlike the previous examples, it does not appear to have a constant average about which it tends to vary. This sequence has a special name, the *random walk*, and it is an example of a nonstationary stochastic sequence in the sense that as the sequence gets longer the variance continues to increase. Whenever the parameter α is greater than or equal to unity, these autoregressive sequences are considered nonstationary. In these cases the magnitude of the autocorrelation will decrease with increasing lag at a slow rate (see Figure 1.16).

An expression for the theoretical or population variance of an autoregressive stochastic sequence can be derived as follows. First, both sides of Equation (1.5) are squared:

$$c_k^2 = \alpha^2 c_{k-1}^2 + w_k^2 + 2\alpha c_{k-1} w_k \tag{1.6}$$

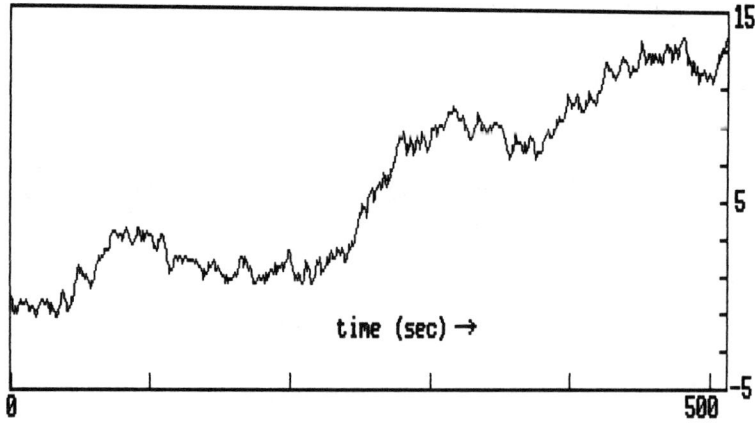

Figure 1.15 Random-walk stochastic sequence.

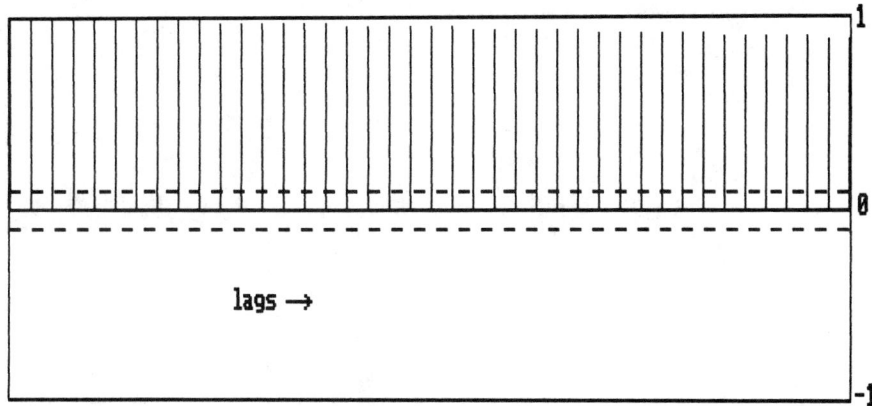

Figure 1.16   Autocorrelation of a random-walk stochastic sequence.

Next, the expected value of each side of the equation is taken. Strictly speaking, this means averaging each term over all possible values at a particular instant of time. Under certain conditions it can also be considered as the average of each term over a large number of samples in time. (The expected value will be discussed in more detail in Section 1.3.8.) Since the mean of c_k is zero, the left side of Equation (1.6) yields the variance of c, which will be denoted as V_c, and the first term on the right side yields $\alpha^2 V_c$. The second term on the right side yields the variance of the white noise sequence V_w. The average of the third term on the right side is proportional to the cross-correlation of c_{k-1} and w_k and therefore is zero, since w_k is completely uncorrelated with itself and with c_{k-1}. Therefore, all that remains is

$$V_c = \alpha^2 V_c + V_w$$

so the variance of the autoregressive sequence is

$$V_c = \frac{V_w}{1 - \alpha^2} \tag{1.7}$$

Equation (1.5) is an example of a first-order autoregressive stochastic sequence. It is a special case of an nth-order autoregressive stochastic sequence that has the form

$$c_k = \alpha_1 c_{k-1} + \alpha_2 c_{k-2} + \cdots + \alpha_n c_{k-n} + w_k$$

For most industrial situations the first-order model suffices. For a discussion of the general case the reader is referred to the text by Box and Jenkins (1970).

1.3.3 Moving Average Stochastic Disturbances

A moving average stochastic sequence c_k can be generated from the following equation:

$$c_k = w_k + \beta w_{k-1} \tag{1.8}$$

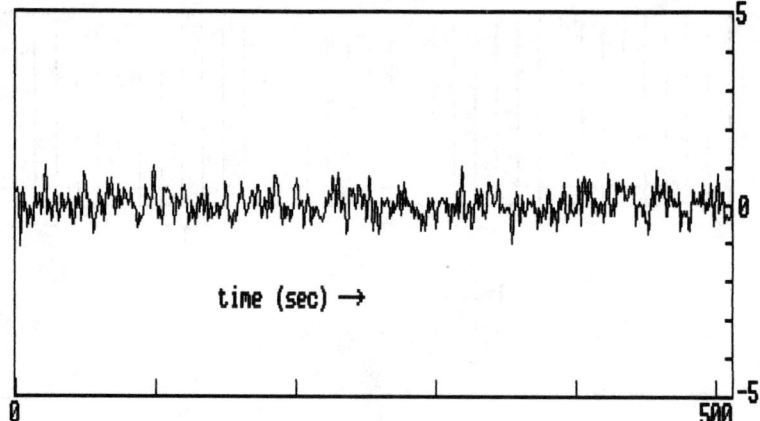

Figure 1.17   Moving average stochastic sequence (coefficient = 0.5).

Note that, unlike autoregressive stochastic sequences, moving average sequences depend only on current and previous values of the unautocorrelated sequence w_k and do not depend on past values of the c_k sequence. Also note that since the white noise sequence has zero mean so does c_k.

Figure 1.17 shows a moving average stochastic sequence generated from the above equation with $\beta = 0.5$, and Figure 1.18 shows the autocorrelation of this sequence. Note that only the first nonzero lag autocorrelation is significantly different than zero, which says that every value of c_k depends on w_{k-1} and therefore c_{k-1}, but not on c_{k-2} or c_{k-3}, and so on. This dependence on just the previous value of the generating white noise sequence follows from the defining equation. In general, the autocorrelations of moving average stochastic sequences do not damp out slowly as do those of autoregressive sequences. On the contrary, they

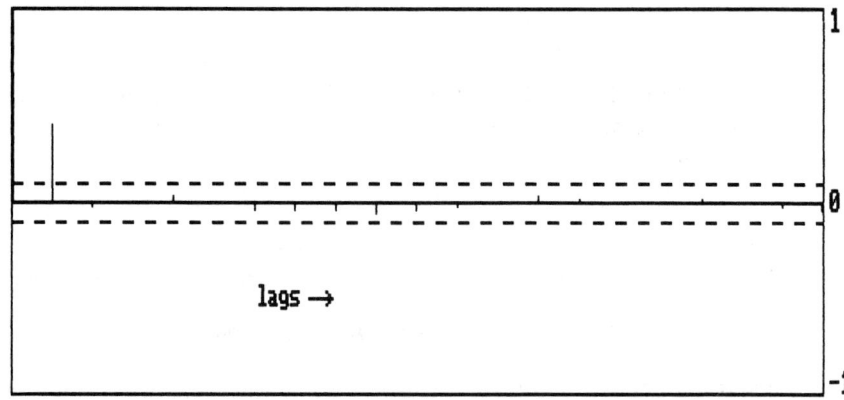

Figure 1.18   Autocorrelation of a moving average stochastic sequence (coefficient = 0.5).

cut off abruptly at a lag equal to the number of appearances of w_k in the generating equation.

By squaring both sides of the defining equation and averaging the result over a large number of samples, an expression for the variance of the moving average stochastic sequence can be derived, just as was done for the autoregressive stochastic sequence. The result of the squaring Equation (1.8) is

$$c_k^2 = w_k^2 + 2\beta w_{k-1} w_k + \beta^2 w_{k-1}^2 \tag{1.9}$$

Taking the expected value or the average of each term in Equation (1.9) over a large number of samples gives

$$V_c = V_w + \beta^2 V_w = (1 + \beta^2) V_w$$

where the independence of the white noise sequence has been used to remove the cross term. No matter what the value of β is, a moving average stochastic sequence will always be stationary.

Equation (1.8) is an example of a first-order moving average stochastic sequence. It is a special case of an nth-order moving average stochastic sequence that has the form

$$c_k = w_k + \beta_1 w_{k-1} + \beta_2 w_{k-2} + \cdots + \beta_n w_{k-n}$$

For most industrial situations, the first-order model suffices. For a discussion of the general case the reader is referred to the text by Box and Jenkins (1970).

1.3.4 Combinations and Extensions

The stochastic sequence c_k generated by the following algorithm has both an autoregressive term and a moving average term:

$$c_k = \alpha c_{k-1} + w_k + \beta w_{k-1} \tag{1.10}$$

and is referred to as an autoregressive moving average (ARMA) sequence. Figure 1.19 shows an example of such a sequence where $\alpha = 0.9$ and $\beta = 0.5$. Figure 1.20 shows the autocorrelation of this sequence. Note that the autocorrelation of this sequence is not significantly different from that of the autoregressive sequence shown in Figure 1.12.

If the sequence defined by Equation (1.10) is designated as the input to a second sequence x_k, where

$$x_k = x_{k-1} + c_k \tag{1.11}$$

then the resulting sequence becomes

$$x_k = (1 + \alpha) x_{k-1} + \alpha x_{k-2} + w_k + \beta w_{k-1} \tag{1.12}$$

which, because Equation (1.11) sums the input c_k, is nonstationary even when the magnitude of α is less than unity. Because of the presence of the summing action, which is the discrete analog of integration, this sequence is often referred

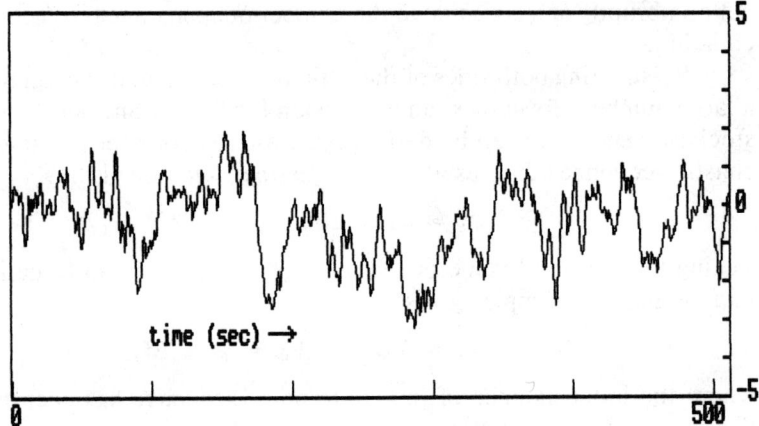

Figure 1.19 Autoregressive moving average stochastic sequence.

to as an autoregressive integrated moving average (ARIMA) sequence. Figure 1.21 compares a random walk with a sequence generated from Equation (1.12) using $\alpha = 0.2$ and $\beta = 0$. For both sequences the white noise sequence w_k has the same standard deviation. Since each of these graphed sequences is only one realization, it is difficult to visually decipher any difference.

1.3.5 Analysis of Stochastic Disturbances

To analyze a stochastic disturbance, one could calculate and plot the autocorrelation. If all the autocorrelations for nonzero lags lie within the boundary denoted by the dashed horizontal lines in the autocorrelation graphs, then there is a strong chance that the disturbance is unautocorrelated; that is, it is a white noise sequence. If, after a few lags, the autocorrelations abruptly drop away to nearly

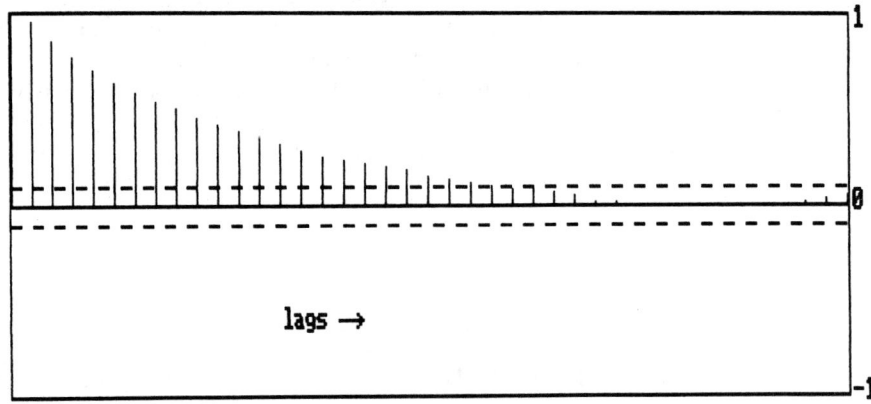

Figure 1.20 Autocorrelation of an autoregressive moving average sequence.

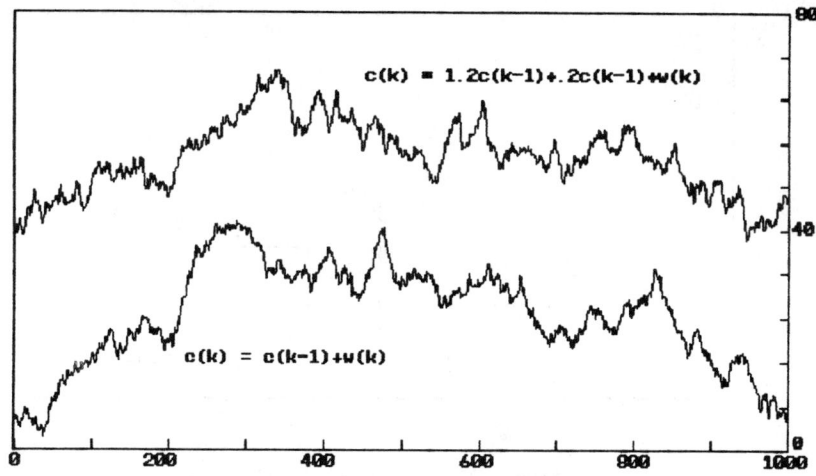

Figure 1.21 Two nonstationary sequences.

zero, then the stochastic sequence is probably a moving average. Autoregressive sequences will have autocorrelations that drop off slowly. The analysis of mixed autoregressive–moving average sequences is more difficult to generalize about and will not be pursued here. For an in-depth analysis of these stochastic sequences the reader is referred to the text by Box and Jenkins (1970).

Nonstationary sequences are characterized by slowly decreasing or non-decreasing autocorrelations. Often in these cases one can difference the data to generate a new sequence that is stationary. (Differencing will be discussed again in Section 3.7). For example, if c_k is the sequence that appears to be nonstationary, then the new sequence d_k constructed from

$$d_k = c_k - c_{k-1}$$

will likely have an autocorrelation that drops off quickly.

1.3.6 The Histogram

The autocorrelation deals with how the elements making up a stochastic sequence are related to each other in time. The histogram, on the other hand, is another way of characterizing stochastic sequences, which ignores how the elements are related to each other in time but instead shows how frequently the elements fall into discrete ranges. For example, Figure 1.22 shows a histogram of a white noise stochastic sequence. The x-axis of the histogram shows how the range, over which the values in the sequence vary, is broken up into discrete cells. The y-axis gives the fraction of the total number of elements that fall into each cell. The histogram shows how the elements are distributed over the range of values, independent of how the elements are ordered in time. The histogram could also be looked at as

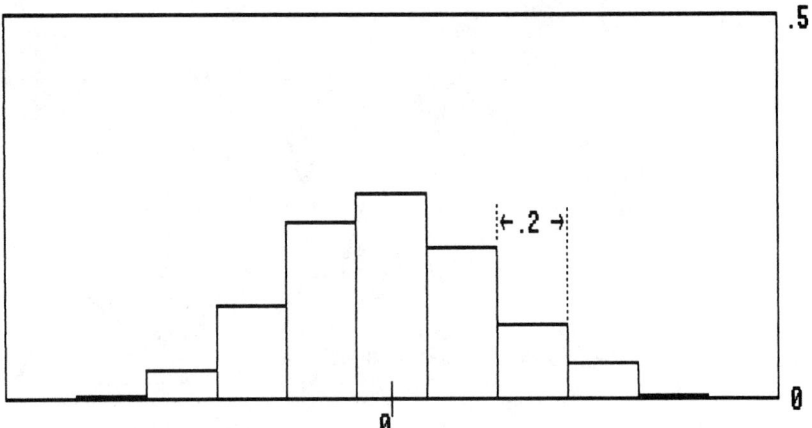

Figure 1.22   Histogram of a white noise sequence.

an extension of the standard deviation that gives a numerical measure of how the elements of a stochastic sequence are distributed about the mean.

One could start with a stochastic sequence, compute its autocorrelation and histogram, and then rearrange the order of the elements in the sequence such that the autocorrelation of the rearranged sequence would be entirely different from that of the original sequence. However, the histogram of the new rearranged sequence would be identical to that of the original.

The histogram can be useful in characterizing controlled variables. A peak would be expected at the cell that contains the target, with the heights of adjacent cells dropping off dramatically.

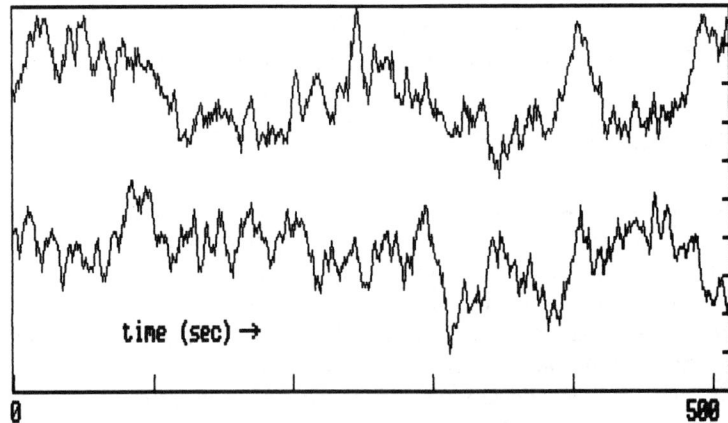

Figure 1.23   Two realizations of a stochastic sequence.

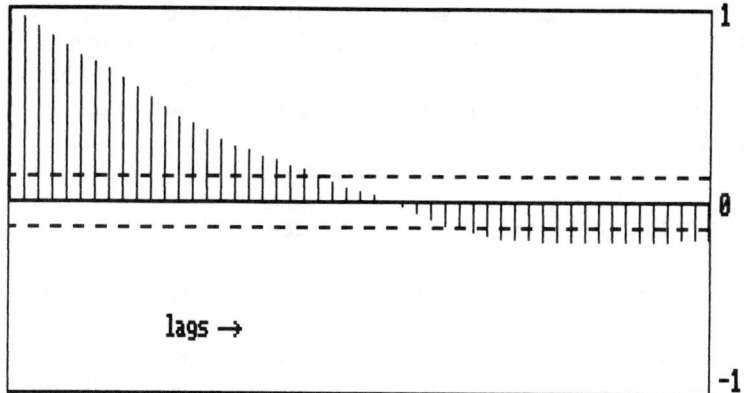

Figure 1.24   Autocorrelation of the first realization.

1.3.7 Concept of Realization and Its Application to the Autocorrelation

Consider Figure 1.23, where two realizations of an autoregressive stochastic sequence having a coefficient of 0.95 are graphed (with different y-axis offsets so as to keep them easily discernible). Each takes a slightly different path in time and each would have slightly different sample variances, averages, and autocorrelations. Figures 1.24 and 1.25 show the autocorrelations for the two realizations depicted in Figure 1.23. Note that, although both autocorrelations have the same characteristic of dropping off slowly with increasing lag, they are different.

The two stochastic sequences shown in Figure 1.23 are but two realizations chosen from an infinite family of realizations that make up the population for the

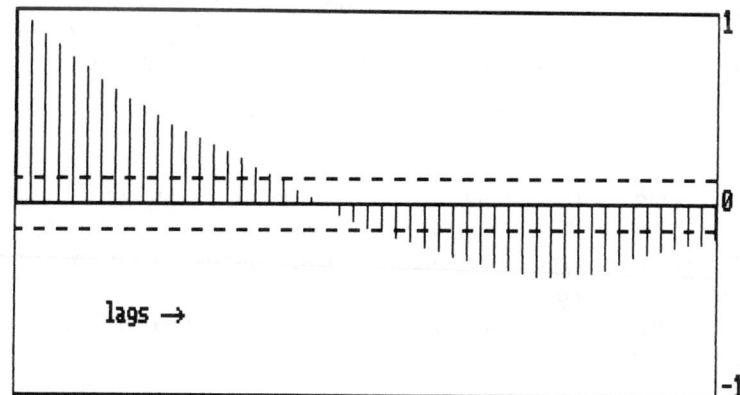

Figure 1.25   Autocorrelation of the second realization.

autoregressive stochastic sequence with a coefficient of 0.95. Conclusions drawn from simulations using stochastic sequences generated from the equations introduced above are only good for that particular realization. Other simulations should be run using other realizations of the stochastic sequence before making general conclusions.

1.3.8 Concept of the Expected Value

Earlier in this section the idea of "taking the expected value" of the terms on both sides of various equations was used to develop expressions for the variance of a stochastic sequence. This idea will now be put on a little firmer footing. Once this idea has been rooted, it will be used to further clarify the concept of a stationary sequence.

The population mean and the population variance of a stochastic sequence are characteristics of the sequence derived from looking at *all* the realizations of that sequence. Let x_i, $i = 1, 2, \ldots$, be a stochastic sequence that is infinite in extent. At any instant of time t, there is a probability density for the quantity x, called $p(x', t)$, which gives the probability that x will have a value between x' and $x' + dx$. By definition, the integral of $p(x, t)$ over all the possible values that x can take on will be unity:

$$\int_{-\infty}^{\infty} p(x, t) \, dx = 1$$

The *expected value* of x at time t, $E\{x\}$, or equivalently the *mean* of x at time t, μ_x, is given by

$$E\{x\} = \mu_x(t) = \int_{-\infty}^{\infty} p(x, t)x \, dx \tag{1.13}$$

In effect, this is a weighted average over the whole ensemble of realizations associated with the stochastic sequence x at time t. This is not to be confused with the *average* of x over N samples in time, which is given by

$$\bar{x} = \frac{1}{N} \sum_{i=1}^{N} x_i$$

The mean of a sequence is often estimated by computing the average of that sequence over N samples. This is acceptable as long as N is large enough so that the elements of the finite length sequence suitably represent most of the possible values that x can take on.

The expected value of $(x - \mu_x)^2$ is the *variance of x around the mean* and is given by

$$E\{(x - \mu_x)^2\} = \sigma_x^2(t) = V_x(t) = \int_{-\infty}^{\infty} p(x, t)(x - \mu_x)^2 \, dx \tag{1.14}$$

where both $\sigma_x^2(t)$ and $V_x(t)$ will be used in this book to denote the population variance of x. The population standard deviation is denoted by σ_x. For most of the sequences in this book, the average will be subtracted off first, so the difference between the variance and the variance about the mean will not be emphasized. Equation (1.14) is, in effect, a weighted average of $(x - \mu_x)^2$ over the whole ensemble of realizations associated with the stochastic sequence x at time t. As was pointed out in Section 1.3.1, the variance of a stochastic sequence is often estimated by using Equations (1.1) or (1.2).

The autocorrelation function of a stochastic sequence x, denoted as $A_x(i)$, is the expected value of $x_k x_{k+i}$ or $E\{x_k x_{k+i}\}$. Note that the index i can have positive or negative values (as well as zero). The autocovariance function of a stochastic sequence x is

$$E\{(x_k - \mu_x)(x_{k+i} - \mu_x)\}$$

and is often estimated by

$$r_x(i) = \frac{1}{N} \sum_{k=1}^{N-i} \frac{[x_k - \bar{x}][x_{k+i} - \bar{x}]}{s_x^2}$$

In Section 1.3.1, this quantity was called the sample estimate of the autocorrelation or just the autocorrelation. Note that the autocorrelation is normalized by the sample estimate of the variance, while the autocorrelation function and the autocovariance function are unnormalized. Also note that, for sequences having zero means, the autocorrelation function and the autocovariance function are the same. Finally, note that for zero lag the autocovariance function equals the variance, and for a zero mean stochastic sequence the autocorrelation function also equals the variance; that is,

$$A_x(0) = V_x, \qquad \text{for } \mu_x = 0$$

The cross-correlation function associated with two stochastic sequences x_k and y_k is $E\{x_k y_{k+i}\}$. The cross-covariance function is

$$E\{(x_k - \mu_x)(y_{k+i} - \mu_y)\}$$

and is often estimated by

$$r_{xy}(i) = \frac{1}{N} \sum_{k=1}^{N-i} \frac{[x_k - \bar{x}][y_{k+i} - \bar{y}]}{s_x s_y}$$

which was referred to earlier as the cross-correlation. Note again the normalization in the cross-correlation, which does not occur in the cross-covariance function.

At this point, a review of this barrage of new functions, the autocorrelation function, the autocovariance, the cross-correlation function, and the cross-co-variance function, should be attempted. The autocovariance function is the extension of the autocorrelation $r_x(i)$ for an infinite sequence. Or, put another way, the autocovariance function is the population characteristic associated with the autocorrelation whose calculation is based on a finite sample of the population. The autocorrelation function is the unnormalized version of the autocovariance function. Using similar verbiage, the cross-correlation function is the extension of the cross-correlation $r_{xy}(i)$ for an infinite sequence. That is, the cross-correlation function is the population characteristic associated with the cross-correlation whose calculation is based on a finite sample of the population. Whenever the word function is appended to either the autocorrelation or the cross-correlation, a population characteristic rather than an estimate derived from a finite sample is being referred to. The autocorrelation function and cross-correlation function are to the autocorrelation and cross-correlation as the population mean μ and the population standard deviation σ are to the sample average $\bar{x}$ and the sample estimate of the standard deviation s_x. In Chapter Five the autocorrelation function will be used in the derivation of the expressions for the power spectral density.

A *stationary* sequence has a population variance that is finite, a population mean that is constant (that is, independent of time), and an autocorrelation function that depends only on the displacement i. One can picture a stationary stochastic sequence as being the result of plucking a sample x_k from a distribution of values described by $p(x)$. The probability density $p(x)$ governing the distribution is constant with time. The relation between samples x_k and x_{k+i} that have been plucked from the distribution at different times can be characterized by the autocorrelation function $E\{x_k x_{k+i}\}$ or the autocovariance function

$$E\{(x_k - \mu_x)(x_{k+i} - \mu_x)\}$$

which in turn can be estimated by calculating $r_x(i)$.

For the case of two stationary sequences x and y, the cross-correlation functions $E\{x_k y_{k+i}\}$ and $E\{y_k x_{k+i}\}$ would be identical.

A sequence is considered to be *ergodic* if its characteristics such as the mean, variance, and autocorrelation function can be determined from calculations applied to an infinitely long time-domain realization. For example, a sequence is said to be ergodic if the limiting case of the average taken over an infinitely long time-domain realization equals the mean. (Remember that the mean is a weighted average taken over all the realizations at a particular moment in time.) In Section 1.3.2, the assumed ergodicity of the autoregressive stochastic sequence was used when an expression was derived for its variance.

1.3.9 Method of Simulating Disturbances and Evaluating the Effects

Frequently, in the remainder of this text, control algorithms will be tested by subjecting various process models to stochastic disturbances, and this is a good place to explain how they are generated and how the results of the tests are evaluated. The starting point is a random number generator invoked by a function call from a BASIC program. (All the simulations are coded in BASIC.) Let r be the number so generated, which has a range from 0.0 to 1.0. Since it is uniformly distributed, its probability density is

$$p(x) = 1, \qquad 0 \le x \le 1$$

which satisfies the requirement that

$$1 = \int_0^1 p(x) \, dx$$

Since the random numbers generated are unautocorrelated, the variable x is an example of rectangularly distributed white noise. These numbers can be centered about zero by defining a new number y as

$$y = x - 0.5$$

so that y ranges from -0.5 to 0.5 but has the same distribution shape as x. A third number, z, is derived from y by multiplying it by a gain g_r in order to give it a range from $-0.5g_r$ to $0.5g_r$. The probability distribution of z is

$$p(z) = \frac{1}{g_r}, \qquad -0.5g_r \le z \le 0.5g_r$$

and the variance of z is

$$V_z = \int_{-0.5g_r}^{0.5g_r} \frac{1}{g_r} z^2 \, dz = \frac{g_r^2}{12} \tag{1.15}$$

Six values of z are averaged to form a fourth quantity w, where the variance of w is given by

$$V_w = \frac{V_z}{6} = \frac{g_r^2}{72}$$

The variable w will tend to be normally distributed about zero and therefore can be used to simulate approximately normally distributed white noise. The histogram in Figure 1.22 verifies this contention.

To generate a realization of an autocorrelated stochastic disturbance, this white noise is fed into the appropriate generating equation. For example, if an autoregressive stochastic sequence, c_k, is desired, then the generating equation is

$$c_k = \alpha c_{k-1} + w_k$$

where w_k is the above-mentioned white noise sequence resulting from choosing a gain g_r.

Simulation will be used in Chapters Two and Three to test control algorithms by applying them to model processes subjected to selected stochastic disturbances that are generated as described above. There are three ways to compare control algorithms using this approach. First, everytime a control algorithm is changed and a new simulation is run, the random number generator would be started with the same seed. Second, control algorithms can be compared by making single but long simulation runs, say, runs that are at least 100 times longer than the effective time constant and deadtime. These runs would have to be long enough to desensitize the results from the choice of a random number seed. Third, several runs can be made, each with different seeds. Then, at the completion of these runs, averages and standard deviations can be estimated over the runs as well as within the runs. For the most part, method 2 is used in this book. No matter what method is used, some uncertainty will always be associated with the results.

It is important to keep the empirical standard deviations calculated from these simulation samples separate from the theoretical or population standard deviations that will occasionally be derived from first principles and that are not subject to the sampling uncertainty. The distinction should be apparent from the context.

Before leaving this section, it should be mentioned that Equation (1.15), which gives the variance for a uniformly distributed stochastic variable over the interval $[-0.5g_r, 0.5g_r]$, can be used to describe quantization noise resulting from an analog-to-digital conversion. If the full-scale range of a bipolar converter is $-R$ to R and if it has B bits, then the quantization step size q is

$$q = \frac{R}{2^B}$$

For small q, Oppenheim and Schafer (1989) show that the A/D converter generates white noise that is uniformly distributed over the interval $[-q/2, q/2]$. Therefore, Equation (1.15) shows that the variance of the quantization noise is

$$V_q = \frac{q^2}{12}$$

This will be useful in Chapters Three and Five when tools are developed to study the response of controlled systems to white noise disturbances.

1.3.10 Intermediate Summary III

The autocorrelation was introduced as a tool for characterizing stochastic disturbances. Using this tool, another look was given unautocorrelated disturbances, autocorrelated disturbances, and nonstationarity. Two kinds of autocorrelated disturbances were considered: autoregressive and moving average. The histogram was also discussed as another way of characterizing a stochastic sequence without taking into account how the elements of the sequence are ordered in time.

There has been an attempt to put stochastic disturbances on a little more quantitative footing without being rigorous. The reader interested in a significantly more thorough and elegant approach is directed to the text by Box and Jenkins (1970). In Chapter Three, the characterization of empirical stochastic disturbances will be revisited using the line spectrum. In Chapter Five, the power spectral density will be used to characterize theoretical stochastic disturbances.

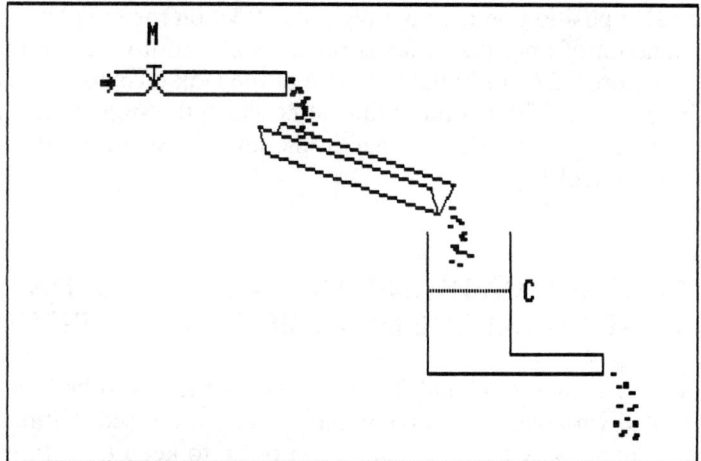

Figure 1.26   Process with deadtime.

1.4 PROCESS DEADTIME

The third example process, shown in Figure 1.26, consists of another tank of water, but this time the inlet stream must flow down an inclined trough after it leaves the inlet pipe and before it reaches the tank. If a step change is applied to

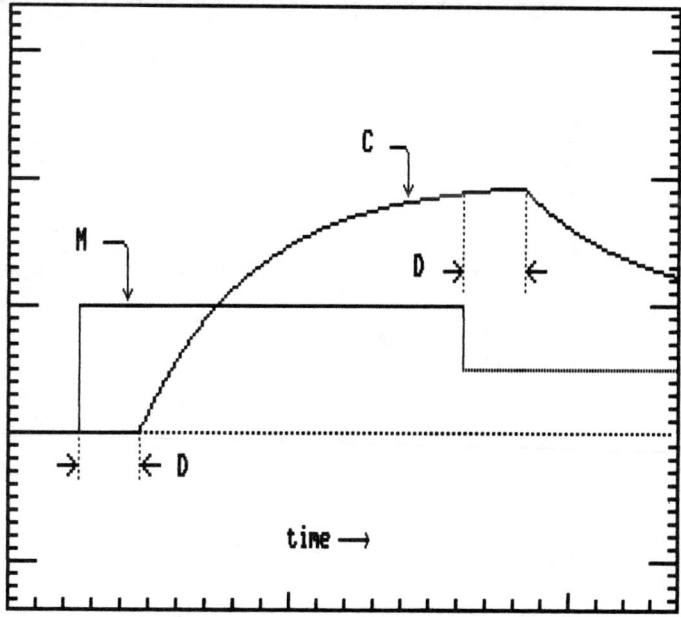

Figure 1.27   Illustration of deadtime.

the valve position on the inlet pipe, its effect on the tank level will be delayed by the amount of time that it takes for the water to flow down the trough.

Figure 1.27 shows the response of the tank level to a step change in the inlet valve position. The period of time after which the step in the valve was made and during which there is no change in the level is called the *deadtime*, represented by the symbol D.

1.5 A PARTIAL SUMMARY AND AN EXAMPLE PROCESS ILLUSTRATING ALL THE DYNAMIC CHARACTERISTICS

Two new symbols, C and M, representing the controlled variable and the manipulated variable, respectively, have been introduced. Ultimately, the goal is to find a simple way to manipulate M in order to keep C on target.

Three concepts of process dynamics have been introduced. The process gain G is the ratio of the ultimate change in the controlled variable to the applied change in the manipulated variable when the latter has undergone a step change. The deadtime D is the time elapsed after the step change in M is made but before C starts to respond. The time constant T is the time after the deadtime has elapsed that it takes for C to reach 63% of its final value after a step change in the manipulated variable. Knowing estimates of these characteristics for a process will

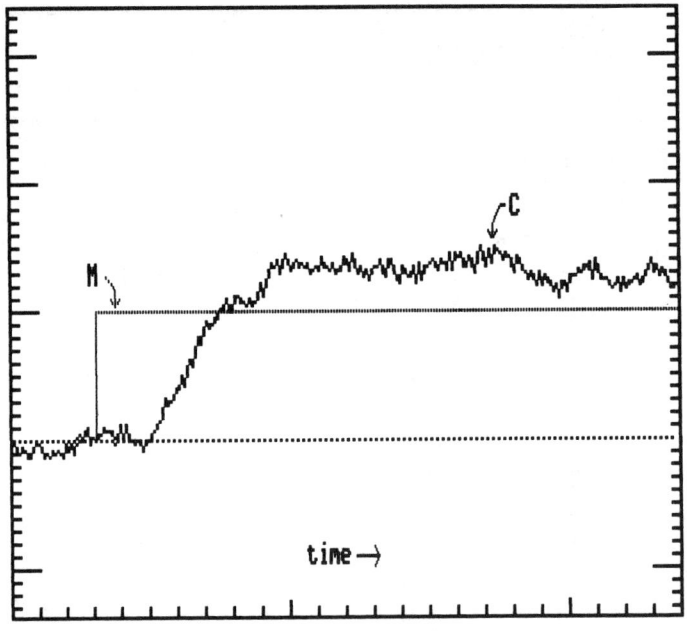

Figure 1.28 Process with T, G, D, and autocorrelated stochastic disturbance.

provide a means of assessing the difficulty of a pending control problem and evaluating the quality of an existing control system.

Figure 1.28 shows the controlled variable and the manipulated variable for a process having a process time constant roughly equal to its process deadtime and where the controlled variable is subject to stationary autocorrelated stochastic disturbances. Note that the visual estimation of the process time constant, deadtime, and gain is made more difficult by the presence of the autocorrelated stochastic disturbances.

1.6 A MATHEMATICAL MODEL OF A PROCESS HAVING A TIME CONSTANT, GAIN, AND A DEADTIME (THE FOWDT MODEL)

In this section some of the foregoing material will be put in perspective by developing a mathematical model for the example tanks that have been used to illustrate concepts in process dynamics. There will be some elementary use of calculus in this section and none of this material is absolutely essential to what follows in this chapter, but if the reader is properly equipped, reading this section will probably add insight into what follows in subsequent chapters.

Start with the tank in Figure 1.1. To develop an equation that describes the dynamic behavior of the tank's level, set the rate at which the water accumulates equal to the difference between the rate at which water flows into the tank and the rate at which water leaves. In engineering terms, this is simply a dynamic mass balance on the tank; that is, the conservation of mass is being invoked. If the flow rate (in pounds per hour) into the tank is F_i, the flow rate out of the tank is F_o, and the amount of water (in pounds) in the tank at any instant is rAC, where r is the density, A is the tank's planar cross-sectional area, and C is the level, then

$$\frac{d}{dt}(rAC) = F_i - F_o$$

$$(\text{acc}) = (\text{in}) - (\text{out})$$

(1.16)

In Section 1.1 it was assumed that, for small changes, the outlet flow rate was proportional to the level in the tank, so

$$F_o = \frac{C}{R}$$

(1.17)

where R is the proportionality constant and represents the resistance to flow through the outlet pipe. By the way, Equation (1.17) requires that for zero level there will be zero outlet flow. If the level C is considered the potential for flow, Equation (1.17) can be considered an example of an extremely simple constitutive equation that relates a flow to a potential. Therefore, the process of deriving a describing equation for the model consists of coupling a conservation law with constitutive equations.

The inlet flow rate can be considered proportional to the valve position M as in

$$F_i = g_v M \tag{1.18}$$

where g_v is the valve's gain in pounds per hour per inch of valve movement.

Combining Equations (1.16), (1.17), and (1.18) gives the following first-order differential equation:

$$(rAR)\frac{dC}{dt} + C = (Rg_v)M \tag{1.19}$$

Temporarily, the following associations will be made (proof will come later):

$$T = rAR$$

$$G = Rg_v$$

In other words, it is suggested that the process time constant T is rAR and that the process gain G is Rg_v. Having done this, Equation (1.19) becomes

$$T\frac{dC}{dt} + C = GM \tag{1.20}$$

This is a general form for a first-order differential equation ("first" order because the highest derivative occurring has order 1).

Now introduce two new sets of variables:

$$C = C_{ss} + c \tag{1.21}$$

$$M = M_{ss} + m \tag{1.22}$$

where M_{ss} and C_{ss} are the values of M and C at steady states; that is, when $dC/dt = 0$, they satisfy

$$C_{ss} = GM_{ss}$$

and they are constants. The lowercase variables m and c are the time-varying parts of M and C. Introducing these variables into Equation (1.20) gives

$$T\frac{d(C_{ss} + c)}{dt} + C_{ss} + c = G(M_{ss} + m)$$

which, taking into account that $dC_{ss}/dt = 0$ and that $C_{ss} = GM_{ss}$, simplifies to

$$T\frac{dc}{dt} + c = Gm \tag{1.23}$$

For the case where m is zero for $t < 0$ and m takes on the constant value M_c for $t \geq 0$ (that is, where a step change is made in m at time zero), Equation

(1.23) has the following solution:

$$c(t) = c(0) \exp(-t/T) + G[1 - \exp(-t/T)]M_c \qquad (1.24)$$

Equation (1.24) can be obtained by a variety of mathematical methods, none of which are pertinent to our discussion. [In Chapter Four, Equation (1.24) will be derived using the Laplace transform.] Because of the definition of $c(t)$ and $m(t)$, the quantity $c(0)$ would also be zero, but for future reference it will be left in the expression.

Graphing C versus t would give a curve just like that in Figure 1.2. Therefore, this equation is the response of C to a step in M of amount M_c at time zero.

If, in Equation (1.24), t goes to infinity so that the exponential terms go to zero, then the final value of $c(t)$ will be GM_c. Remembering the definition of the process gain in Section 1.1 as the ratio of the ultimate change in the controlled variable to the change in the manipulated variable confirms that G is in fact the process gain.

If $t = T$, then, since $c(0) = 0$, Equation (1.24) gives

$$c(T) = 0.6321206GM_c$$

which indicates that C reaches 63% of its final value when $t = T$. Hence T is the process time constant.

Now break time into discrete intervals of length h and let t_i denote the time at the start of the ith interval:

$$t_i = t_{i-1} + h$$

Next, specify that every h seconds M changes its value and that it holds this value over that interval. Furthermore, denote the value of M during the time t_i to t_{i+1} as $m(t_i)$. Since the variation of M has been broken up into a sequence of constant values, Equation (1.24) can be applied over the period of time during which M is constant:

$$c(t_i) = c(t_{i-1}) \exp(-h/T) + G[1 - \exp(-h/T)] m(t_{i-1}) \qquad (1.25)$$

where it is seen why $c(0)$ was left in Equation (1.24) even though it was zero. This equation simply says that, if M is given the value $m(t_{i-1})$ at time t_{i-1} and kept constant at that value until just before time t_i, the value of C at time t_i can be calculated if $c(t_{i-1})$ is known. If the following associations are made

$$\alpha = \exp(-h/T)$$

$$\beta = 1 - \alpha$$

$$c_i = c(t_i)$$

$$m_i = m(t_i)$$

then c_i can be written as a weighted sum of c_{i-1} and m_{i-1}:

$$c_i = \alpha c_{i-1} + \beta G m_{i-1} \qquad (1.26)$$

If the process time constant T is large relative to h, then the parameter α will be nearly unity, the parameter β will be nearly zero, and c_i will depend strongly on c_{i-1}; that is, it will change relatively slowly even though M may change significantly. Conversely, if the time constant is small, then the parameter α will be small relative to β and c_i will be strongly dependent on $G m_{i-1}$, and every change in M will show up quickly in C.

Strictly speaking, all the quantities appearing in Equation (1.23) should be written with their time arguments, giving

$$T \frac{dc(t)}{dt} + c(t) = Gm(t) \qquad (1.27)$$

For a process that has deadtime, the argument [in Equation (1.27)] of m is changed from (t) to $(t - D)$, indicating that the effect of a change in M does not have an effect on C until the time D has elapsed:

$$T \frac{dc(t)}{dt} + c(t) = Gm(t - D) \qquad (1.28)$$

If Equation (1.28) is solved, just as Equation (1.20) was, for the case when there is a step in M from M_{ss} to $M_{ss} + M_c$ at time zero, then

$$c(t) = c(0) \exp(-t/T) + G\{1 - \exp[-(t - D)/T]\}M_c U(t - D) \qquad (1.29)$$

where $U(t - D)$ is the unit step function that by definition equals zero whenever its time argument is negative and equals unity otherwise. Equation (1.29) is the same as Equation (1.24) for the case of no deadtime except that the step in M of amount M_c is delayed D seconds.

As with the case without deadtime, time can be broken up into discrete steps, and subscripts can be used to shorten the notation. The deadtime D is also assumed to be an integer multiple of the time interval length:

$$D = dh$$

where d is the integer. Again it is assumed that M can be divided into a sequence of constant values, $m_1, m_2, \ldots, m_i, \ldots$. Instead of applying this discretization to Equation (1.29), the subscript of M in Equation (1.25) is simply changed from $i - 1$ to $i - 1 - d$, giving

$$c_i = c_{i-1} \exp(-h/T) + G[1 - \exp(-h/T)] m_{i-1-d} \qquad (1.30)$$

Therefore, the controlled variable can still be looked upon as a weighted sum of the previous controlled variable and the previous manipulated variable, where the latter quantity is delayed an extra d samples if there is a deadtime of hd.

In this section, continuous and discrete mathematical models for a process exhibiting a time constant, a gain, and a deadtime have been presented. Since the models can be described by first-order differential or difference equations, this model is referred to as the FOWDT or first-order-with-deadtime model. With this model in mind, the reader might want to reread Section 1.1.1, where the effects of the example tank's dimensions on the time constant and gain were discussed.

1.7 DEVIATIONS FROM THE FOWDT MODEL

1.7.1 Nonlinearity

In the first example, the outlet flow rate was assumed to be proportional to the level in the tank, and the transient behavior shown in Figure 1.2 was constructed from a mathematical model having this characteristic. For small changes in the level, the *linear* relationship between the level and the outlet flow rate was a realistic assumption.

However, for large changes in the level, elementary physics says that the outlet flow rate is proportional to the square root of height of liquid above the outlet pipe; that is, the level and the outlet flow rate have a nonlinear relationship. As a consequence of this, there will be a difference in the process gain and the process time constant depending on the sign of the step in the manipulated variable.

This can be explained by the following chain of logic. After a step change in the manipulated variable, a new steady state will be arrived at when the level changes enough to cause the outlet flow rate to match the inlet flow rate. When the inlet flow step change is positive, the level will have to rise until the outlet flow rate, which is proportional to the square root of the level, is large enough to equal the inlet flow rate. The converse will happen when the step change is negative. Because of this square-root relationship, the amount by which the level will have to rise, in the case of a positive step change in the inlet flow rate, will be greater than the amount by which the level will have to drop in the case of a negative step change in the inlet flow rate. Therefore, the process gain will be greater for a positive step change than for a negative step change.

A similar argument leads to the conclusion that the process time constant will depend on the sign of the step change. However, since most dynamic analyses for control purposes deal in small changes, the dependence of the time constant and gain on the sign of the step change will often be negligible.

An interesting analogy can be made between mass transfer processes, like the first example, and high-temperature energy transfer processes. The temperature takes the place of the level, and the rate of energy loss replaces the outlet mass flow rate. Instead of the outlet flow rate being proportional to the square root of the level, the energy loss rate is proportional to a greater-than-unity power, say 2, of the temperature, because at high temperatures energy can be transferred by radiative as well as convective means.

After a positive step change in the manipulated variable, which is assumed to be related linearly to energy input rate, the process temperature will rise until the energy loss rate matches the increased energy input rate. Because of the quadratic relationship between energy loss rate and temperature, the temperature will not have to increase as much as it would have had to decrease for a negative step of equal magnitude in the energy input rate. Therefore, the process gain is less for positive steps in the manipulated variable than for negative steps.

Another example of a common nonlinearity is the valve position–flow characteristic. When the FOWDT model was derived in Section 1.6, the flow through the valve was related to the valve position by a linear valve gain, g_v. In reality, most valves have nonlinear valve characteristics so that over the whole range of valve movement from completely open to completely closed a single gain will not suffice. A good discussion of control valve characteristics is given in Chapter Seven in the book by Eckman (1962).

1.7.2 Inflection Points

Consider Figure 1.29, which shows a process consisting of three tanks connected to each other instead of just one tank, as was the case in Figure 1.1. Figure 1.30 shows the response of the level in the third tank (which will be the controlled variable for this example) when a step change is applied to the valve in the inlet pipeline to the first tank. Upon comparing Figures 1.2 and 1.30, one sees that the transient response of C in the latter has an inflection point.

This behavior can be explained if it is assumed that all changes are small and that the outlet flow rate from the third tank is proportional to its level. Furthermore, it is assumed that the flows between the tanks are proportional to the differences in the levels; that is, the flow from the first tank to the second tank is proportional to the difference between the level in first tank and that in the second tank.

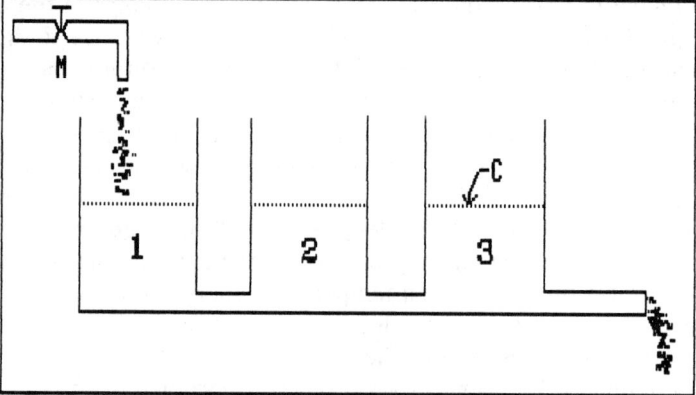

Figure 1.29   Three-tank process.

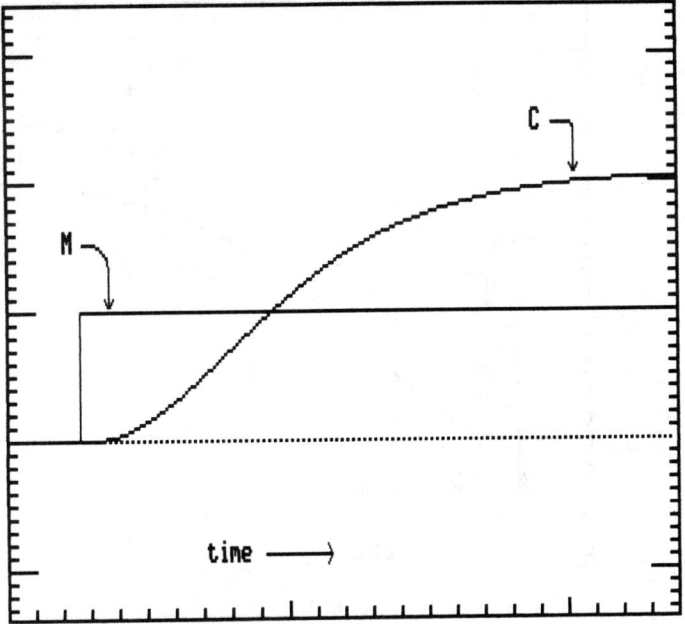

Figure 1.30   Illustration of inflection point.

The main reason why the third tank's level behaves differently than the single tank's level of the first example is that the time behavior of the flow into the third tank differs significantly from that of the flow into the single tank in the first example. In the first example, the inlet flow rate was given a step change and then was held constant at that value indefinitely. The level began to increase, but at a decreasing rate, as the outlet flow rate approached the new inlet flow rate. In the three-tank example, the inlet flow rate to tank 3 does not initially see the step in the first tank's inlet flow rate. In fact, it grows only as the level in the second tank rises. Unlike the process in the first example, the third tank's level increases slowly. The third tank's *rate of change* of level also increases as the second tank's level increases, pushing more water into the third tank.

Later, when the third tank's level has increased to the point where it is pushing a significant amount of water out through the exit pipe, its rate of change slows down because the third tank's outlet flow rate approaches its inlet flow rate. Consequently, there is an inflection point where the rate of change of the third tank's level stops increasing and starts to decrease.

The step change responses of many industrial processes have inflection points, but in these cases an approximation consisting of a deadtime and time constant, as shown in Figure 1.31, is usually sufficient for control purposes. Also, the inflection point is often obscured by the presence of stochastic disturbances. This is demonstrated in Figure 1.32, where stochastic disturbances are added to the process shown in Figure 1.30. Here, the best one could do is to estimate the

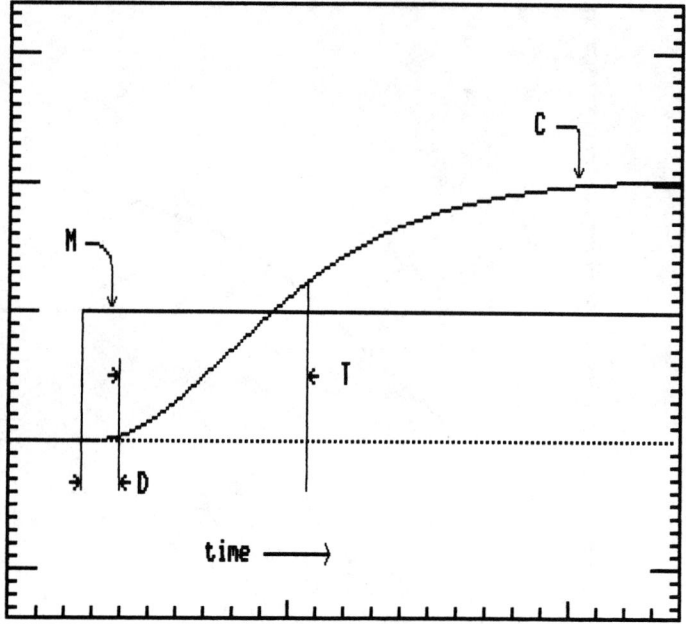

Figure 1.31 Deadtime/time constant approximation to inflection point.

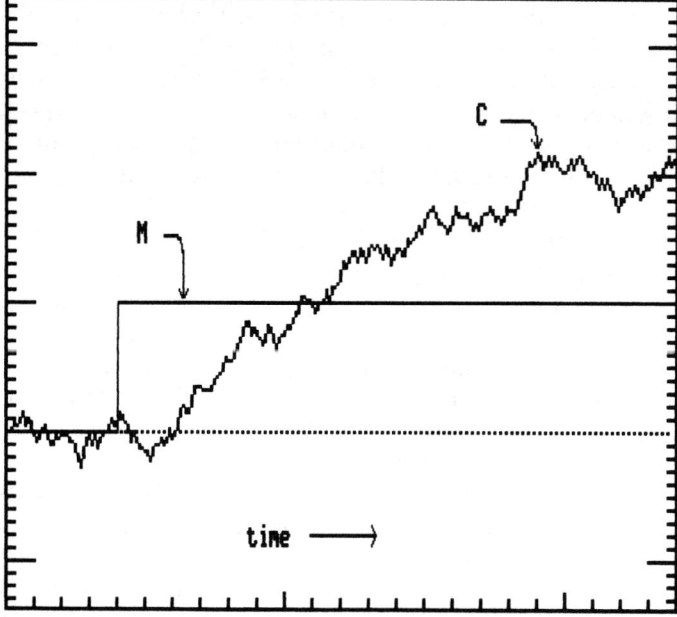

Figure 1.32 Inflection point with noise.

deadtime, the time constant, and the gain without attempting to discern an inflection point.

1.7.3 Integrators

In Section 1.6 a first-order differential equation describing the dynamic behavior of the level in a tank was derived. The mass balance gave

$$\frac{d}{dt}(rAC) = F_i - F_o \tag{1.31}$$

$$(\text{acc}) = (\text{in}) - (\text{out}).$$

If the difference in flows is written as $\Delta F_c = F_i - F_o$, then integration of Equation (1.31) gives

$$C(t) = C(t_0) + \frac{1}{rA}\int_{t_0}^{t}\Delta F_c\,du \tag{1.32}$$

In other words, the level is the integral of the difference of the flow rates.

Now, consider a modification to Figure 1.1 where a constant flow pump is placed in the outlet stream so that the outlet flow rate is no longer a function of the level in the tank, but instead is effectively constant. The level is therefore the integral of the inlet flow rate:

$$C(t) = C(t_0) - \frac{1}{rA}F_o(t - t_0) + \frac{1}{rA}\int_{t_0}^{t}F_i(u)\,du \tag{1.33}$$

Equation (1.33) represents a process that is called an *integrator,* and it appears to be a special case not covered by the FOWDT family of models. However, from a mathematical point of view it could be looked upon as a limiting case of a first-order model with an infinite gain and an infinite time constant, where the ratio of the gain to the time constant is finite.

These comments can be supported by some simple mathematics. Consider Equation (1.23) derived in Section 1.6 for the first-order model:

$$T\frac{dC}{dt} + C = GM \tag{1.34}$$

If the time constant and process gain are large, then the second term on the left side is swamped by the derivative term and by the term on the right side, giving

$$T\frac{dC}{dt} \approx GM$$

which can be integrated to give

$$C(t) = C(t_0) + \frac{G}{T}\int_{t_0}^{t}M(u)\,du$$

which is the equation of an integrator where the ratio of two large quantities, G/T, is finite.

On the macroscopic time scale, the controlled variable of a first-order process will ultimately approach a new steady state after a step in the manipulated variable. However, just after the step, before the effect of the first small change in the controlled variable has been felt, the controlled variable is well approximated by the integral of the step change. This is because immediately after a step change is applied to a FOWDT model, for a short period of time, C is overwhelmed by $T\ dC/dt$ and GM. Consequently, immediately after a positive step change in M, the quantity C will integrate up until it becomes significant relative to $T\ dC/dt$ and GM, at which time the process no longer acts as an integrator and starts to show the FOWDT tendency of driving toward a new steady state.

1.7.4 Underdamped and Inverse Responding Processes

The step response of an underdamped process (model 12) is shown in Figure 1.33. Note that, even though there is no feedback control activity in the manipulated variable, damped oscillations appear in the controlled variable.

In another relatively rare type of process behavior, the step change response of the controlled variable first moves in a direction opposite to that of the ultimate change. Figure 1.34 shows such a type of process whose behavior can be qualitatively described as the sum of two first-order processes: one with a negative

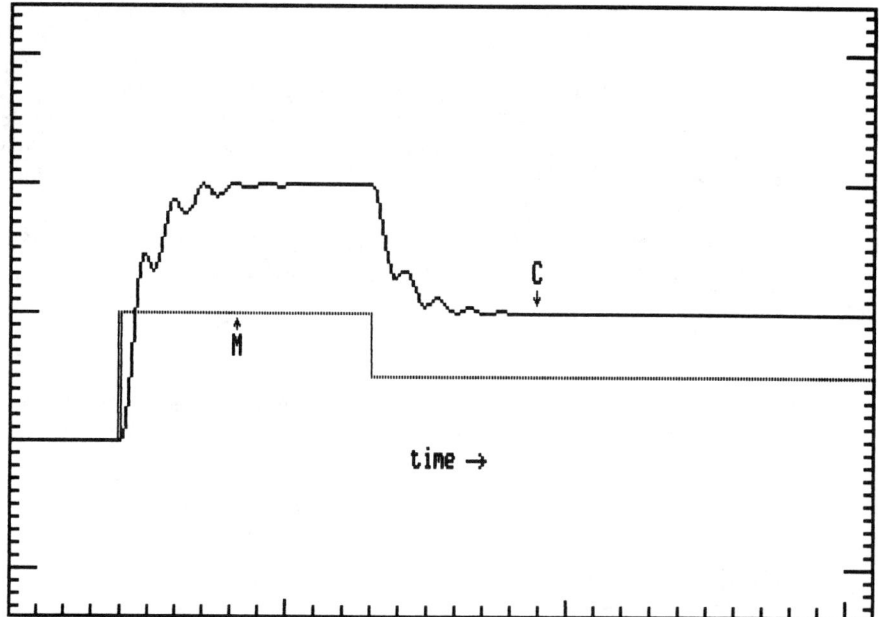

Figure 1.33 Underdamped process.

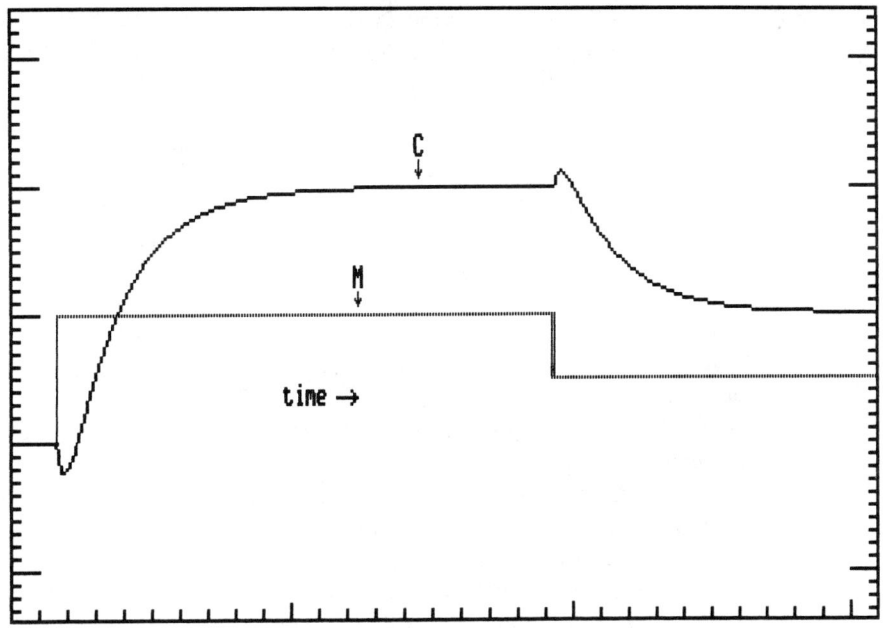

Figure 1.34   Inverse responding process.

gain and a short time constant, and one with a larger positive gain and a longer time constant. After the step in the manipulated variable is made, the first component reacts quickly, but, as time progresses, the second component with its larger positive gain and longer time constant dominates.

Since both of these example processes are relatively rare, little time will be spent on them in the balance of the text.

1.8 DYNAMIC CHARACTERISTICS AND THE DIFFICULTY OF CONTROL

Even though a specific control algorithm has yet to be discussed, general comments can be made about the degree of difficulty posed by a process having large or small values of the dynamic characteristics that have been discussed above.

1.8.1 Process Time Constant and Process Gain

Start with a process having no deadtime and not subject to stochastic or deterministic disturbances. In general, as long as there is no deadtime, the degree of control difficulty is independent of the process gain and the process time constant. A process with a large process gain (meaning that small changes in the manipulated variable cause large ultimate changes in the controlled variable) will require a conservative control strategy; that is, the sizes of the steps in the manipulated variable will have to be relatively small. The converse is true for a process with a small process gain; that is, whatever the control strategy is, it will have to be aggressive, meaning that the sizes of the steps will have to be relatively large. Whether the strategy is aggressive or conservative, the degree of control difficulty is the same in either case.

A process with a small process gain and a restricted range for the manipulated variable to move in will be more difficult to control than one with the same range for the manipulated variable but with a larger process gain. In the former case the manipulated variable may spend quite a bit of time limited at its minimum or maximum allowed values. The converse is true for a large gain process and a crudely resolvable manipulated variable (which might be the consequence of a low-resolution D/A converter). Here the smallest allowable change in the manipulated variable will still cause an unacceptably large response in the controlled variable because the process gain is too large.

As long as there is no deadtime and no disturbances, the length of the time constant will have no impact on the degree of difficulty of control. A process having a long time constant may require more patience, but even though it will take a long time to see the *ultimate* effect of a change in the manipulated variable, there will be *some* detectable response immediately. It is this immediate response, however little, that will provide the basis for the next control move and hence allow for good control. Furthermore, the sluggishness of a large-time-constant process can be beneficial should one make a few incorrect control moves. In other words, a large-time-constant process has a smoothing, filtering effect on changes in the manipulated variable, as well as on disturbances, a fact that can be taken advantage of in a well-designed control strategy.

Should the controlled variable be subject to stochastic disturbances, auto-correlated or not, then, in the case of the process with a long time constant and/or a small gain, the immediate response to a control move may be indistinguishable from the stochastic disturbances. In this case it will be difficult to evaluate the effect of the last control move, and the degree of control difficulty will be higher. Conversely, if the process has a shorter time constant or a higher gain, then the result of a control move may be more clearly discernible and good control may be easier to obtain.

1.8.2 Deadtime

If a process has deadtime, then one must wait until that deadtime has elapsed before seeing *any* effect of a control move; therefore, deadtime can have a significant impact on the degree of control difficulty. However, it is the ratio of the deadtime to the time constant that really should be used as the measure of difficulty. This can be seen by considering the third example shown in Figure 1.26. Assume that the tank cross-sectional area is large so that the time constant will be large relative to the deadtime resulting from the time required for the water to flow down the trough. Now, if one is making adjustments to the inlet flow rate in order to keep the tank level on target, previous comments suggested that patience should be used first because of the large time constant and, second, because of the deadtime. This patience, coupled with the smoothing effect of the large time constant, will effectively decrease the control difficulty's sensitivity to the deadtime. For example, if the tank's time constant were 40 minutes and the dead time were 3 minutes, a wait of 3 minutes after each control move would be a small price to pay since it would take 40 minutes more before 63% of the final effect of each move would be seen. If an incorrect control move were made, the smoothing effect of the large time constant would so attenuate the effect of the move on the level that a corrective move could easily be made 3 minutes later that would help bring the level back toward the target.

Conversely, should the process time constant be small relative to the deadtime, then there would be little of the forgiving smoothing available. Consequently, the effect of an incorrect control move would not be detected until the deadtime had elapsed. Furthermore, once the deadtime had elapsed, the small time constant would not protect the incorrect control move with smoothing.

1.8.3 Stochastic Disturbances

The *variation* in the controlled variable due to *un*autocorrelated stochastic disturbances cannot be decreased by adjusting the manipulated variable. The reason for this is as follows. Since, by definition, the unautocorrelated disturbance at any instant of time is independent of past disturbances and since, unless the nature of the stochastic disturbances changes, future disturbances will be independent of the present disturbance, it follows that any attempt to adjust the manipulated variable in order to respond to such an observed disturbance will be fruitless. In other words, it does no good to react to a disturbance that has no dependence on the past and no influence on the future.

However, the *mean* of a controlled variable subject to unautocorrelated stochastic disturbances can be controlled by adjusting the manipulated variable. The failure to distinguish between controlling the mean and the variation of a process variable subject to unautocorrelated stochastic disturbances has led many workers in the field of process control astray.

Should the stochastic disturbances be autocorrelated, then the variation of the controlled variable can often be decreased by adjusting the manipulated variable, especially if the autocorrelated stochastic disturbances are nonstationary and if the ratio of the deadtime to the time constant is small. In this case, the goal of any pattern of control moves would be to transform the autocorrelated nonstationary stochastic disturbance into one that is at least stationary, with the mean equal to the target. Also, the degree of autocorrelation should be decreased as much as possible as a result of the control activity. However, because of earlier comments, one must realize that no sequence of control moves would be able to decrease the variation once the stochastic disturbances become unautocorrelated.

Based on the foregoing discussion, one might be tempted to say that *optimal control* has been achieved when the mean of the controlled variable is on target and the variations about the mean are unautocorrelated. Note that this definition of optimal control does not require that the variations be zero, just that they be unautocorrelated. Therefore, within reason, a process variable can be considered to be well controlled even though it appears to be noisy as long as the noise is unautocorrelated and the mean is on target.

Before leaving this definition of optimal control, it should be noted that, in the discrete time domain, whether a noisy signal is white or colored often depends on the length of the sampling interval. A noisy sequence that is autocorrelated when sampled at a small time interval can be made to appear to be white if the sampling interval is lengthened enough. This is a consequence of the simple fact that as the sampling interval increases there is less chance that a disturbance at the moment of sampling will be dependent on disturbances at previous instants of sampling. Therefore, it would be unreasonable to expect a control strategy to transform an autocorrelated disturbance into a white one when the process has a large process time constant unless the sampling and control interval were approximately the same size as the process time constant.

In general, the control engineer is concerned with keeping the mean of the controlled variable on target such that the residuals between the controlled variable and its target are stationary and as nearly unautocorrelated as possible. Once the control engineer has accomplished this, the statistician would be concerned with finding ways to decrease the variation of the controlled variable about its mean. Furthermore, the statistician would also be concerned with searching for an assignable cause should the nature of the residuals change significantly. This interplay is discussed in somewhat more detail in Section 2.7, where feedback control is compared with statistical process control.

1.9 THE FILTER AS A UNITY-GAIN, NO-DEADTIME, FIRST-ORDER PROCESS

On-line filters can be used to decrease the perceived noise riding on a process variable. However, these filters also add lag to the process variable, so if the filtered variable is to be a controlled variable, one must be careful that the increased lag does not significantly degrade the control.

There are many kinds of filters, but one of the simplest and most used consists of a weighted sum of the input value and the previous filtered value. For example, if $c(k)$ represents the value of a process variable at time t_k, and if $c_f(k)$ represents the value of the filtered process variable at time t_k, then this filter looks like

$$c_f(k) = \alpha c(k) + (1 - \alpha)c_f(k - 1) \qquad (1.35)$$

or

$$c_f(k) = \alpha[c(k) - c_f(k - 1)] + c_f(k - 1) \qquad (1.36)$$

where the filtering coefficient is α. To make any sense, the parameter α must lie between zero and unity, with more filtering taking place as the parameter α decreases. The filter described in Equations (1.35) and (1.36) is variously referred to as a discrete first-order filter, an exponentially weighted moving average, an autoregressive filter, and many other names. This variety of names is a consequence of the many areas in science and engineering where this filter occurs.

If the input to the filter, $c(k)$, were to experience a step change, the output of the filter, $c_f(k)$, would respond just like the first process example with a time constant, T_f, given by

$$T_f = -\frac{h}{\ln(1 - \alpha)} \tag{1.37}$$

If the input to the filter stops changing, the filter output will ultimately reach the value of the input; therefore, this filter is just like the first process example except that its gain is unity.

If the reader has read Section 1.6, then he or she will recognize the mathematics associated with this filter as being similar to that of the first-order (FO) model (without deadtime) and will be able to derive Equation (1.37) by inspection. This is especially easy if the defining equations are written side by side.

Filter: $c_f(k) = (1 - \alpha)c_f(k - 1) + \alpha c(k)$

FO model: $C(t_i) = \exp(-h/T)\, C(t_{i-1}) + G[1 - \exp(-h/T)]M(t_{i-1})$

Not only does this comparison show how α can be related to the filtering time constant, but it also suggests that processes behaving like the first-order model act as filters if one considers the manipulated variable as the filter input and the controlled variable as the output. This interpretation of a process as a filter may help explain some of the comments made in Section 1.8.2 about the effect of smoothing on the degree of control difficulty.

To get a quantitative feel for the reduction of noise between the input and the output of this filter, consider the case where the input to the filter, $c(k)$, is zero-mean white noise; that is, each value of the input sequence is independent of any of the previous values of the sequence. If both sides of Equation (1.35) are squared and the average or expected value is taken of the result, an expression can be derived for the ratio of the root mean square noise in the filter output to that of the input. The reader should note the similarity of this approach with what was done in Section 1.3 when stochastic disturbances were studied. First, squaring both sides of Equation (1.35) gives

$$c_f(k)^2 = \alpha^2 c(k)^2 + 2\alpha(1 - \alpha)c(k)c_f(k - 1) + (1 - \alpha)^2 c_f(k - 1)^2 \tag{1.38}$$

Since $c(k)$ has a zero mean, the expected value of the left side of Equation (1.38) gives the variance of the filter output, which will be signified as V_o. The expected value of the first term on the right side gives α^2 times the variance of the filter input, which will be signified as V_i. The expected value of the second term on the right side is zero since $c(k)$ is assumed to be white noise, which is therefore not correlated with the previous value of the filter output. Finally, the average of the last term on the right side gives $(1 - \alpha)^2$ times the variance of the filter output, which is V_o. Therefore, the result of averaging the square of the filter equation gives

$$V_o = \alpha^2 V_i + (1 - \alpha)^2 V_o \qquad (1.39)$$

Solving Equation (1.39) for V_o and taking the square root gives the following expression:

$$\frac{f_o}{f_i} = \sqrt{\frac{\alpha}{2 - \alpha}} \qquad (1.40)$$

where f_o is the root mean square of the filter output, while f_i is the same thing for the filter input; that is, $f_o = \sqrt{V_o}$ and $f_i = \sqrt{V_i}$.

Equation (1.40) shows how the white noise riding on a process variable is reduced as a function of the filter coefficient α. Thus there are two equations that can be used to determine the proper value of the coefficient α and the sampling interval h. In one scenario, the user might specify the desired reduction on the noise, f_o/f_i. Equation (1.40) could then be used to solve for a value of the parameter α. Next, the user would use Equation (1.37) to solve for a sampling interval that would give the desired time constant.

To illustrate the use of this filter, refer to Figure 1.35 where a filter with $\alpha = 0.1$ is applied to a noisy signal that is responding to a step change in the

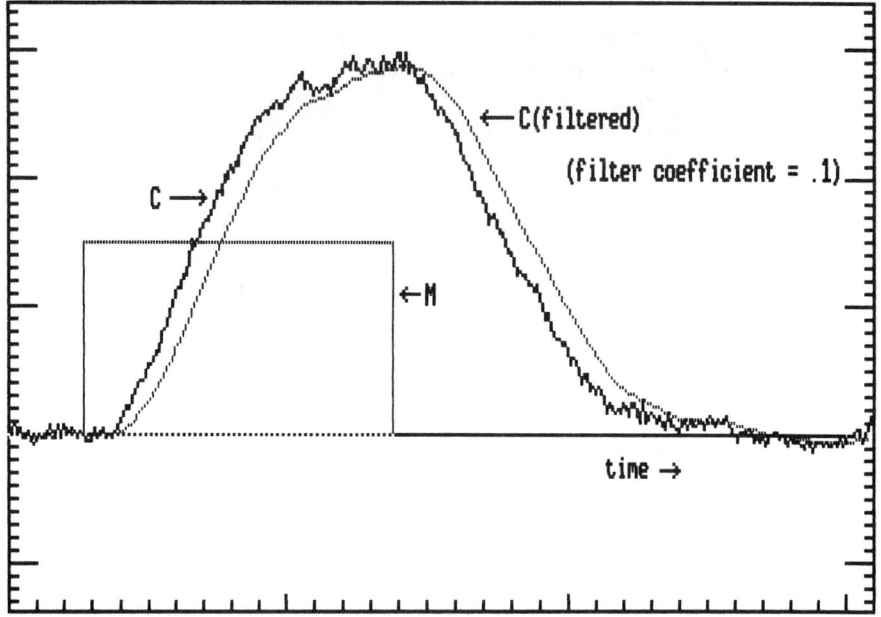

Figure 1.35 Effect of filter.

manipulated variable. Note that the filtered signal is smoother than and lags the unfiltered controlled variable. Using Equation (1.37), the filter time constant is found to be

$$T_f = \frac{-1}{\ln(1 - 0.1)} = 9.49 \text{ s}$$

Equation (1.40) gives the ratio of the rms of the filter output to the input as

$$\frac{f_o}{f_i} = \sqrt{\frac{0.1}{2 - 0.1}} = 0.229$$

Thus this filter has a time constant of approximately 10 seconds and reduces the rms of the white noise content by a factor of 4.36.

1.10 DETERMINATION OF THE DYNAMIC PROCESS PARAMETERS FROM STEP CHANGE RESPONSE DATA

To paraphrase R. W. Hamming, the object of parameter determination is insight into process dynamics rather than numbers. A simple method that is based on reformulating the FOWDT model as a linear least-squares regression equation will be presented first. Although this model does not include autocorrelated noise, it will give the reader insight into the relationship between regression coefficients and FOWDT model parameters. With knowledge of the FOWDT model parameters, the control designer can make effective decisions on control strategy. Later, a more complicated approach that determines noise model parameters in addition to FOWDT parameters will be discussed.

1.10.1 A Simple Method Based on the FOWDT Model

Section 1.6 showed that the FOWDT model can have the following representation:

$$C_i = \alpha C_{i-1} + \beta M_{i-1-d} \tag{1.41}$$

where $\alpha = \exp(-h/T)$ and $\beta = G(1 - \alpha)$. On the other hand, a linear regression equation relating the independent variables X_1 and X_2 to the dependent variable Y can be written as

$$Y_i = A_0 + A_1 X_{1i} + A_2 X_{2i} + e_i \tag{1.42}$$

where A_0, A_1, and A_2 are the linear least-squares regression coefficients to be determined by minimizing the sum of e_i^2 over the data set:

$$S_d = \sum_{i=1+2d}^{N} e_i^2$$

The linear least-squares coefficients can be used to find values of the process time constant T, the process gain G, and the process deadtime D if the following pairings between Equations (1.41) and (1.42) are made:

$$X_{1i} = C_{i-1}$$

$$X_{2i} = M_{i-1-d}$$

$$Y_i = C_i$$

$$A_1 = \exp(-h/T)$$

$$A_2 = G[1 - \exp(-h/T)] = G(1 - A_1)$$

$$h = \text{sampling interval}$$

$$T = \text{process time constant}$$

$$G = \text{process gain}$$

$$D = dh = \text{process deadtime}$$

Note that A_0 is not used in these pairings, but it appears in the linear regression equation to account for the fact that, in general, C_i will not be zero when M_i is zero. Many authors dispense with A_0 in their model equations but, empirically, it is necessary, as can be seen in the following development.

Start with

$$C_i = A_0 + A_1 C_{i-1} + A_2 M_{i-1-d} + e_i, \quad i = d + 2, d + 3, \ldots \quad (1.43)$$

and rewrite Equation (1.43) using the new variables that were introduced in Section 1.6:

$$C_i = C_{ss} + c_i, \quad M_i = M_{ss} + m_i$$

where C_{ss} and M_{ss} are constants that at this point are arbitrary:

$$c_i + C_{ss} = A_0 + A_1(c_{i-1} + C_{ss}) + A_2(m_{i-1-d} + M_{ss}) + e_i \quad (1.44)$$

If it is required that

$$c_i = A_1 c_{i-1} + A_2 m_{i-1-d} + e_i$$

that is, if it is required that there be no offset, then

$$
\begin{aligned}
A_0 &= C_{ss}(1 - A_1) - A_2 M_{ss} \\
&= C_{ss}(1 - A_1) - G(1 - A_1)M_{ss} \quad (1.45) \\
&= (1 - A_1)(C_{ss} - GM_{ss})
\end{aligned}
$$

Therefore, it follows that for the data c_i, m_i, $i = d + 2, d + 3, \ldots$, to fit an equation with no constant term, C_{ss} and M_{ss} must be chosen to satisfy Equation

(1.45). Unfortunately, to use this equation requires the knowledge of the parameters A_0, A_1, and A_2, which at the outset are unknown. However, Equation (1.45) suggests that the problem can be ameliorated by assigning averages of C and M to C_{ss} and M_{ss}, respectively.

Some authors avoid the constant term by differencing the data so that the model equation contains only changes in C and M. In this case the model equation would look like

$$\Delta C_i = A_1 \, \Delta C_{i-1} + A_2 \, \Delta M_{i-1-d} + e_i \qquad (1.46)$$

where the residuals e_i are not the same as those in Equation (1.43). As will be seen in Chapters Three and Four, the act of differencing data can significantly affect its harmonic content such that the parameters obtained from fitting Equation (1.46) may be quite different from those obtained from fitting Equation (1.43).

Only the case where the manipulated variable has been given a series of step changes of alternating signs, where the duration of the step is on the order of the process time constant, will be considered. In another popular method, the manipulated variable takes on the form of pseudorandom binary noise (PRBN). The chemical process industries make little use of sinusoidal excitation. Prett and Morari (1987) compare these three approaches.

Assume that there is a data set consisting of C_j, M_j, $j = 1, 2, \ldots, n$, where M has been given a series of steps of varying size, duration, and sign, and where, as a result of the activity of M, the process variable C also shows significant activity. To determine the parameters occurring in Equation (1.43), one would start by assuming that the deadtime is zero so that $d = 0$ and carry out a linear least-squares regression. For this value of the deadtime index d, the regression would yield values of A_0, A_1, and A_2, as well as a sum of the squares of the error S_d. Next, the deadtime index d would be increased by 1 and the analysis repeated. This cycle would be continued until the deadtime index exceeded the largest expected value. At this point, there would be a sequence of values for S_d, A_0, A_1, and A_2, that is, a set of four values for each deadtime index. To find the deadtime that best fits the data, one would search for that value of d that was associated with the smallest S_d. Associated with this best value of d would be the best values of A_0, A_1, and A_2. The values of the process time constant, process gain, and deadtime could be calculated from

$$T = -\frac{h}{\ln(A_1)} \qquad \text{(process time constant)}$$

$$G = \frac{A_2}{1 - A_1} \qquad \text{(process gain)}$$

$$D = hd \qquad \text{(process deadtime)}$$

If a recursive linear least-squares algorithm were available, this approach could be carried out on line. To determine the deadtime, using this approach on line, several of these analyses would be carried out in parallel, each with a different deadtime, and the one with the smallest sum of squares of errors would determine the best deadtime. For further information on recursive least squares, the reader can consult the texts by Astrom (1970), Franklin and Powell (1980), and Ljung and Soderstrom (1983).

Several caveats are associated with this method whether it is carried out on line or off line. First, since this model does not take into account the effect of autocorrelated stochastic disturbances, one must be careful to choose a portion of the data where the manipulated variable is given sizable steps and where the response variable (the controlled variable) is not experiencing a significant disturbance; that is, the data can be noisy but the stochastic disturbances should be stationary. If this cannot be avoided, then one must make the portion of the data to be analyzed long enough so that the effects of the disturbances can be averaged out. Alternatively, one could search for small portions of the data where it is clear that the disturbances are nearly white noise.

The best results using this method are usually obtained by making several steps in the manipulated variable, each time waiting until most of the response has taken place before making the next change. By alternating the sign of the changes in the manipulated variable, a long series of steps can be applied without driving the process too far away from the nominal conditions. Also, the size of the step can be varied depending on how far the controlled variable is from its target, thereby disturbing the process as little as possible.

1.10.2 Model Verification

To verify the parameter values obtained, one should always plot the extrapolated model along with the experimental data versus time. To generate extrapolated model data, initial conditions would be chosen for C and M based on actual data and then, using only subsequent experimental values for M and previous model output values for C, the model output would be computed and plotted along with the experimental values of C. In other words, once the coefficients have been found, the following equation would be used:

$$C_i' = A_0 + A_1 C_{i-1}' + A_2 M_{i-1-d}, \qquad i = d + 2, d + 3, \ldots \qquad (1.47)$$

where C_i' represents a value of C calculated from the model and where $C_{d+1}' = C_{d+1}$; that is, the model calculations would use only one experimental value of C, namely C_{d+1}, to get the calculations started. After that, only model values of C would be used.

One can be misled by the quality of the model if experimental values are used for the C_{i-1}' term in the right side of Equation (1.47) because now the model is being used to predict only the next value based on observed previous values. Almost any model, for example,

$$C_i' = C_{i-1}$$

would do a visually good job of fitting the data in a graphical sense.

The time-domain plot should also contain the experimental values of M as well as C in order to visually verify that the deadtime was chosen correctly. One advantage of using step change disturbances is that the visual verification of the time constant and deadtime of the model is relatively straightforward.

1.10.3 Examples

To illustrate the method, consider an example where pseudoexperimental data are generated from the following equations:

$$C_i' = 0.9C_{i-1}' + 0.1M_{i-1-5}$$

$$C_i = C_i' + N_i$$

$$N_i = 0.95N_{i-1} + w_i$$

where M_i is given steps every 20 time units and where the white noise w_i, which is driving the autocorrelated disturbance, has a standard deviation of approximately 0.1. Figure 1.36 shows 200 samples of the data. Although the noise contaminating these data is stationary, it is heavily autocorrelated. Applying the method of this section, values of A_1, A_2, and A_3 were generated for deadtimes ranging from 0 to 7. The sum of the squares of the residuals

$$\sum_{i=7}^{200} e_i^2$$

was minimum for a deadtime of 5, and for this deadtime the coefficients were

$$A_0 = 0.0190508, \qquad A_1 = 0.9049481, \qquad A_2 = 0.1002699$$

Thus it appears that values of d, A_1, and A_2 have been found that come reasonably

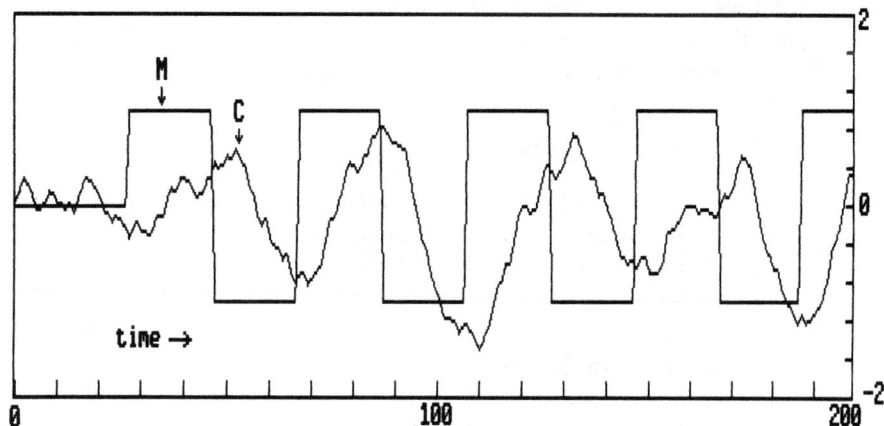

Figure 1.36   Pseudoexperimental data from step changes: $C'(t) = 0.9C'(t-1) + 0.1M(t-1-5)$, $C(t) = C'(t) + N(t)$.

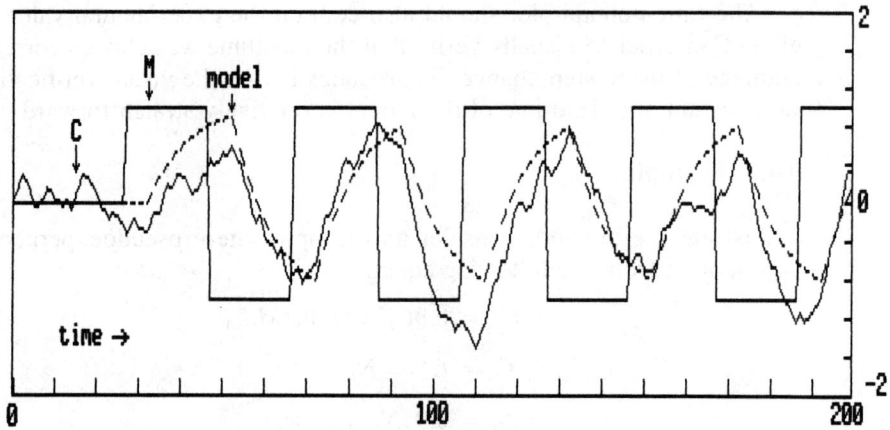

Figure 1.37 Extrapolated model and pseudoexperimental data: $A(1) = 0.01905$, $A(2) = 0.9049$, $A(3) = 0.1003$.

close to the true values, even though the model does not include the autocorrelated noise.

To verify the model, a known initial value was given to C and then the actual values of M along with previous model output values of C were fed to the model to generate values of C for the whole experimental run. Figure 1.37 shows how this extrapolation of the model fits the data.

This experiment was repeated with the same model except that the noise was generated from

$$N_i = N_{i-1} + w_i$$

where the standard deviation of the white noise that drives this random-walk model was approximately 0.13. Unlike the previous example, the data here are nonstationary. The same linear least-squares procedure as before was applied to 200 data points, and a model having a deadtime of 5 time units and coefficient values of

$$A_0 = -0.00487, \qquad A_1 = 0.9592838, \qquad A_2 = 0.105368$$

was determined. As expected, the estimates of the coefficients are significantly different from the true values, although the deadtime is correctly determined. Figure 1.38 shows the experimental data used for the regression, and Figure 1.39 shows how the extrapolated model fitted the data.

1.10.4 A Model Including Noise Parameters

More sophisticated methods (see the texts by Astrom, 1970, Box and Jenkins, 1970, Franklin and Powell, 1980, Ljung and Soderstrom, 1983, and Ljung, 1987) that attempt to determine parameters in a disturbance model as well as a process

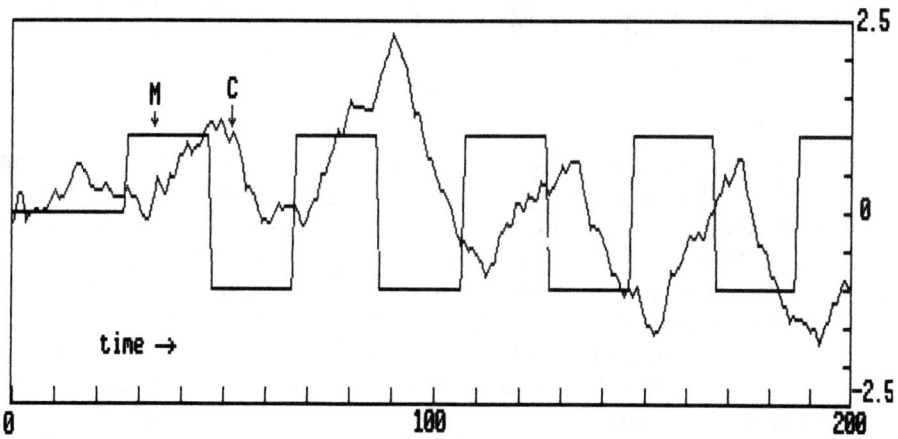

Figure 1.38 Pseudoexperimental data from step changes (second example).

model can be used. However, the computational effort required is much greater and, although the fit is statistically better, sometimes it is not clear that more insight has been gained. Since these methods usually require initial guesses for the parameters, the method of this section could be used to generate them.

As an example of a more sophisticated method, consider the following approach. Assume that the deadtime index d is known from some other method, say the one from this section. Values of the parameters A_0, A_1, A_2, and α in

$$C_i' = A_0 + A_1 C_{i-1}' + A_2 M_{i-1-d}$$

$$C_i = C_i' + N_i$$

$$N_i = \alpha N_{i-1} + w_i$$

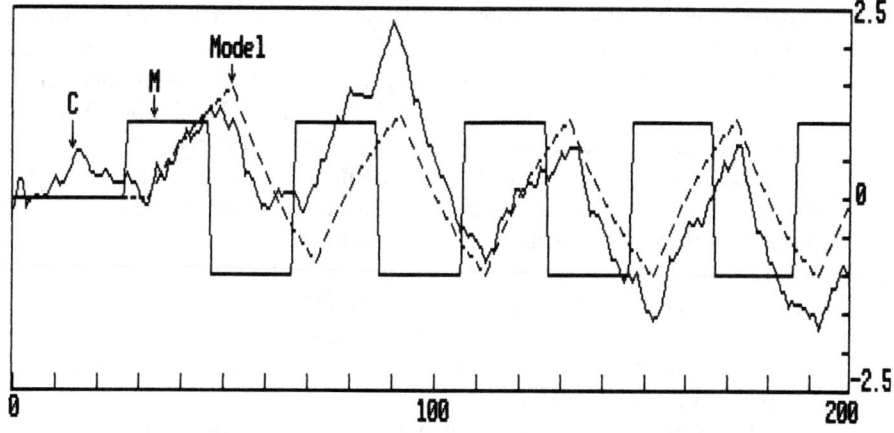

Figure 1.39 Extrapolated model and pseudoexperimental data (second example).

will be searched for that would make

$$S = \sum_{i=d+2}^{N} w_i^2$$

minimum. Note that C_i' and C_{i-1}' are model values, while C_i represents the experimental values that are contaminated by the autocorrelated noise, N_i. On the surface the only difference between this approach and the method presented earlier in this section is the presence of the one additional parameter, α. Unfortunately, the inclusion of this one extra parameter makes the problem an order of magnitude more difficult, in that the methods of linear least squares no longer are applicable. There are many different methods by which the search for the minimizing parameter values could be carried out, and the reader is directed to the references for details.

If the simple method of Section 1.10.1 is not sufficient, there are several reasonably priced commercial packages, for example, MATLAB™, available for both mainframes and personal computers that are capable of carrying out parameter determination using methods far more sophisticated than those mentioned here. One should be aware that experienced process analysts (for example, Prett and Morari, 1987) agree that the choice of the identification technique is not nearly as important as the quality of the experimental data. High-quality data will make almost any identification technique look good. Conversely, there are no identification techniques that will be able to extract good estimates of model parameters if the experimental data are heavily corrupted by nonstationary disturbances.

1.11 SUMMARY

At this point the introduction to the fundamentals of process dynamics has been completed. Using three simple physical models, the concepts of process gain, process time constant, process deadtime, and stochastic disturbances have been introduced. The basic process model has been the first-order-with-deadtime, or FOWDT, model which describes many industrial processes, especially when those processes are subjected to stochastic disturbances. Deviations from the FOWDT concepts have been discussed: processes with inflection points in their step change response, nonlinear processes, underdamped processes, and inverse responding processes. It has been suggested that, for most purposes, the FOWDT model does a remarkable job at describing the dynamics of industrial processes. In those cases where it fails, it can serve as a benchmark, allowing the analyst to describe the process in terms of deviations from FOWDT behavior.

The basic model for stochastic disturbances has been the first-order autoregressive sequence driven by white noise. To characterize disturbances, the autocorrelation was introduced.

Even though a specific control algorithm has not been discussed, it has been suggested how knowledge of the above concepts can allow the reader to anticipate the difficulty he or she will have in trying to control the process. Some attention was also given filters and methods of estimating values of the three parameters that make up the FOWDT model, and the scene is now set for the discussion of control algorithms, which will be the subject of the next chapter.

1.12 DISCUSSION OF THE REFERENCES

The following text was mentioned in reference to regressive least squares, but it has many other attractive features, among them a totally different approach to feedback control than taken in our text.

ASTROM, KARL J. *Introduction to Stochastic Control Theory,* Academic Press, New York, 1970.

A classic text that has become the standard reference for autoregressive–moving average (ARMA) processes is

BOX, G. E. P., and JENKINS, G. M., *Time Series Analysis, Forecasting and Control,* Holden-Day, San Francisco, 1970.

For control engineers, the theory of nonlinear phenomena has long been a curious but usually neglected field of study. Engineers have gotten into the habit of linearizing their process models because they assume the control system will keep the process near the point of linearization. This allows them to separate the stochastic phenomena from the deterministic process response. The science of chaos suggests that there are some situations where apparently stochastic behavior is the result of nonlinear deterministic behavior. An interesting qualitative introduction to the field of chaos is

GLEICH, J., *Chaos, Making a New Science,* Penguin Books, New York, 1987.

As with the book by Astrom, the following book was cited for regressive least squares. Also, like the Astrom book, it contains a readable introduction to modern control theory and discusses methods that in many ways are significantly different from the approaches taken in this book.

FRANKLIN, G. F., and POWELL, J. DAVID, *Digital Control of Dynamic Systems,* Addison-Wesley, Reading, Mass., 1980.

A comprehensive reference for on-line methods of identifying the dynamic parameters of a process is

LJUNG, L., and SODERSTROM, T., *Theory and Practice of Recursive Identification,* MIT Press, Cambridge, Mass., 1983.

A book dealing with off-line system identification methods, and which is matched with the MATLAB™ software package, is

LJUNG, L., *System Identification—Theory for the User,* Prentice-Hall, Englewood Cliffs, N.J., 1987.

The following reference is an interesting and useful compendium of papers by researchers from academia and by practicing control engineers from industry. It is cited in Section 1.10 for one of its papers, "Process Identification—Past, Present, Future," by D. M. Prett, T. A. Skrovanek, and J. F. Pollard.

PRETT, DAVID M., and MORARI, MANFRED, *The Shell Process Control Workshop,* Butterworths, Boston, 1987.

Although rather dated compared to the other books referenced here, the following text is one of the clearest written. It also is the basis for our choice of the symbols C and M to represent the controlled variable and the manipulated variable.

ECKMAN, DONALD P., *Automatic Process Control,* John Wiley & Sons, New York, 1962.

In Section 1.3.9, a result from Section 3.7 of the following text regarding rectangular white noise generated by the quantization of A/D converters was quoted.

OPPENHEIM, A. V., AND SCHAFER, R. W., *Discrete-time Signal Processing,* Prentice-Hall, Englewood Cliffs, N.J., 1989.

An advanced text on process control using mathematics far beyond the scope of this text, but which contains some interesting comments on disturbances and measurement errors in the first chapter, is

MACIEJOWSKI, J. M., *Multivariable Feedback Design,* Addison-Wesley, Reading, Mass., 1989.

CHAPTER 2

CONTROL ALGORITHMS

Although the digital computer is widely used and accepted in the field of process control, it does encourage users to occasionally stray from simplicity because of its awesome speed, power, and most importantly its flexibility. Engineers will sometimes bypass a simple control algorithm and attempt a control strategy that is more sophisticated or even original in the belief that superior performance will be obtained. This usually results from a lack of understanding of simple control algorithms and the temptation that the digital computer presents to an engineer anxious to flex his or her mathematical muscles and create something "just a little bit better."

In the long run, the simple control algorithm, when applied to carefully chosen controlled variables and manipulated variables that in turn result from correctly installed sensors and actuators, will provide the best process performance. This is especially true when the in-plant technology receiver is considered. This person is usually a "fire fighter" rather than a control theory expert and simply does not have the time to baby a sophisticated control strategy. He or she wants a "robust" control system that effectively "runs itself" once it is commissioned. In any case, a progression from a simple but robust control algorithm to something more complicated should only take place if it is clear why the simple approach was unsatisfactory.

In this chapter the emphasis will be mostly on just one control algorithm: the proportional-integral-derivative (PID) control algorithm. It has been widely used in the process control field for decades for a good reason—it works . . .

when properly installed. Its performance can often be improved by the addition of the Smith predictor when the process has a large deadtime to time constant ratio.

The reader will note that although the main interest is in noisy processes the estimates for values of the parameters occurring in the control algorithms come only from the process gain, time constant, and deadtime. A brief mention will be made of the Box–Jenkins control algorithms since they take into account the nature of the stochastic disturbance, as well as the process gain, time constant, and deadtime. These kinds of algorithms have some severe problems and the reader is cautioned to proceed with care.

Attention will be given an important issue arising from the increased participation of statisticians in the control of processes, which appears to be the result of the widespread use of statistical process control or SPC. Although SPC techniques were originally designed to determine changes in distributions and thereby signal an *out of control* condition, they are starting to appear as part of feedback control algorithms. Examples of this approach compared to conventional PI control will be given.

Finally, a simple application of the Kalman filter to the case where there is a process model and where the measurement of the controlled variable is noisy will be discussed.

2.1 A SIMPLE CONTROL PROBLEM

Figure 2.1 shows the single-tank process that was introduced in Section 1.1. A new quantity is introduced: the target or set point for the controlled variable, symbolized by R and represented by the dashed line.

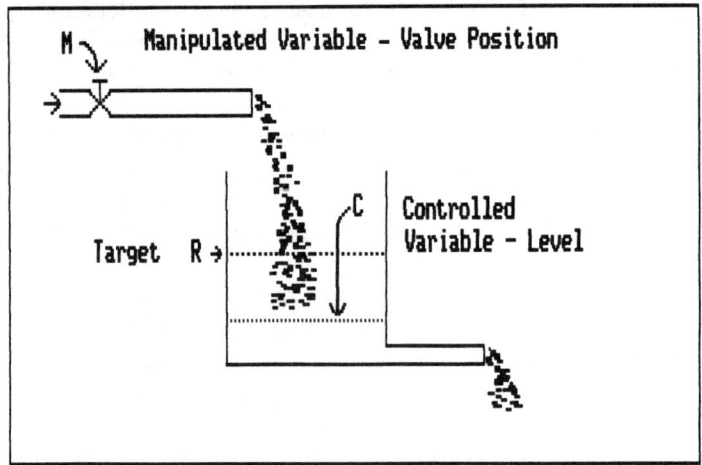

Figure 2.1 Tank with a target.

The problem is to somehow adjust the manipulated variable, the valve position on the inlet pipeline, in order to drive the controlled variable, the tank level, up to the target.

2.2 A SIMPLE CONTROL STRATEGY

A simple control strategy to get the level on target is the following:

1. Measure the level C.
2. Calculate the error E between the target R and the controlled variable C: $E = R - C$.
3. Make a *change* in the manipulated variable, ΔM, proportional to the error: $\Delta M = kE$, where k is the proportionality constant.
4. Wait h seconds (h is the control interval).
5. Repeat step 1.

Figure 2.2 shows how the manipulated variable and the controlled variable behave when the above algorithm is applied in response to a step change in the set point. Note that, as advertised, the changes in the manipulated variable become smaller as the controlled variable approaches the target, that is, as the error gets smaller.

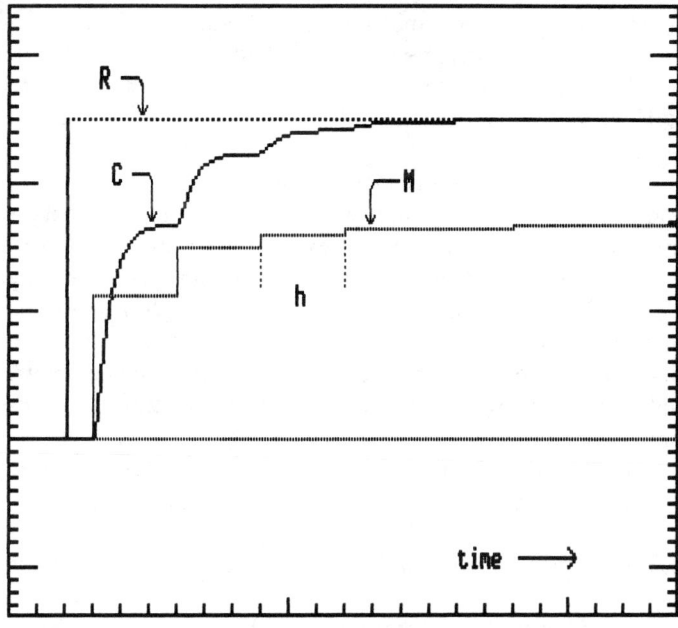

Figure 2.2 Simple control strategy.

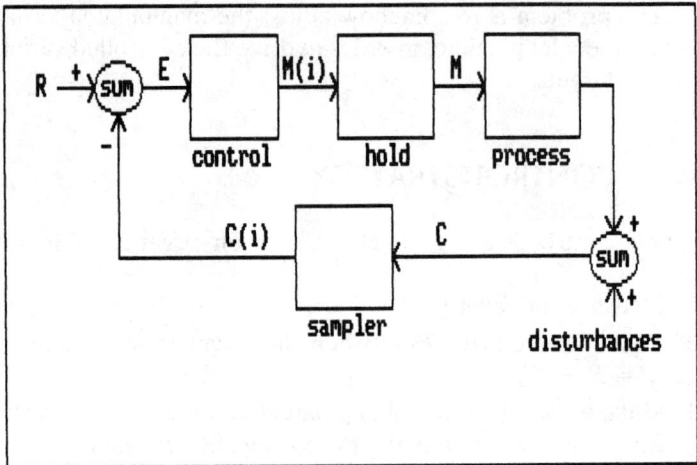

Figure 2.3 Block diagram concept.

2.2.1 Block Diagram Concept for Feedback Control Loops

Figure 2.3 shows a picture of the logical steps that were outlined in the previous paragraphs. Each box in the figure has an input and an output. The box representing the process has the manipulated variable as the input and the controlled variable as the output. The process box is assumed to contain not only the process dynamics but the sensor/measurement dynamics as well.

The sampler box has the controlled variable in continuous form as its input and the sampled value of the controlled variable, denoted as $C(i)$, as its output. Here $C(i)$ is a snapshot of the controlled variable at time t_i, where $t_i = t_{i-1} + h$.

The control algorithm box has the error, available every h seconds because of the sampler box, as its input. As its output, the control algorithm box has the manipulated variable, computed every h seconds when the control algorithm is activated. The output of the control algorithm box is denoted as $M(i)$ since it is not a continuous variable. The hold box has $M(i)$ as its input and a piecewise continuous (staircase) version of the manipulated variable as its output. The value of the manipulated variable, computed each time the control algorithm is activated, is held at this value during the period of time of length h between control actions.

This block diagram concept will prove useful in subsequent discussions of other control strategies and statistical process control.

2.2.2 Review of and Label for the Simple Control Strategy

In Sections 2.1 and 2.2, three new symbols were introduced: R for the set point or target, E for the error between the target and the controlled variable, and h

for the control interval. Then a control strategy was described where, every *h* seconds, a change, ΔM, in the manipulated variable was made proportional to the error *E*. Note that the control interval was longer than the process time constant.

It is suggested to the reader that this control strategy is a commonsense, everyday approach that one uses whether one is adjusting the temperature of the shower water before stepping into the stall or adjusting the steering wheel of an automobile in order to keep it on the road.

This strategy is called the *integral* control algorithm, and the proportionality constant *k* is replaced by the product *Ih*, where *I* is the integral control gain. Now the control algorithm has the form

$$\Delta M = IhE$$

The control gain *I* is a measure of the aggressiveness of the control algorithm. Figure 2.4 shows how the control algorithm drives the tank level to the target when the value of *I* is doubled relative to that in Figure 2.2.

2.2.3 Why Is the Control Algorithm Called Integral?

To answer this question requires some elementary algebra and some calculus. If the reader is not mathematically inclined and is willing to accept the argument that the algorithm should be called integral, then he or she may want to pass

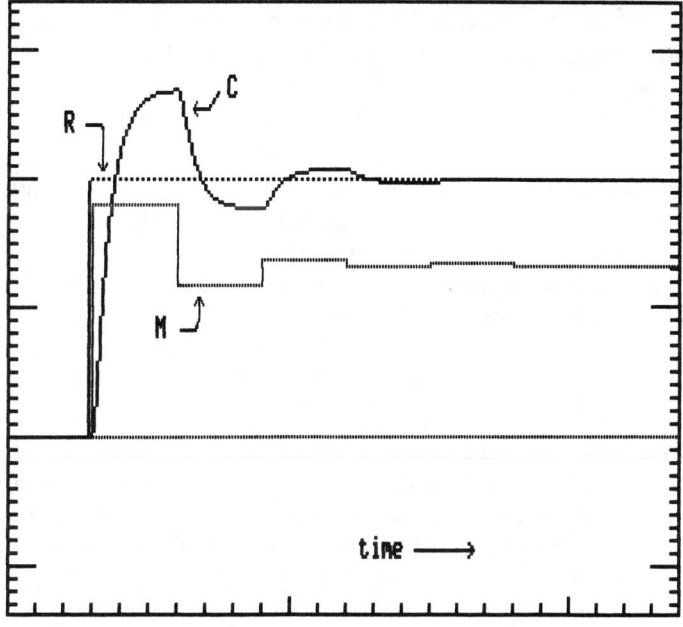

Figure 2.4 Simple strategy with doubled gain.

quickly through the following derivation to the last paragraphs in this section where the results are discussed.

First, the basic control equation is rewritten with time arguments and is augmented by an equation representing the fact that the change ΔM has to be added to the previous value of the manipulated variable to get the new value:

$$\Delta M(t_i) = IhE(t_i) \tag{2.1}$$

$$M(t_i) = M(t_{i-1}) + \Delta M(t_i) \tag{2.2}$$

where $t_i = t_0 + ih$. If $\Delta M(t_i)$ is eliminated between Equations (2.1) and (2.2), the result is

$$M(t_i) = M(t_{i-1}) + IhE(t_i) \tag{2.3}$$

Changing the subscripts in Equation (2.3) from i to $i - 1$ gives

$$M(t_{i-1}) = M(t_{i-2}) + IhE(t_{i-1}) \tag{2.4}$$

Using Equation (2.4) to eliminate $M(t_{i-1})$ from Equation (2.3) gives

$$M(t_i) = M(t_{i-2}) + IhE(t_i) + IhE(t_{i-1})$$

If this elimination procedure is continued n times, the result will be

$$M(t_i) = M(t_{i-n}) + Ih[E(t_i) + E(t_{i-1}) + \cdots + E(t_{i-n-1})] \tag{2.5}$$

which says that the manipulated variable is equal to some initial value, $M(t_{i-n})$, plus Ih times the sum of the errors at all control instants since that initial time to the present. After exchanging the subscripts i and n, Equation (2.5) can be rewritten using the summation symbol, Σ, as follows:

$$M(t_n) = M(t_0) + Ih \sum_{i=1}^{n} E(t_i) \tag{2.6}$$

This form of the control algorithm suggests that an appropriate name might be "summation control" rather than "integral control."

If the control interval h decreases toward an infinitesimal value while the number of control intervals n increases so that the product nh remains constant, the sum in Equation (2.6) becomes an integral:

$$M(t) = M(t_0) + I \int_{t_0}^{t} E(u) \, du \tag{2.7}$$

This form of the control algorithm says that the manipulated variable at time t is equal to some initial value $M(t_0)$ plus a term consisting of the product of the integral control gain and the integral of the error over the period of time from t_0 to t. This is the basis for the name integral. It also should explain why the product Ih was used as the proportionality constant in Equation (2.1).

In this simple control strategy there is a wait of h seconds between control moves to allow the process to respond fully to each move. This type of control strategy could be called a *discrete* control strategy in the sense that discrete control moves were made at discrete instants in time. Therefore, it is ideally suited for implementation on a digital computer. However, when the control interval h is several orders of magnitude less than the process time constant, the control changes effectively take place continuously. Also, when the integral control algorithm is implemented on electronic or pneumatic equipment in analog form, the control moves do indeed take place continuously.

Historically, feedback controllers were first implemented in continuous form using pneumatic and then later electronic means. In the early 1950s, some chemical process industry companies began to use digital computers and discrete versions of conventional feedback control algorithms. By the early 1980s, single-loop controllers were routinely based on microprocessors.

2.2.4 What Can Go Wrong with the Integral Control Algorithm?

There are several reasons why the integral control algorithm worked in the example. First, it was acceptable to wait for the total response of the controlled variable to each of the moves in the manipulated variable; that is, the control interval h was greater than $T + D$. In most industrial situations, such patience is an unallowable luxury.

Second, there were no disturbances, deterministic or stochastic, during the long control interval. It is unrealistic and in fact dangerous to predicate a control algorithm on the assumption that there will be no disturbances over an interval of time that long.

Third, in the example the process time constant was short and the process deadtime was zero, two characteristics allowing the integral control algorithm to be successful.

Now it will be shown what happens when there is no patience with a process (model 2) that has a small but significant deadtime. Figure 2.5 shows how the integral control algorithm, with an aggressive control gain and a short control interval, drives the controlled variable to a new set point. Note how the manipulated variable "integrates" up just after the set point is changed but before the deadtime has elapsed. This is because the error, $R - C$, is constant since C has not changed due to the deadtime.

Also note how the controlled variable passes right through the target and overshoots. Before that happened, the controlled variable was approaching the target in a desirable fashion but, because C is below R, the control algorithm was still asking the manipulated variable to increase and as a result there was overshoot. What is really needed here is some sort of braking action to prevent some of the overshoot. This braking action will be a characteristic of the control algorithm to be discussed in the next section.

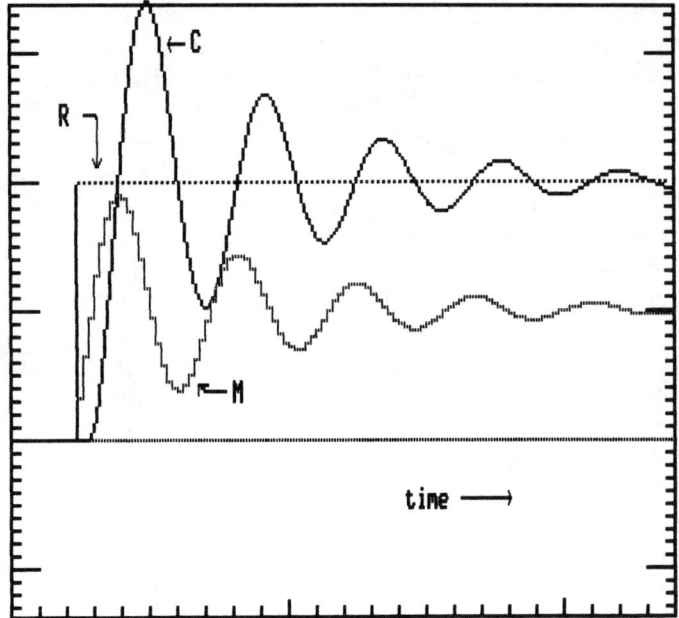

Figure 2.5   Aggressive integral-only control.

2.3 THE PROPORTIONAL–INTEGRAL CONTROL ALGORITHM

To arrive at the proportional–integral control algorithm, one term is added to the integral control algorithm [Equation (2.1)], giving

$$\Delta M = IhE + P\,\Delta E \tag{2.8}$$

Now the *change* in the manipulated variable equals the sum of the integral component, which is proportional to the error, plus the proportional component, which is proportional to the *change* in the error. The reader must be careful when the word proportional is used (it was used three times in the last sentence). The control gain P for the proportional component has the same function as the control gain I for the integral term; it is a measure of the aggressiveness of the control action.

The addition of the proportional term will allow the control interval h to be significantly smaller than the sum of the process time constant and the deadtime. But before going into the reasons why the addition of one term provides so many benefits, this control algorithm will be applied to the example process to which integral-only control was applied in Figure 2.5.

Figure 2.6 shows a couple of significant differences in the way the manipulated variable behaves. At the first control instant after the set point is changed, the manipulated variable makes a large step and then makes a few smaller positive steps before it begins to decrease, well before the controlled variable reaches the set point for the first time. This is to be compared to the behavior of the manipulated variable in Figure 2.5.

The first large step after the set point change is the result of that one time change in the set point, and most of it comes from the proportional component:

$$P \, \Delta E = P \, \Delta(R - C) \qquad (2.9)$$
$$= P \, \Delta R - P \, \Delta C$$

The first term on the right side of Equation (2.9) is nonzero only when the set point R is given the step change. This term provides the large step in the manipulated variable shown in Figure 2.6. The second term on the right side of Equation (2.9) is initially zero since the controlled variable, C, is not changing because of the deadtime.

During the period of time after the set point change but before the deadtime has elapsed, the only contribution to ΔM is from the integral component:

$$IhE = Ih(R - C)$$

Once the deadtime has elapsed and C starts to respond, both the integral term and the proportional term make contributions to the change in the manipulated variable.

To illustrate the proportional component, consider Figure 2.7, where the

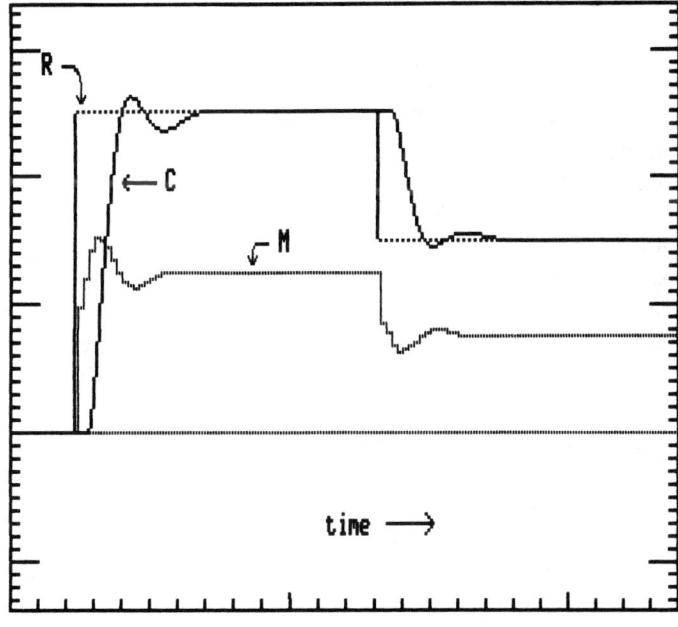

Figure 2.6   Proportional–integral control.

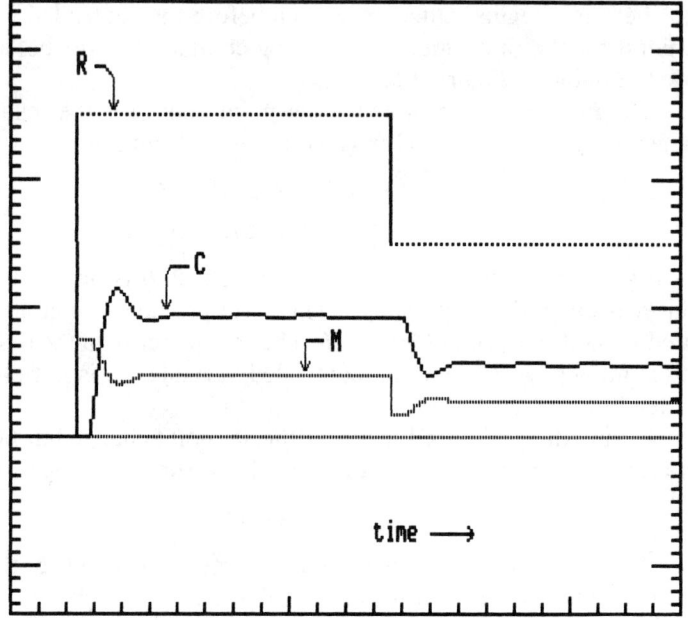

Figure 2.7 Proportional-only control.

same process used in Figures 2.5 and 2.6 responds to proportional-only control, which is

$$\Delta M = P \Delta E$$

The same one time jump at the first control instant after the set point is changed occurs just as with the PI control. Then, since there is no integral component, no change takes place during the period after the set point was changed but before the deadtime has elapsed. When the controlled variable starts to respond after the deadtime has elapsed, the change in the error, $\Delta E = -\Delta C$, is negative, and the manipulated variable decreases, soon settling out at a value that does not cause the controlled variable to be on target. Note that without the integral term present there is nothing to keep moving the manipulated variable until the error is zero. As a result, there is an offset, sometimes called *proportional droop*.

2.3.1 Why Is the New Component Called Proportional?

This section is similar to Section 2.2.3. There will be some use of algebra and calculus, so the mathematically uninclined reader will probably want to pass through the following derivation quickly.

As before, the derivation starts with the basic working equations, rewritten with time arguments:

$$\Delta M(t_i) = IhE(t_i) + P \Delta E(t_i) \tag{2.10}$$

$$M(t_i) = M(t_{i-1}) + \Delta M(t_i) \tag{2.11}$$

$\Delta M(t_i)$ is eliminated between Equations (2.10) and (2.11), and then, carrying out the same kind of somewhat tedious algebra as in Section 2.2.3, the following expression results:

$$M(t_n) = M(t_0) + Ih \sum_{i=1}^{n} E(t_i) + PE(t_n) \tag{2.12}$$

If the control interval h decreases toward zero while the number of control intervals n increases, so that the product nh remains constant, Equation (2.12) becomes

$$M(t) = M(t_0) + I \int_{t_0}^{t} E(u)\, du + PE(t) \tag{2.13}$$

This form of the control algorithm says that the manipulated variable is equal to its initial value at time t_0, plus a component proportional to the integral of the error over the time interval from t_0 to the present time t, plus a component proportional to the current value of the error.

   This last component is called the proportional component because the manipulated variable is proportional to the error. The other component is called the integral component because the manipulated variable is proportional to the integral of the error.

   The *proportional–integral* control algorithm has been around since the 1920s, but it really became a widely used industrial tool in the 1940s. Until the late 1950s the proportional–integral controllers were mostly pneumatic. With the advent of solid-state technology in the 1960s, many controllers became electronic. In the 1980s, most controllers became microprocessor or digitally based. For situations requiring extensive supervisory activity or many control loops, mainframe computers have been used since the late 1960s.

2.3.2 Overall Gain Version

The most popular form of the proportional–integral control algorithm is slightly different than the one given in Equation (2.13). Here an overall gain is pulled out of the proportional and integral terms, giving the following:

$$M(t) = M(t_0) + K_p \left[E(t) + R_p \int_{t_0}^{t} E(u)\, du \right]$$

where K_p is the overall gain and R_p is the repeats per minute. Another slightly different but widely used form is

$$M(t) = M(t_0) + K_p \left[E(t) + \frac{1}{\tau_i} \int_{t_0}^{t} E(u)\, du \right]$$

where τ_i is the integral time.

Historically, the first controllers were pneumatic and proportional-only. When the integral component was added, the overall gain approach lent itself more easily to pneumatic construction.

Instead of trying to explain and interpret these two new control parameters, K_p and R_p, and τ_i, the reader will be referred to other conventional textbooks on control (such as the one by Astrom, 1988). Note in passing that by comparing the last two control equations one can come up with the following equivalences:

$$K_p = P$$

$$R_p = \frac{I}{P} = \frac{1}{\tau_i}$$

Note also that with the overall gain version it is not possible to have integral-only control.

2.3.3 How the Proportional Term Improves Controllability

There are three cases. First, consider the case where the controlled variable C is approaching the set point R. The proportional term counters or brakes the effect of the integral component so that, when the two control gains, P and I, are properly chosen, the controlled variable will not overshoot the set point as it did in Figure 2.5. Referring to Figure 2.8, the controlled variable approaches the set point from below. Start with the basic proportional–integral control algorithm:

$$\Delta M = IhE + P\,\Delta C$$

Because C is below R, the integral component of ΔM will be positive:

$$IhE = Ih(R - C) > 0$$

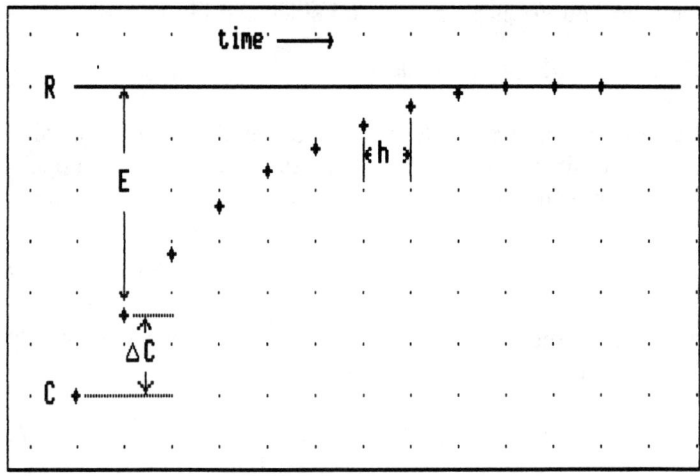

Figure 2.8   Case 1.

Therefore, if the control gains are positive, the integral term will contribute a positive component to the manipulated variable. Referring to the control problem introduced in Section 2.1, these changes in the manipulated variable would be equivalent to increasing the inlet flow rate.

Because C is approaching R from below, the proportional component will be negative, as can be seen in the following discussion. The proportional component is

$$P \, \Delta E = P \, \Delta(R - C)$$

$$= -P \, \Delta C, \quad \text{since } \Delta R = 0$$

Since the changes in C are positive as it approaches R, the above term will be negative (assuming that the control gains are positive). Therefore, the proportional term will contribute a negative component to the manipulated variable in opposition to the integral term.

When C is far below R, the error will be so large that the integral term will overcome the proportional term and the control algorithm will increase the manipulated variable. As C gets close to R, the error becomes small, and the proportional term will overcome the integral term and cause the change in the manipulated variable to turn around and decrease before C equals R. This turning around or braking can prevent the controlled variable from passing through the set point and overshooting, as was the case in Figure 2.5.

In the second case, the controlled variable is diverging from the set point, as shown in Figure 2.9. Since C is greater than R, the integral component will be negative:

$$IhE = Ih(R - C) < 0$$

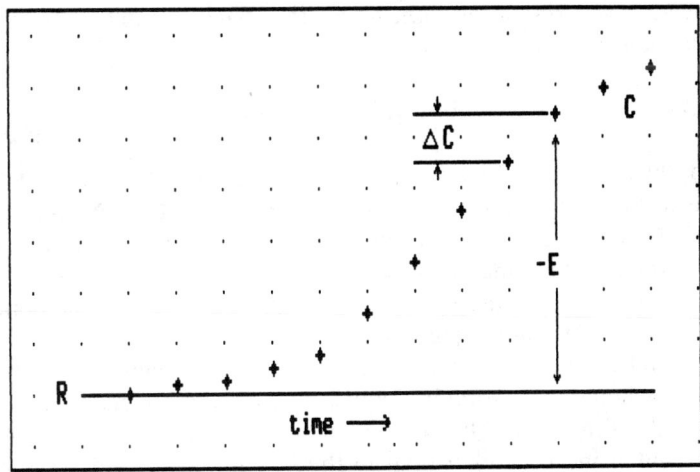

Figure 2.9 Case 2.

However, since the changes in C are still positive, the proportional component will also be negative:

$$P \, \Delta E = P \, \Delta(R - C)$$
$$= -P \, \Delta C, \quad \text{since } \Delta R = 0$$
$$< 0, \quad \text{since } \Delta C > 0$$

With both components being negative, the control algorithm will cause the change in the manipulated variable to decrease; that is, the integral and proportional components will augment each other.

Thus, when the controlled variable approaches the set point, the proportional component will counter the integral component, giving a braking action. When the controlled variable is diverging from the set point, both the proportional and integral components will have the same signs, augmenting each other, in an attempt to turn the controlled variable around. It is this combination of countering and augmenting that makes the proportional–integral control algorithm so powerful. Note that, since the proportional component can brake the controlled variable as it approaches the set point, there is no need to constrain the control interval h to be greater than the sum of the deadtime and the time constant. In fact, the smaller h is, the better; this way the control algorithm can deal with disturbances promptly.

Case 3 covers the situation mentioned in Section 2.3, where the set point is given a step change. There will be an accelerative "kick" in the right direction as a result of the set point change:

$$\Delta M = IhE + P \, \Delta E$$
$$= IhE + P \, \Delta(R - C)$$
$$= IhE + P \, \Delta R - P \, \Delta C$$

The second term on the right side is the "kick" term and occurs only when there is a change in the set point.

2.3.4 Effect of Doubling the Control Gains

To illustrate the effect of changing the control gains, start with model 7, which has a time constant of 40 seconds, a deadtime of 10 seconds, and a process gain of 2.0. The baseline control gains are $P = 0.71$ and $I = 0.0175$. Figure 2.10 shows the effect of doubling the integral gains. Note the higher rate of integration during the deadtime after the set point change, and note that the control variable overshoots the set point significantly.

When the proportional gain is doubled (Figure 2.11), two things happen. First, when the set point is changed, the jump in the manipulated variable is twice as high as in the baseline case. This initial kick causes the controlled variable to rise much more quickly than in the baseline case and a slight overshoot is exhibited. Second, the excessively high proportional gain prevents the controlled

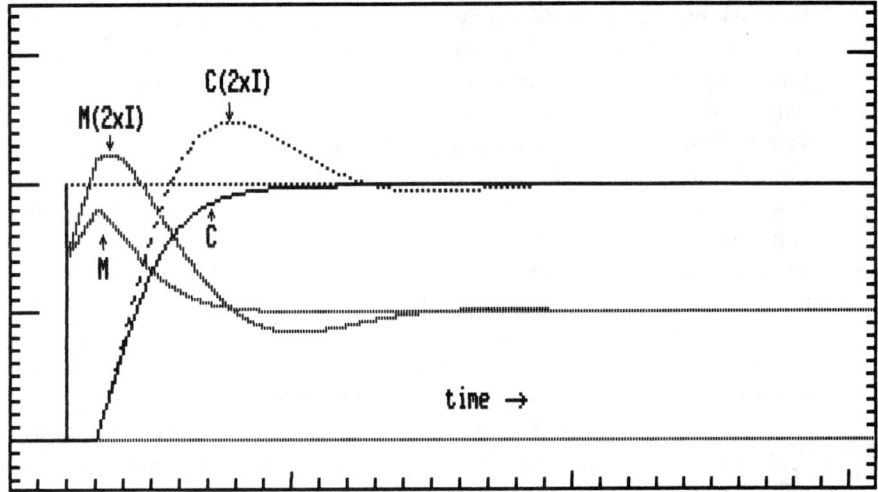

Figure 2.10 Effect of doubling the integral gain.

variable from quickly returning to the set point after experiencing the overshoot and the following undershoot.

2.3.5 Prevention of Integral Windup

Consider the case where the controlled variable stays on one side of the target for long periods of time, say, below the target. The integral component will then, for the case of a process with a positive process gain, increase to the point where

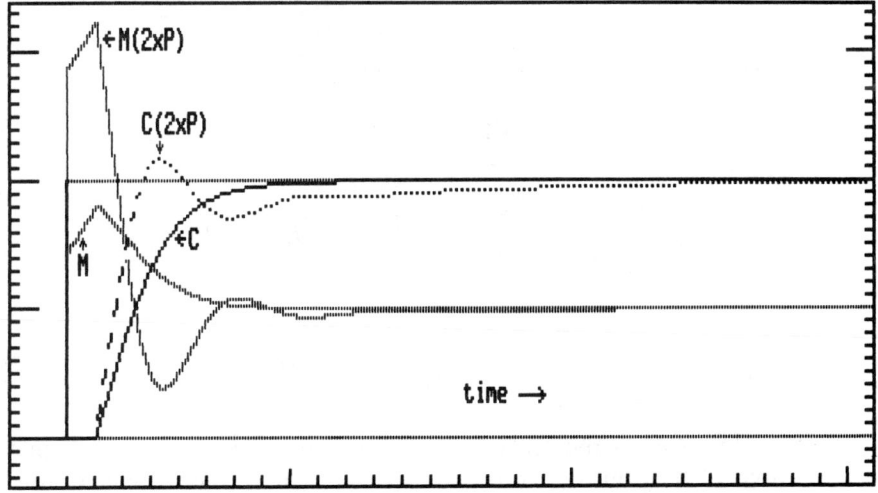

Figure 2.11 Effect of doubling the proportional gain.

it is asking the manipulated variable to take on values that are not physically possible. For example, if the manipulated variable is a valve, the maximum valve position is 100% open. To prevent the control algorithm from asking for unrealistic values of the manipulated variable, it should be designed to check the desired value of the manipulated variable against minimum and maximum limits every time ΔM is calculated. If the desired manipulated variable is out of limits, then ΔM should be recalculated such that the desired value of the manipulated variable is just on the limit. If this logic is followed, then the control algorithm will be able to bring the manipulated variable off the limits quickly when the controlled variable starts to move up toward and perhaps through the target. Otherwise, the control algorithm would have to bring the desired value of the manipulated variable (which would be, in the hypothetical example, above the actual manipulated variable since it is at its allowed maximum value) down into the realizable range first. The time required to get the desired manipulated variable back into the realizable range is lost time in being able to respond to disturbances and therefore will seriously degrade the quality of control.

The following logical steps will prevent integral windup:

1. Calculate the change in the manipulated variable:

$$\Delta M_i = P \, \Delta E_i + IhE_i$$

2. Calculate a tentative manipulated variable:

$$M_i \text{ (tent)} = M_{i-1} + \Delta M_i$$

3. Check the tentative value against minimum and maximum limits and adjust M_i and ΔM_i appropriately:

$$\text{IF } M_i \text{ (tent)} > M_{\max} \text{ THEN}$$

$$M_i = M_{\max}$$

$$\Delta M_i = M_{\max} - M_{i-1}$$

$$\text{ELSE IF } M_i \text{ (tent)} < M_{\min} \text{ THEN}$$

$$M_i = M_{\min}$$

$$\Delta M_i = M_{\min} - M_{i-1}$$

$$\text{ELSE}$$

$$M_i = M_i \text{ (tent)}$$

$$\text{END IF.}$$

Note that ΔM_i is recalculated in those cases where M is on a limit because the output of the control algorithm that is sent to an actuator may not be M_i, but instead may be the increment ΔM_i. In most cases, M_{i-1} will be the last value of the manipulated variable *calculated* by the control algorithm and not the *actual* value. However, in some cases where it is critical that manipulated variable limits not be exceeded, it may be best to read in the present value of the manipulated variable and assign it to M_{i-1} before each calculation of ΔM_i.

2.4 THE DERIVATIVE TERM

Most texts on process control do not stop with the proportional and integral terms but add a derivative term so that the basic control algorithm in the continuous form looks like

$$M(t) = M(t_0) + I \int_{t_0}^{t} E(u)\,du + PE(t) + D_g \frac{dE(t)}{dt} \qquad (2.14)$$

where D_g is the control gain for the derivative term. Many controllers on the market also include the derivative term. Conventional textbooks on process control will demonstrate how the use of derivative control allows a faster, more stable response to both set point changes and load changes. This is because the derivative term counters and augments the integral term in a manner similar to the proportional term. However, because it is based on the rate of change of the controlled variable, rather than on just the value of the controlled variable, there is an anticipatory benefit such that, when there are no high-frequency components in the controlled variable signal, the integral gain can be increased and faster response to set point changes and disturbances can be achieved.

Unfortunately, in realistic situations the controlled variable is subject to stochastic disturbances that have high- as well as low-frequency components. In this case, derivative control almost always creates havoc because it amplifies the high-frequency content of the disturbances and makes the manipulated variable far too active.

For the most part, derivative control is simply not a practical industrial tool, and not much time will be spent on it. However, in Section 2.6.3 it will be shown how using derivative control can be beneficial if filtering is carefully used. Therefore, the following derivation shows how to obtain a discrete form for the derivative term.

Start by noticing that the discrete form of the PID control algorithm could be obtained by evaluating the continuous form, Equation (2.14), at two times, t_i and t_{i-1}, and taking the difference of the two resulting equations:

$$M_i - M_{i-1} = P[E_i - E_{i-1}] + IhE_i + D_g \left[\frac{dE_i}{dt} - \frac{dE_{i-1}}{dt} \right] \qquad (2.15)$$

In Equation (2.15), the notation has been made more compact by using subscripts instead of time arguments, and the last term on the right side contains the difference of the derivatives evaluated at times t_i and t_{i-1}.

Before pursuing the derivatives any further, E should be replaced by $-C$; that is, the target or set point should be removed from the error in the derivative term so that, when steps are made in the set point, pulses in the manipulated variable are not generated:

$$\frac{dE_i}{dt} \rightarrow -\frac{dC_i}{dt}$$

Next, the derivative will be approximated with a finite difference:

$$-\frac{dC_i}{dt} \approx -\frac{C_i - C_{i-1}}{h}$$

After replacing the derivatives with this approximation, Equation (2.15) becomes

$$M_i - M_{i-1} = P[E_i - E_{i-1}] + IhE_i - D_g\left[\frac{C_i - 2C_{i-1} + C_{i-2}}{h}\right] \qquad (2.16)$$

Since the derivative term amplifies high-frequency signals, the control algorithm in this form is relatively useless unless some way can be found to decrease the effect of the high-frequency components in the controlled variable. This can be accomplished by passing the derivative term through the filter described in Section 1.9 and then using the filter output in the control algorithm. For example, let the variable q_i be the difference term:

$$q_i = \frac{C_i - C_{i-1}}{h}$$

and let q_i^f be the filtered difference term such that

$$q_i^f = \alpha q_i + (1 - \alpha)q_{i-1}^f$$

where the filtering coefficient is α.

Now Equation (2.16) becomes

$$M_i - M_{i-1} = P[E_i - E_{i-1}] + IhE_i - D_g(q_i^f - q_{i-1}^f) \qquad (2.17)$$

Note that the price that has been paid for adding the derivative is the added burden of having to estimate two more control parameters, the derivative gain D_g and the filtering coefficient α.

In passing, it should be noted that some control designers prefer to remove the set point from the proportional term as well as the derivative term, yielding a modified PI algorithm that looks like

$$\Delta M = -P\,\Delta C + IhE$$

This control algorithm is identical in performance to the regular PI control algorithm as long as there are no set point changes. If there are, then this latter form will not give the proportional kick. Phelan (1977) has dubbed this the "pseudoderivative feedback'' control algorithm and has written a textbook devoted entirely to it.

2.5 TUNING THE PID CONTROL ALGORITHM

The last two sections should have shown that the basic proportional–integral control algorithm can be an effective industrial tool for feedback control. As a result, the reader should be in a position to at least receive or recommend this technology. To actually use it requires an ability to pick values of the parameters that go into the algorithm.

There are three basic approaches. The simplest and the least effective, especially for novices, is the trial and error method where one tries a little bit of proportional here and a little bit of integral there, and so on, until satisfaction or total confusion is obtained. This is by far the most popular method and, in the hands of an experienced tuner, can be fairly effective.

The best way to converge toward a successful combination of control gains using this first approach is to make a series of step changes in the set point and, after each change, decide whether the proportional or integral gains need to be modified based on a fundamental understanding of how the proportional–integral control algorithm is supposed to work. This method could be made more systematic, but experience shows that each tuner has his or her own idiosyncratic approach, sometimes based on a thorough understanding of how the integral and proportional terms interact with each other and with the process and sometimes based on intelligent intuition.

The second approach is often called the Ziegler–Nichols (1942) frequency method (also see Astrom, 1988, for a more recent discussion). Here one sets the integral gain to zero and increases the proportional gain until the controlled variable experiences sustained oscillations. Sometimes the best way to initiate the oscillations is to change the set point. If the value of the proportional gain at which the sustained oscillations occur is called P_o and the period of the oscillations is called T_o, then the recommended gains are

$$P = \frac{P_o}{2}$$

$$I = \frac{P}{T_o}$$

Since this method was originally developed for continuous pneumatic controllers where the control interval was effectively infinitesimal, it is sometimes not too useful in cases where the control interval is significant relative to the time constant. It also tends to give slightly underdamped response to both set point changes and load changes. A useful compromise between these two methods consists of setting the integral gain to zero, increasing the proportional gain until oscillations occur. The proportional gain is then halved, and the integral gain is carefully increased until satisfactory performance is obtained.

The third method takes a fundamental approach and is in the long term the most effective, but in the short term probably the most expensive. It consists of a well-planned program of *open-loop* (with the control algorithm deactivated) step changes in the manipulated variable. From the resulting data, the dynamic process characteristics (the process gain, process deadtime, and process time constant) are estimated, and from these characteristics, initial estimates of the control gains may be calculated from the following equations:

$$P = \frac{[1 - \exp(-h/T_d)]\exp(-D/T_d)}{G(1 - \exp(-h/T))} \tag{2.18}$$

$$I = \frac{[1 - \exp(-h/T_d)]\exp(-D/T_d)}{hG} \tag{2.19}$$

where

h = control interval
D = process deadtime
G = process gain
T = process time constant

and where T_d is the desired time constant associated with the transient response of the controlled variable when a step change is made in the set point. For example, if one desired the controlled variable to approach the set point with a time constant equal to the process time constant, then T_d would be set equal to T.

These tuning rules are conservative, and in practice it is sometimes useful to choose T_d equal to $T/2$ and multiply the resulting gains by 1.5. These rules are derived in Chapter Four after the Z-transform has been introduced.

There are several other tuning rules in the control engineering literature (for example, see the text by Astrom, 1988), and it is not suggested that the ones presented here are superior; only that they have been used extensively and they appear to work satisfactorily. They also show clearly how the dynamic process characteristics have a direct impact on the control gains. If, for example, there is negligible deadtime and if the desired time constant T_d is set equal to the process time constant T, the above expressions simplify to

$$P = \frac{1}{G}$$

$$I = \frac{1}{GT_d}$$

showing that the proportional gain is the inverse of the process gain and that the integral gain is equal to the proportional gain divided by the time constant of the desired response. These are approximate but handy expressions for use in the field under conditions when a hand calculator is not available.

Although the third approach may not be the most efficient way to obtain satisfactory values for P and I, the side benefits of characterizing the process's dynamics are often more valuable. Usually, one gains a serendipitous insight into the process dynamics during the study of the data resulting from the step change response data.

The reader is probably wondering how one goes about systematically tuning the derivative term if it is present; so do many control engineers. Although most of the accepted tuning rules contain formulas for the derivative gain, they have not enjoyed unequivocal success. The following is a simple but not foolproof alternative method:

1. With $D_g = 0$ and the derivative filtering coefficient equal to unity, find the best P and I gains by any method.

2. Increase the derivative gain until the manipulated variable becomes unacceptably active.

3. Decrease the derivative filter coefficient until the manipulated variable activity level becomes acceptable.

4. Retune the P and I gains by any method, probably trial and error. If the addition of the derivative is going to be beneficial, it will be possible to increase the P and I gains and obtain better performance of the controlled variable.

5. Repeat steps 2, 3, and 4 until no improvements are obtained.

2.6 EXAMPLES

In this section, several features of the PI and PID control algorithms will be illustrated by means of example simulations.

2.6.1 Using the Proportional–Integral Control Algorithm

In Section 1.5, attention was given a process (model 4) having a significant deadtime and that was subject to a random-walk disturbance generated from white noise having a gain g_r of 2.0. From Figure 1.28, assuming that hash marks on the vertical axis represent one controlled variable unit (CVU) and one manipulated variable unit (MVU) and that each hash mark on the horizontal axis represents 10 seconds, the process gain can be estimated, simply by visual inspection, to be approximately 2.0 CVU/MVU, the time constant, 40 seconds, and the deadtime, 20 seconds.

To apply the tuning rules of Section 2.5 to this example, a value of T_d must be chosen. This latter quantity is the time constant that describes the response of the controlled variable to a step change in the target. Assuming a control interval of 1.0 second, a value of 20 seconds is chosen for T_d. (Choosing T_d any smaller will result in more conservative control gains because of the nature of the formulas.) The computed values of P and I are multiplied by 1.5 to get

$$P = 0.54 \text{ MVU/CVU}$$

$$I = 0.013 \text{ MVU/(CVU*sec)}$$

Figure 2.12 shows how PI control, with these control gains, is able to drive the controlled variable up to the new value of the target. Note in Figure 2.13, which is a continuation of Figure 2.12, that because of the presence of autocorrelated stochastic disturbances the controlled variable is not kept exactly on target, in contrast to the case having no disturbance shown in Figure 2.6. Also note that the overshoot in Figure 2.12 suggests that the integral gain may be a little too high.

Study of Figure 2.13 indicates that due to the unpredictable nature of autocorrelated disturbances there are times when the performance of the control algorithm looks relatively satisfactory and other times when there are some significant deviations from target. In practice, it is quite a challenge to resist changing the control gains during these short and unpredictable periods of significant deviation. The reader should be cautioned to exercise patience when tuning controlled systems subject to heavily autocorrelated stochastic disturbances.

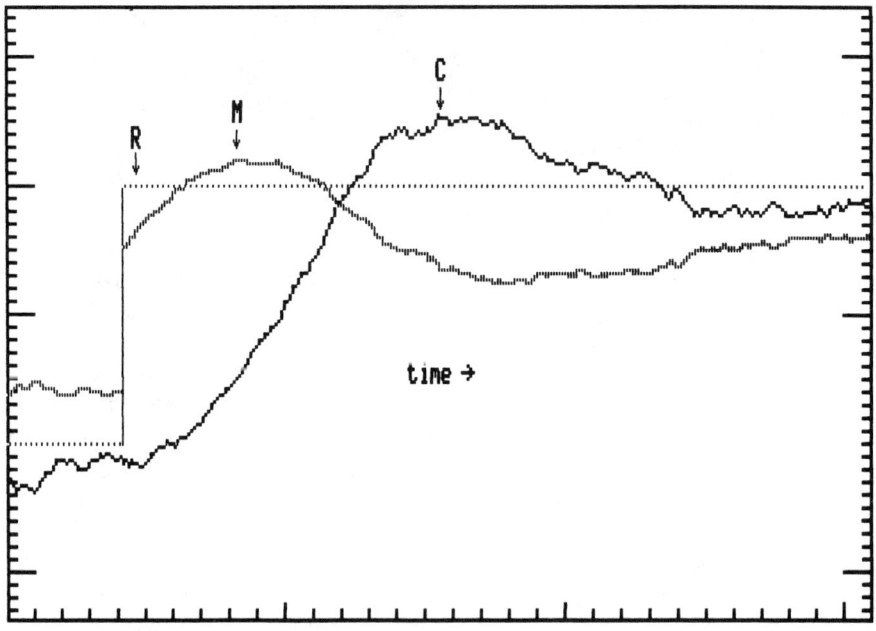

Figure 2.12 Control of a realistic example.

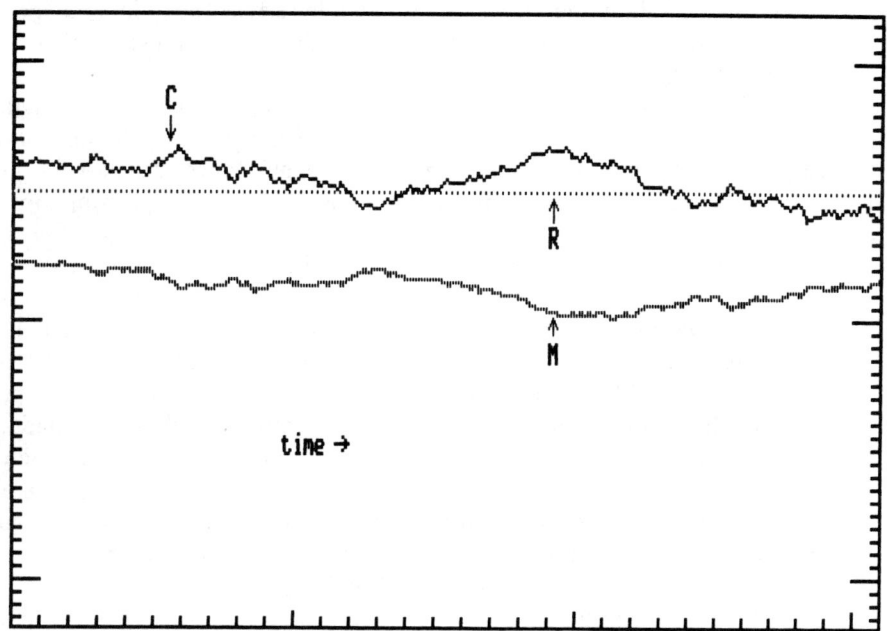

Figure 2.13   Regulation of a realistic example.

2.6.2 Effect of Control Deadband and Resolution Constraints

If a control deadband is used, there is no control action unless the absolute value of the error between the target and the controlled variable, $|E| = |R - C|$, exceeds a quantity, D_b, appropriately called the *deadband*. This is somewhat equivalent to decreasing the resolution on the converter (whether analog-to-digital or serial-to-digital or whatever) that operates on the measurement signal coming into the computer from the controlled variable.

A deadband is sometimes used (not necessarily correctly) when the controlled variable is subject to an appreciable amount of stochastic disturbance and when the control interval is large relative to the process time constant or deadtime. This may be the case where a measurement of the controlled variable is uncertain and only available infrequently, as in the case where the controlled variable is the result of an off-line chemical analysis.

Before demonstrating the effect of a deadband, it should be pointed out that waiting until a certain error is exceeded before making a change in the manipulated variable can only increase the variance of the controlled variable about its target. Without a deadband, the manipulated variable will be more active, and although this may be deemed undesirable under certain circumstances, the absence of a deadband will allow the manipulated variable to react as quickly as possible to a deviation in the controlled variable caused by a deterministic or autocorrelated stochastic disturbance.

Figure 2.14 shows how well the same process appearing in Figure 2.12 is controlled when there is a deadband of 2.0 CVU shown by the dotted lines. The result of the deadband is to increase the root mean square of the error (as estimated from 2048 samples) of the controlled variable about its target from approximately 1.3 to 1.5.

As the reader might have gathered, there really is no good reason to apply a deadband to a controlled variable. However, if there is a significant amount of stochastic variation riding on the controlled variable and if there is a desire to minimize changes in the manipulated variable, the better strategy is to use all the resolution available in the controlled variable measurement and to apply a dead-band to the manipulated variable *after* the control calculation is made. In other words, changes in the manipulated variable would only be made when their ab-solute value exceeded some threshold value. This way, the maximum amount of information is being allowed to proceed around the control loop one more step before stopping it with a deadband. The farther around the loop the information proceeds, the better chance there is of making the best control move.

The foregoing has assumed that the controlled variable could be measured to (and that the manipulated variable could be adjusted to) any resolution desired.

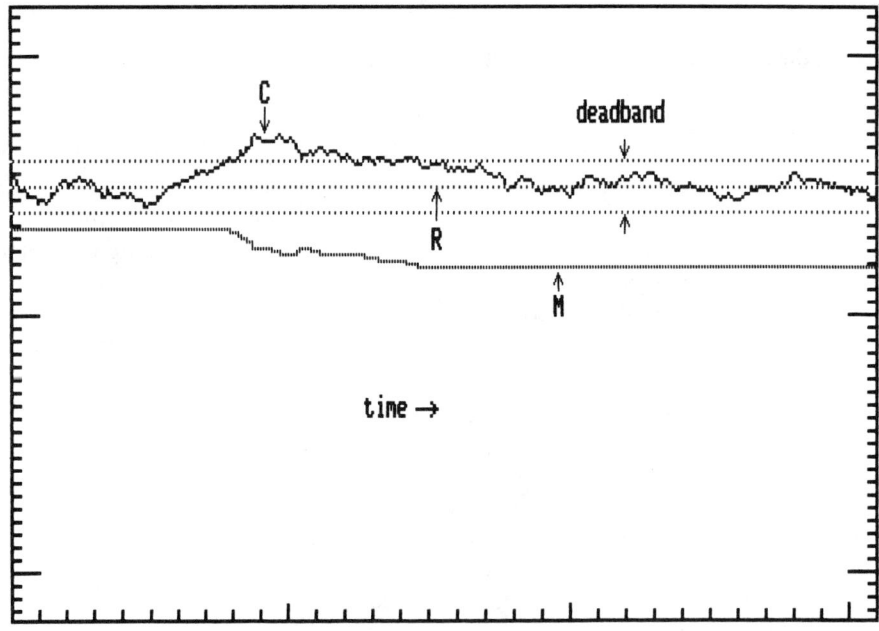

Figure 2.14   Illustration of deadband.

In reality, the resolution is defined by the device that converts the controlled variable's sensor signal into the digital form used by the digitally based control algorithm and by the device that converts the manipulated variable to whatever form used by the associated actuator. Often this resolution is so small as to be insignificant and ignorable; however, sometimes it is not and in these cases *effective deadbands* occur.

For example, take the case where the manipulated variable is associated with a valve that cannot be positioned more finely than 5% of full scale. Therefore, when the control algorithm calls for a change in the valve position of 2.41%, the actual change might be rounded to 0%, resulting in an effective deadband. A similar thing can happen when the resolution of the controlled variable is relatively coarse.

Figure 2.15 shows how the process appearing in Figure 2.12 is controlled when the manipulated variable cannot be resolved any finer than 0.5 MVU. Here the standard deviation of the controlled variable about its target is approximately 1.6.

2.6.3 Effect of Filtering and Adding a Derivative Term

Starting with the example process that was used in Figure 2.12, the stochastic disturbance riding on the controlled variable is removed completely. Second, based on the observations in Section 2.6.1, the integral gain is reduced from 0.013 to 0.009 MVU/CVU∗sec), while the proportional control gain is kept at 0.54 MVU/CVU. The results are shown in Figure 2.16, and they will serve as a basis for comparison with the following PID approach.

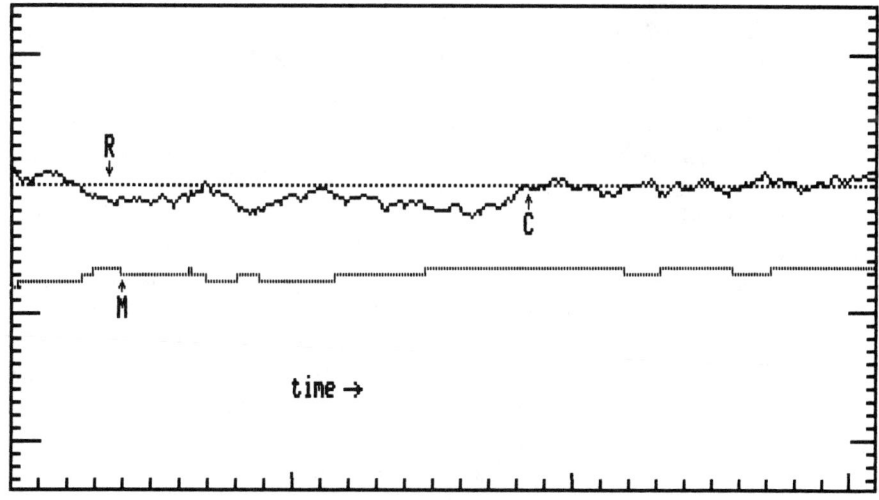

Figure 2.15   Resolution limit on M.

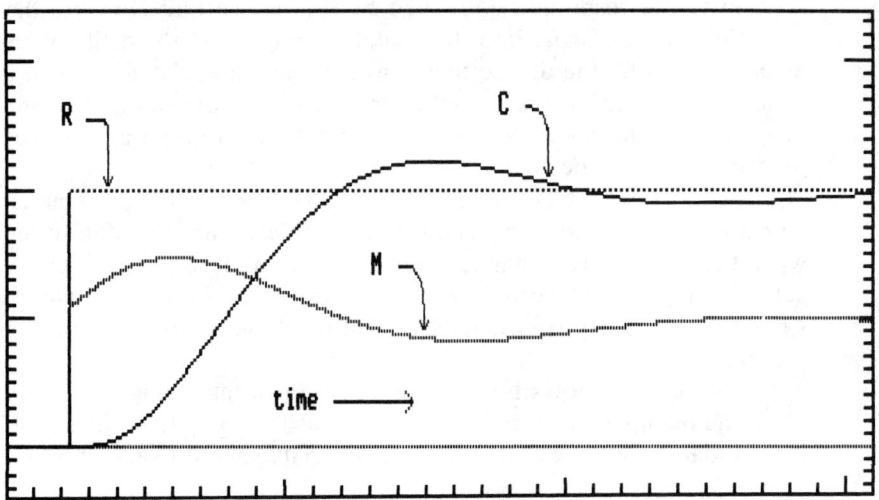

Figure 2.16 PI control of a noiseless process.

Since there is no noise riding on the controlled variable, the derivative filter coefficient is set to unity. Using the strategy outlined in Section 2.5, the derivative gain is iteratively set to nonzero values, and the proportional and integral gains are increased. Somewhat arbitrarily, this process was stopped with a derivative gain of 15 MVU*sec/CVU, a proportional control gain of 1.1 MVU/CVU, and an integral control gain of 0.012 MVU/(CVU*sec). The results are shown in Figure 2.17, where the response of the controlled variable is seen to be better than that in Figure 2.16, with no unacceptable activity in the manipulated variable. This example shows how the addition of the derivative term allows the proportional and integral control gains to be higher and provides quicker response to set point changes.

Without changing the control gains, a small amount of white noise is added to the controlled variable. This noise shows up in Figure 2.18 as a slight amount of hash on the trace of the controlled variable, but manifests itself as a significantly higher level of activity in the trace of the manipulated variable. In an attempt to cut down on the activity in the manipulated variable, one could apply a filter to the noisy controlled variable and then use the filtered controlled variable in every term of the PID control algorithm. The result of this approach (using a filter coefficient of 0.1) is shown in Figure 2.19. Note that, although the activity of the manipulated variable has been decreased, the performance of the controlled variable is clearly worse than that in Figure 2.18. This is because of the increased lag in the feedback path due to the filter. In effect, the PID control algorithm is now looking at a new process: the original process plus a first-order unity-gain element (see Figure 2.20). This new process has a larger effective time constant and a larger effective deadtime; hence it will be more difficult to control.

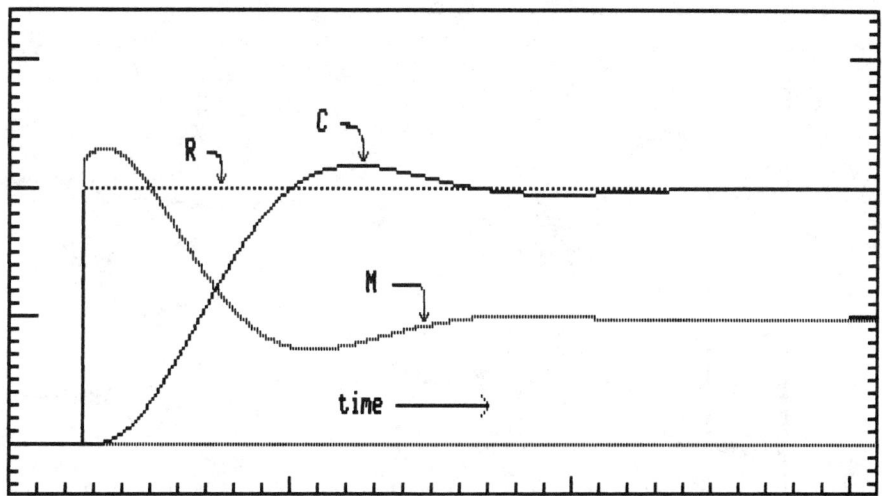

Figure 2.17   PID control of a noiseless process.

If only the derivative term is filtered (using a filter coefficient of 0.1), as advocated in Section 2.4, not only is the manipulated variable activity decreased to what would probably be considered acceptable, but the performance of the controlled variable, as shown in Figure 2.21, is effectively as good as that in Figure

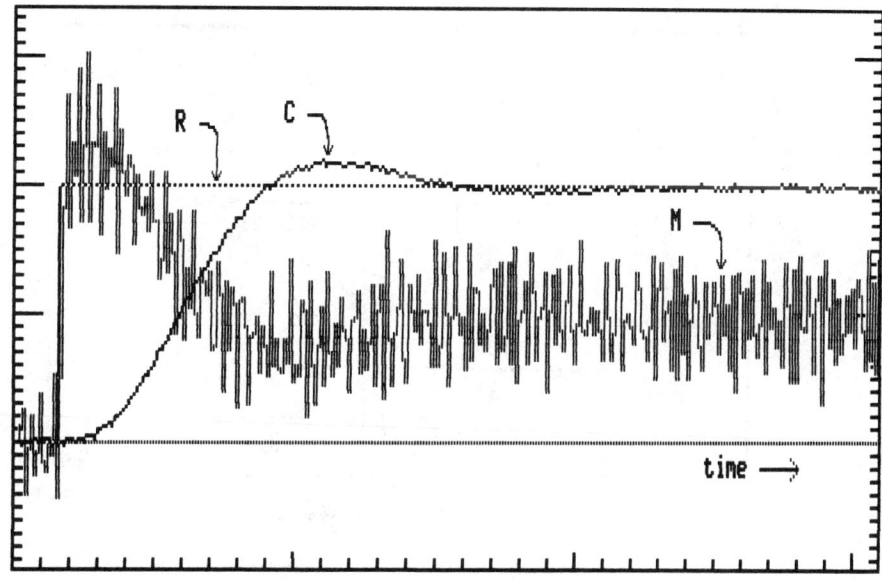

Figure 2.18   PID control of a slightly noisy process.

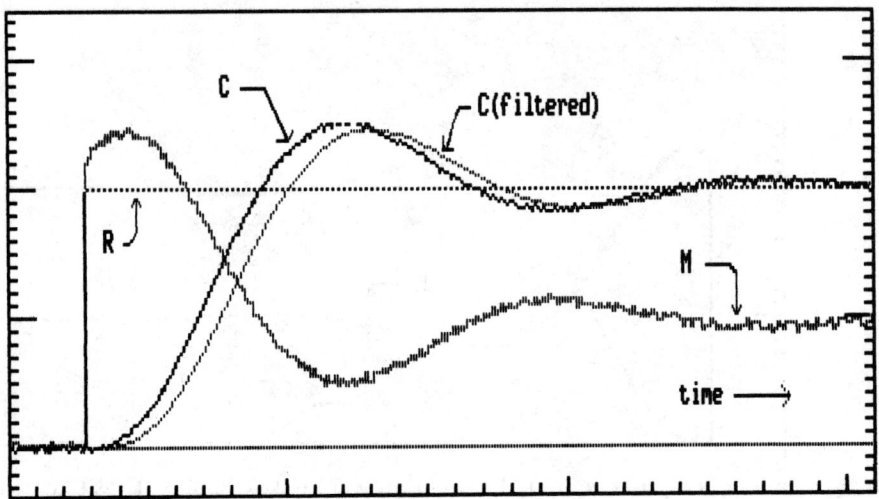

Figure 2.19 PID control of a filtered noisy process.

2.17 or 2.18. Figures 2.22 and 2.23 compare the traces of the controlled variable for PI, PID, and PI with a filtered derivative (PIfD). Although the difference between PI and PID is apparent to the naked eye, the differences between PID and PIfD are more elusive. However, previous figures have shown that the activity of the manipulated variable in the former case is significantly higher.

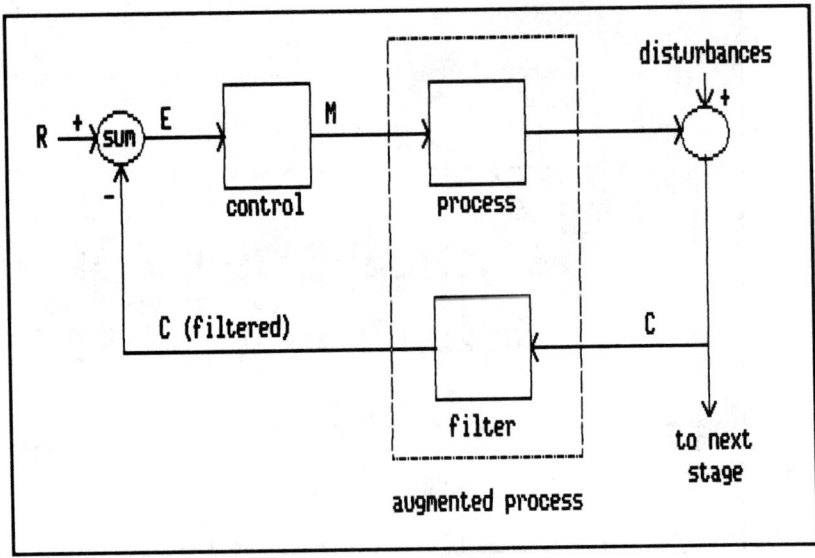

Figure 2.20 Augmented process after adding a filter.

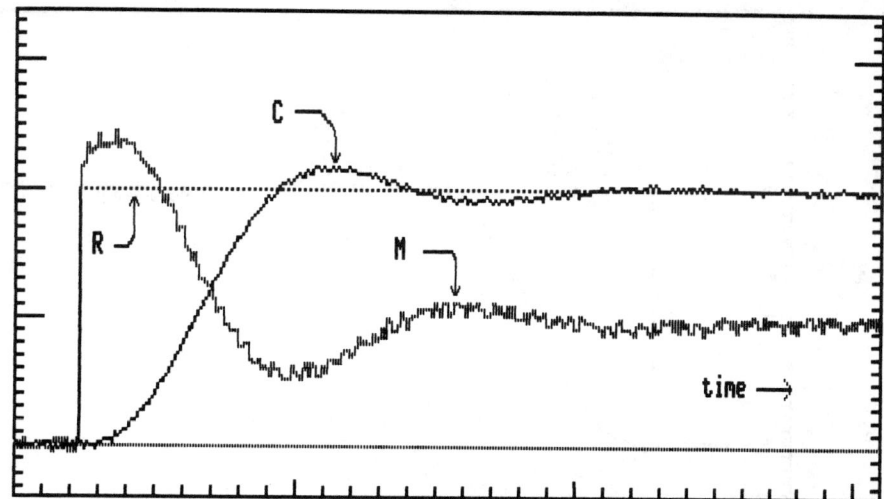

Figure 2.21 PID control with a filtered derivative.

Therefore, one might conclude that at least for this example process, where the stochastic disturbances are white, the extra effort required to estimate two more parameters is worth it. Note, however, that tuning a simulated process such as this, where one can turn things on and off at will, is quite different from tuning

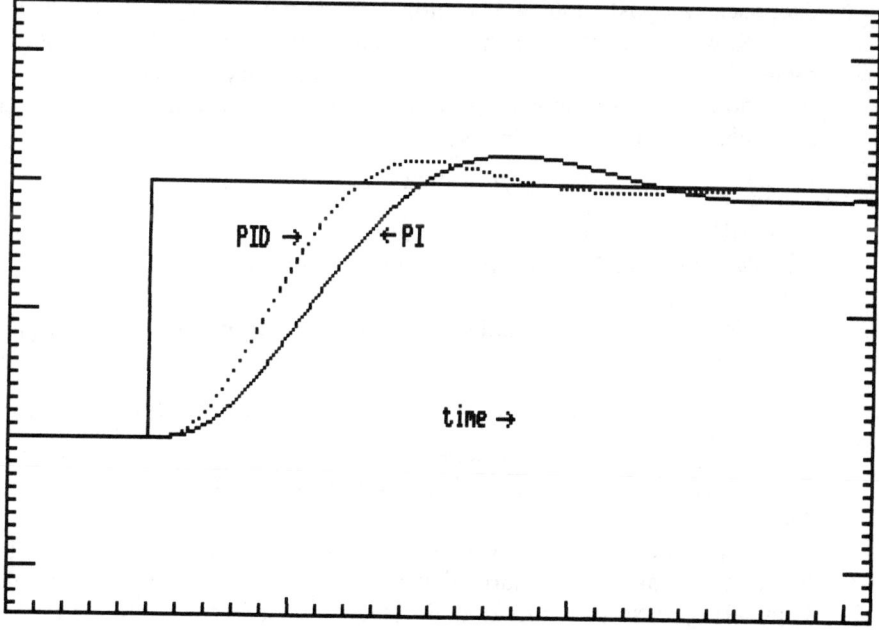

Figure 2.22 Effect of derivative on *C*.

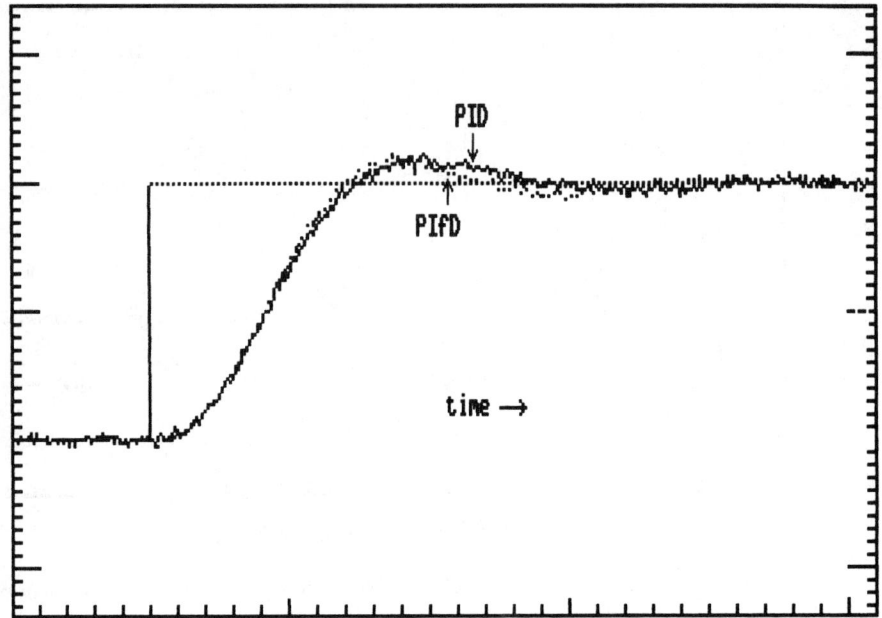

Figure 2.23   Effect of filtering the derivative.

a real industrial process where off-specification product resulting from tuning exercises can cost thousands of dollars per minute.

Now the structure of the stochastic disturbance in the example is modified by making it a random walk, that is, a nonstationary stochastic disturbance. Consider how well the control algorithm regulates the process in the absence of set point changes. Control gains of

$$P = 0.54 \quad \text{and} \quad I = 0.009$$

are used. The time-domain trace of the controlled variable under these conditions is shown in Figure 2.24, where the root mean square of the error for one simulation run of 1024 samples is 1.66.

Next, derivative is added and the proportional and integral gains are increased to

$$P = 1.1, \quad I = 0.012, \quad D_g = 15.0$$

while using a derivative filter coefficient of 0.1. Note that the values for P, I, and D_g are the same as those used to generate the performance depicted in Figure 2.17.

The time-domain trace of the controlled variable is shown in Figure 2.25 where the root mean square of the error for one run of 1024 samples is 1.31. Comparing these two time-domain plots, it is seen that, although the root mean square of the error of the controlled variable may be less when the derivative is

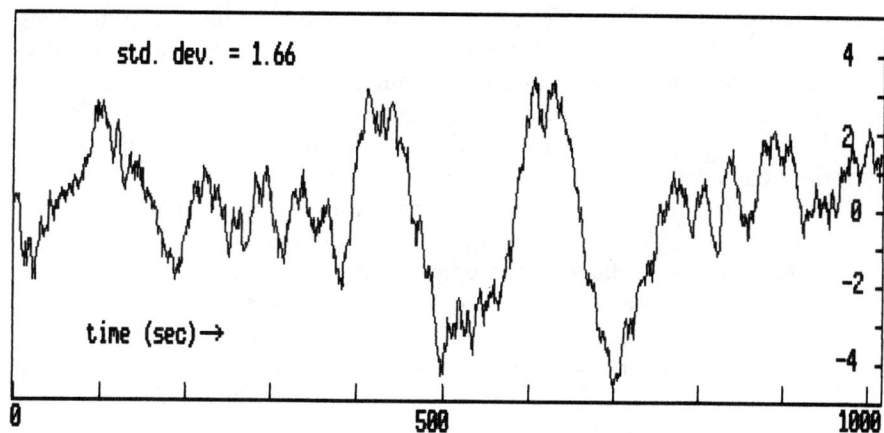

Figure 2.24 Test process subject to random walk using PI control.

used, the difference between the two rather noisy erratic curves may not warrant the additional trouble. In Chapter Five, this problem will be given another look after the powerful tools of Z-transforms and power spectral density have been developed.

2.6.4 Effect of Filtering and Controlling on a Different Interval

In all the examples so far, the sampling and control intervals have been the same. The previous section showed that, in this situation, filtering of the controlled variable can only degrade the controllability. There are some situations where it

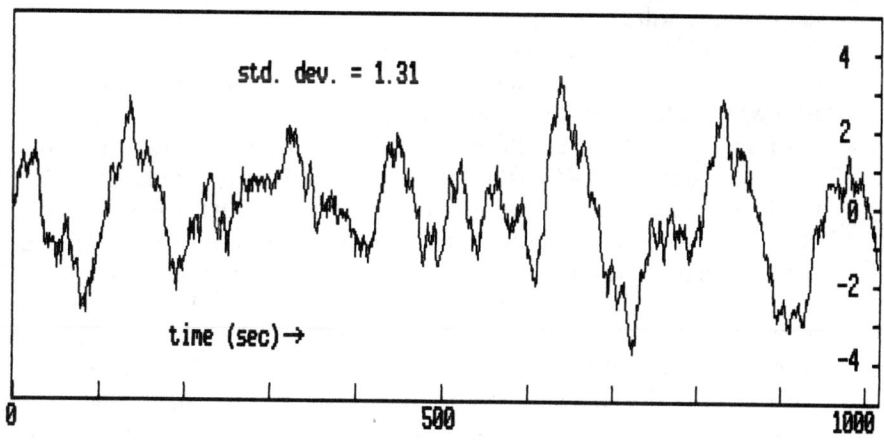

Figure 2.25 Test process subject to random walk using PIfD.

is feasible for the sampling interval to be shorter than the control interval. For example, the dynamics of the actuator may be such that it does not make sense to calculate control moves more frequently than h_c seconds, but the sensor for the controlled variable may allow sampling every h_s seconds, where $h_s = h_c/n$, $n > 1$. In this case, filtering can sometimes be applied effectively as long as the time constant of the filter is not greater than half of the control interval. This is not a hard and fast rule, but common sense suggests that keeping the filtering time constant, less than or equal to the control interval, will likely introduce enough additional lag so as to degrade the controllability.

For example, consider a process with a time constant of 0.5 second, no deadtime, and a gain of 2.0 (model 6), which is subjected to two noise streams in parallel. The first stream X_k consists of autocorrelated noise driven by white noise:

$$X_k = \alpha X_{k-1} + w_k$$

where the standard deviation of the white noise w_k is 0.236 and the autocorrelation coefficient α is 0.99. The theoretical standard deviation of X_k, calculated from the formula derived in section 1.3.2, is

$$0.236 \sqrt{\frac{1}{1 - (0.99)^2}} = 1.67$$

The second stream Y_k consists of unautocorrelated white noise with a standard deviation of 0.707. Therefore, the controlled variable is contaminated by the sum of Y_k and X_k, which has a standard deviation of 1.81 (the root mean square of the standard deviations of the two components). Sampling will take place at 1-second intervals and control will take place at 10-second intervals. Because of the short time constant relative to the control interval, integral-only control with a control gain of 0.05 will be used.

Applying integral-only control to the unfiltered controlled variable yields standard deviations on the order of 1.2. If a moving average of the last five samples of the controlled variable is used in the integral-only control algorithm, the standard deviation of the unfiltered controlled variable is reduced to approximately 1.1. Alternatively, if a first-order filter with a filtering coefficient of 0.2 (resulting in a filtering time constant of approximately 5 seconds) is applied to the controlled variable and the result used in the control algorithm, the standard deviation of the unfiltered controlled variable is also approximately 1.1.

Therefore, it appears that applying a filter slightly improves the controllability. This is only true when the white noise is significant relative to the autocorrelated noise. When the white noise in parallel with the autocorrelated noise has less intensity, the addition of the moving average and/or the first-order filter may worsen control. Conversely, when the white noise in parallel with the autocorrelated noise has more intensity, the addition of filtering or averaging may improve control.

Next, consider the process (model 4) used in the previous section. Here the effective time constant is significantly larger than that of the previous example in this section. The process model is subjected to the same two disturbance streams used above, the control interval is 10 seconds, the sampling interval is 1.0 second, and the control gains are $P = 0.3$ and $I = 0.005$. Under these conditions, the addition of the moving average or the first-order filter does not improve control. In fact, even when the standard deviation of the white noise component is doubled, the use of the filtered or averaged controlled variable in place of the raw controlled variable does not improve the controllability.

This makes sense, because this example process with its long time constant is effectively a low-pass filter and therefore attenuates the extra activity of the manipulated variable caused by the presence of the unautocorrelated white noise. On the contrary, with the first example process, where the time constant is small compared with the control interval, there is much less process filtering of the extra activity of the manipulated variable caused by the presence of the white noise.

Therefore, one must conclude that filtering of the controlled variable can only be effective when the white noise component of the disturbances is quite high relative to the autocorrelated component and when the process itself affords little filtering.

2.6.5 Effect of the Control Interval

To illustrate the effect of changing the control interval, consider the same example process (model 4) that has been used in the last several sections. When subjected to a random-walk disturbance using $P = 0.54$, $I = 0.013$, and $h = 1.0$, the standard deviation of the controlled variable about its target for one simulation run based on 1024 samples is 1.29. If the control interval is increased to 10.0 and the root mean square of the error recomputed, the result is 1.35, an increase of almost 5%. This is to be expected, since a shorter control interval means the control algorithm has a better jump on each disturbance. In general, the shorter the control interval, the better the control. In the limit when h is so small as to cause the control actions to appear to be continuous, the variation of the controlled variable about its target should be the least. Therefore, if the control strategy is PI or PID, an analog controller has the potential of outperforming a discrete controller.

2.6.6 Application of PI to Two Unusual Processes

In Section 1.7.4, two examples of deviations from FOWDT behavior were discussed. In this section, PI will be applied to those two processes in order to give the reader a feel for the degree of difficulty presented. Fortunately, one is not faced with these types of processes often.

The response to set point changes is shown in Figures 2.26 and 2.27. For

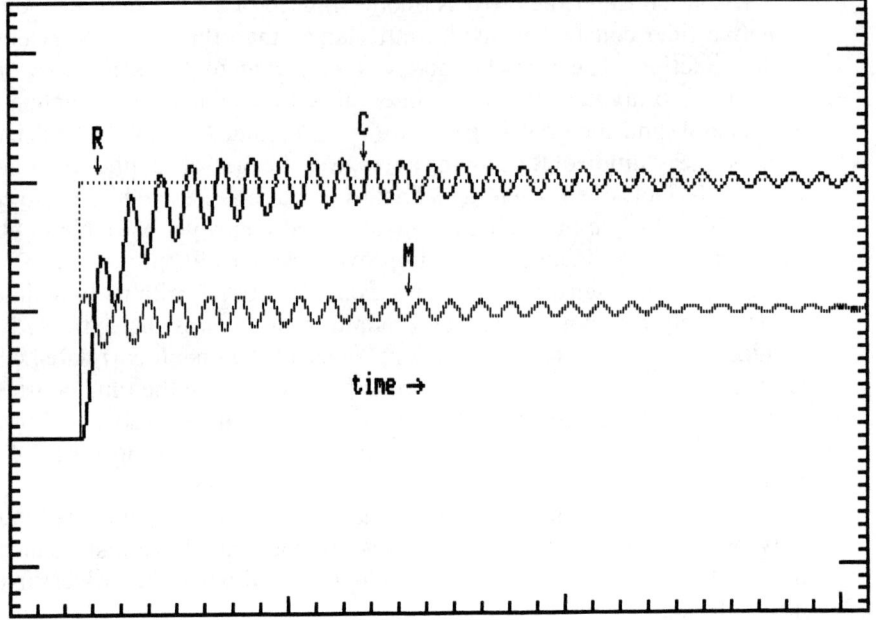

Figure 2.26   PI control of an underdamped process.

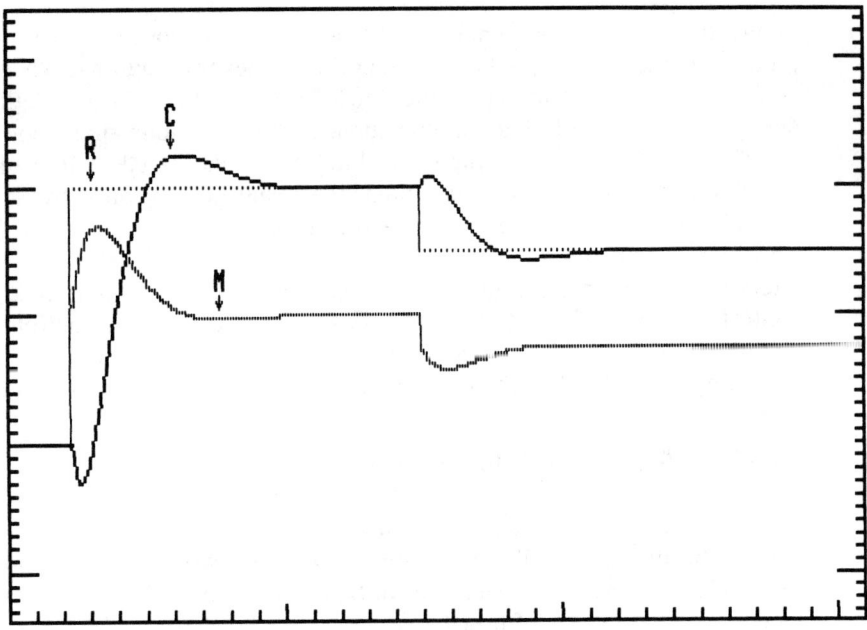

Figure 2.27   PI control of an inverse responding process.

the underdamped process (model 12), which appears to have a process gain of 2.0 and an overall time constant of approximately 30 seconds, control gains of $P = 0.5$ and $I = 0.04$ are used. Note that the only thing one can do about the short-term damped oscillation is to wait for it to die out. Comparing Figure 2.27 with Figure 1.33, it appears that, without control, the short-term oscillations damp out more quickly. In Chapter Five, this process will be studied again in the frequency domain.

For the inverse responding process (model 13), which is considered to be the sum of two first-order processes, the control gains are based approximately on the dynamics of the process with the larger positive gain and the longer time constant; hence the choice of $P = 0.5$ and $I = 0.035$. Note that the inverse response challenges the control algorithm in somewhat the same way as a process with a deadtime.

2.7 PID CONTROL ALGORITHM AUGMENTED BY THE SMITH PREDICTOR

It has been pointed out previously that the bane of control algorithms is deadtime. In an attempt to overcome this problem and still maintain some of the robust simplicity of the PID control algorithm, Smith (1957) developed a predictor structure that is outlined in this section. Figure 2.28 shows the step response of a process (model 5), with a significant deadtime depicted schematically in Figure 2.29. The former figure also shows the response of the fictitious unmeasurable process variable denoted by C', which represents how the process would respond if there were no deadtime. In other words, the example process is considered to be made up of two subprocesses, one that contains a gain G and a time constant T and one that consists of only a deadtime D. The output of the first subprocess, C', has a step change response containing no deadtime, and it is the input to the second subprocess whose output is the controlled variable C.

Figure 2.30 shows how this process would be controlled with conventional PID control, where the controlled variable is fed back as the input to the control algorithm. Because of the deadtime, the control gains would have to be conservative, and the variance of the controlled variable might be unacceptable. Figure 2.31 shows the ideal way to control the process; that is, it would be preferable to feed back the fictitious process variable C' since it would respond to any changes in the manipulated variable with only a time constant effect and no deadtime. Therefore, higher control gains could be used, and a smaller variance in the controlled variable could be expected.

Unfortunately, information on the process variable C' is not available. But, if one knew the process gain and time constant, the response of C' to changes in the manipulated variable could be predicted using the expressions derived in Section 1.6. This predicted value will be called C_p' and appears as the output of a model having a model gain g_m and a model time constant T_m. The input to this

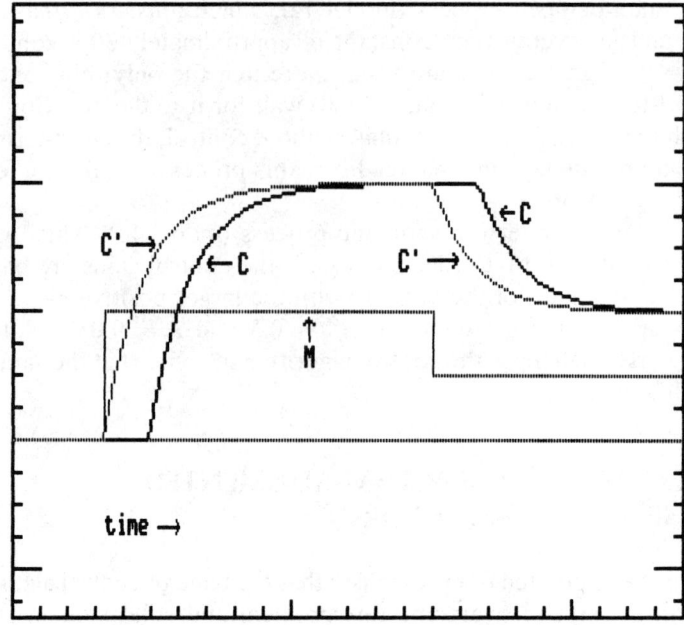

Figure 2.28   Step response showing the fictitious C'.

model is the manipulated variable, which would be a known quantity since it is the output of the control algorithm.

This control concept, shown in Figure 2.32, would work except for the fact that it has no way of responding to process disturbances symbolized by N. To overcome this problem, the output of the model described in the previous paragraph, C'_p, is fed into another model consisting of only a deadtime D_m, which is an estimate of the process deadtime D. The output of the second model, C_p, should match the actual process output, C, except for disturbances. Therefore, if C_p were subtracted from C, that difference would be a measure of how much

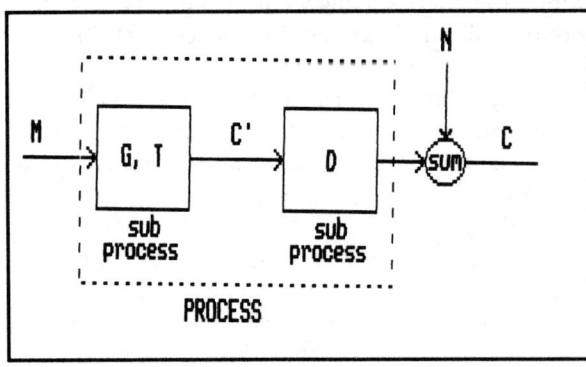

Figure 2.29   Breaking up the process into two subprocesses.

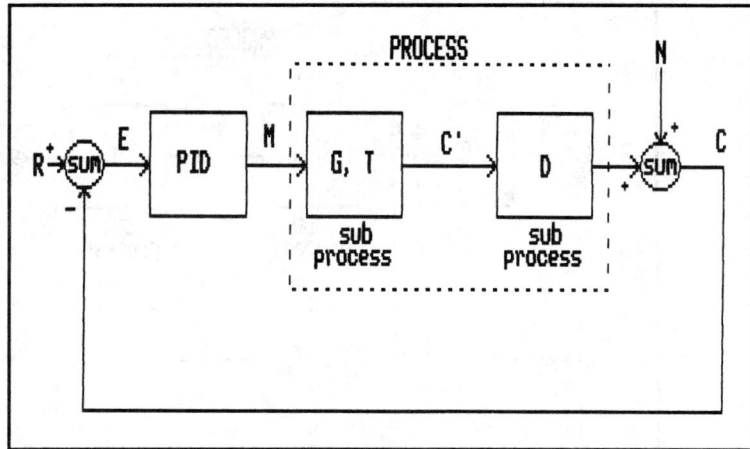

Figure 2.30   Conventional PID control structure.

the combined models misestimate the true controlled variable. If that difference were added to C_p', the result would be an estimate of C' and could be used as the input to the control algorithm. This structure is shown in Figure 2.33, and it represents the famous Smith predictor developed by O. J. M. Smith in 1957. This approach can be considered an example of a *model-reference* control strategy since it predicates the changes in the manipulated variable on not only the controlled variable but also quantities derived from a model of the process.

If the model parameters g_m, T_m, and D_m closely match the process, then high control gains can be used, and the root mean square of the control error can be significantly decreased relative to that resulting when a regular PID control

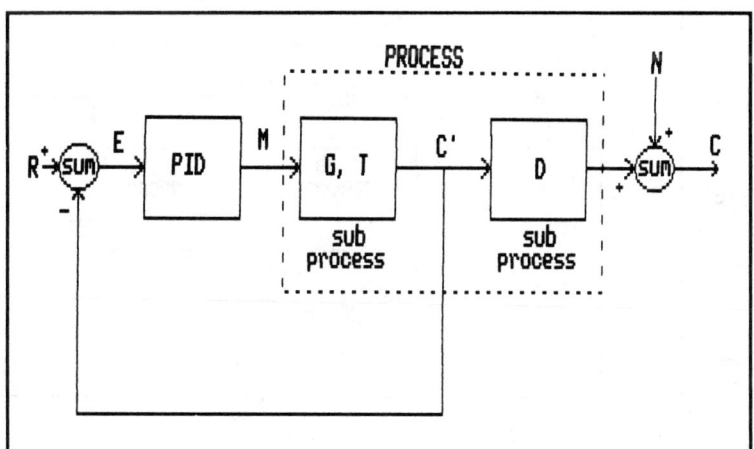

Figure 2.31   PID control using the fictitious C' variable.

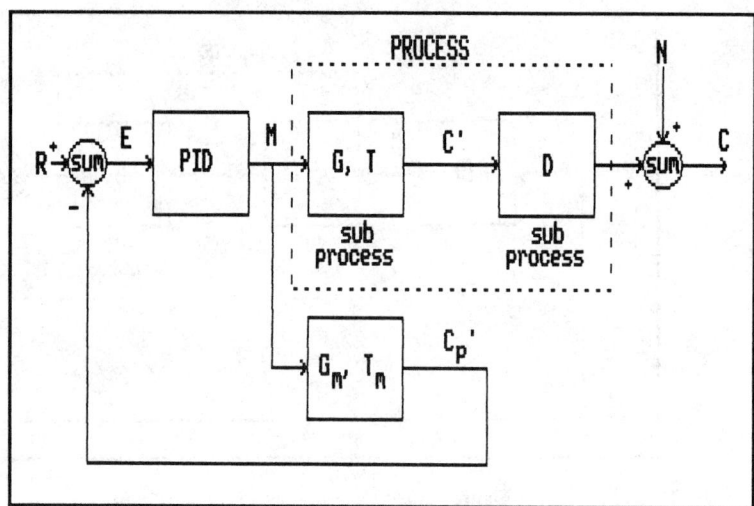

Figure 2.32   PID control using the predicted C_p' variable.

strategy is employed. The performance of the Smith predictor–PID combination, like all model-reference control strategies, is strongly dependent on how well the model actually simulates the real process, so one must proceed with caution.

Drawing on the results of Secton 1.6, the equations for the Smith predictor–PID control strategy can be developed as follows. First, using Figure 2.33 as a guide, the error that is used in the PID algorithm is slightly different from the regular expression for PI control:

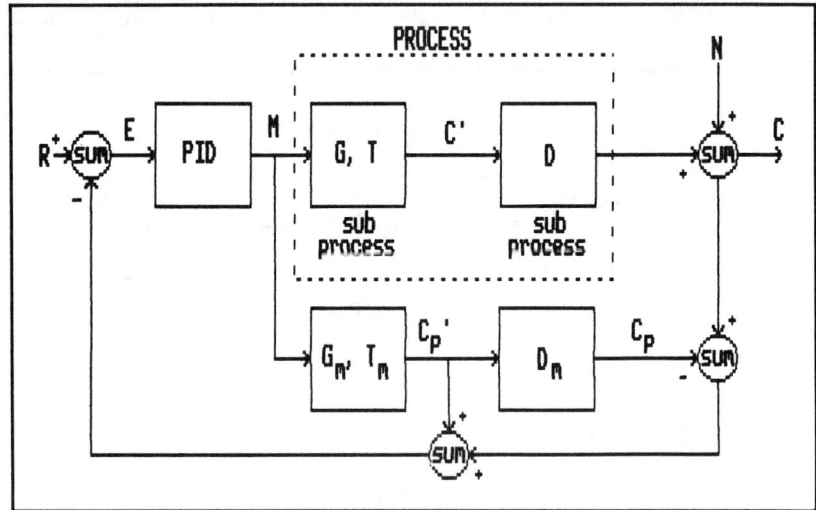

Figure 2.33   PID using the Smith predictor.

$$E = R - (C - C_p + C'_p)$$

where

R = set point
C = controlled variable
C'_p = model without deadtime
C_p = model with deadtime

The predicted values can be obtained from the following expressions:

$$C'_p(i) = AC'_p(i - 1) + BM(i - 1)$$

$$C_p(i) = C'_p(i - d_m)$$

where

$A = \exp(-h/T_m)$
$B = g_m(1 - A)$
$D_m = d_m h$

From the expression for the error, the reader can see that if the model fits the process perfectly, that is, there are no disturbances and the model gain, time constant, and deadtime match those of the process, then $C = C_p$, and the error will depend only on C'_p, which is the output of the model without deadtime. Since the process under control would be without deadtime, aggressive control could be used and the variance would be relatively small.

If there are disturbances and/or model–process mismatches, then the error expression can be looked upon as

$$E = R - C'_p - (C - C_p)$$

where $C - C_p$, which represents the difference between the real process and the reference model with deadtime, will act as a correction to the error of

$$R - C'_p$$

To illustrate the Smith predictor, consider an example process having a deadtime of 15 seconds and a time constant of 15 seconds, plus a slight inflection point (model 5). Figure 2.34 shows how the process responds to a step change in the set point with the Smith predictor in effect. The model gain was set at 2.0, exactly matching that of the process. However, the model deadtime was set to 17 seconds rather than the 15 seconds of the example process in order to compensate for the small inflection point. The initial control gains were calculated from the equations in Section 2.5, using the time constant and gain from the example process but with zero deadtime and a desired response time, T_d, equal to 10 seconds:

$$P = 1, \quad I = 0.1$$

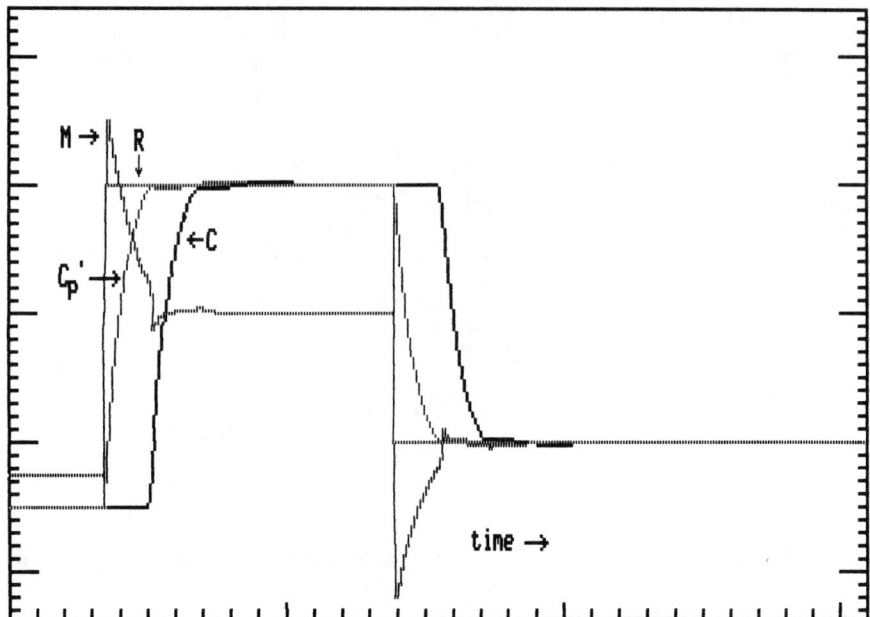

Figure 2.34 Application of the Smith predictor.

Note how the manipulated variable immediately starts to back off after the one-time kick due to the change in the set point. This is because the control algorithm is primarily seeing C_p'. About the time that the deadtime has elapsed and the controlled variable starts to respond, there is some "jerky" behavior in the manipulated variable resulting from the slight mismatch between the model and the example process.

Figure 2.35 shows how the process responds to a step change in the set point with only PI in effect. The initial control gains were again computed from the Section 2.5 equations and had to be more conservative because of the deadtime:

$$P = 0.3, \quad I = 0.016$$

Note how the manipulated variable integrates up during the period after the one-time kick due to the set point change and before the deadtime has elapsed.

A comparison of these two graphs suggests that, although PI does not fair too badly, the performance of the PI–Smith predictor combination is clearly better. In general, aside from the questions of model mismatch, the improvement resulting from adding the Smith predictor will be better as the deadtime/time constant ratio increases.

A random-walk stochastic disturbance is now added to the example process so that C and C_p will not be equal. The control gains are left unchanged. Figures

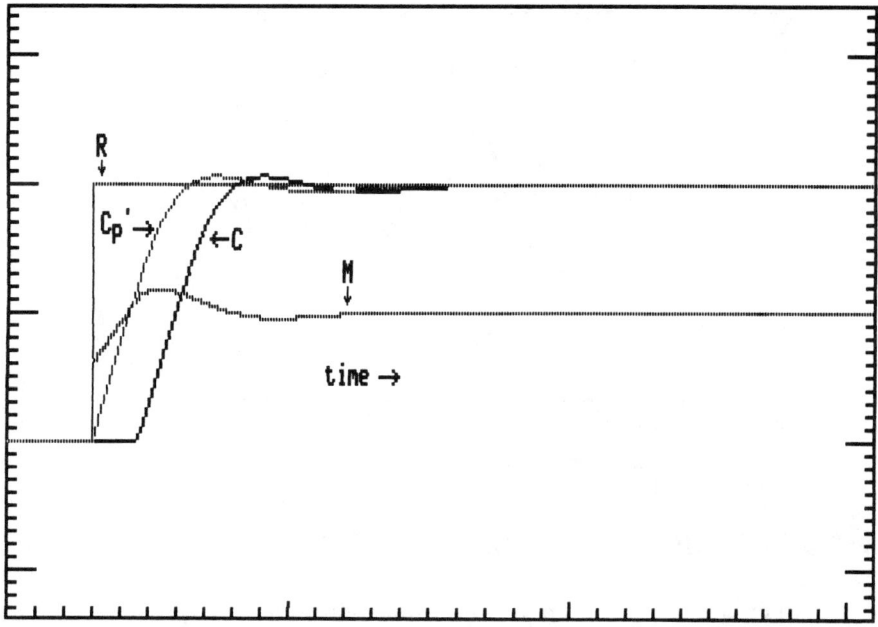

Figure 2.35 Application of PI control.

2.36, for PI, and 2.37, for the Smith predictor, show that the time-domain behavior of the two control algorithms is somewhat similar, with the Smith predictor having a slightly lower root mean square of the error (based on 1024 samples). In Chapter Five the Smith predictor will be studied again.

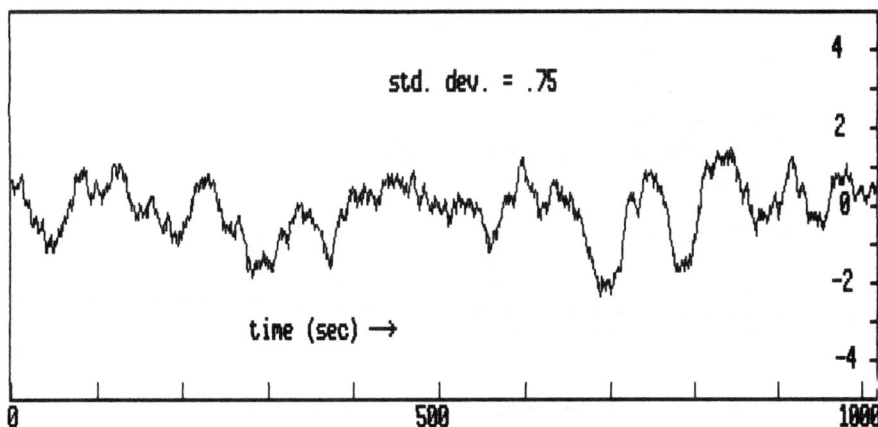

Figure 2.36 PI control applied to a random-walk disturbance.

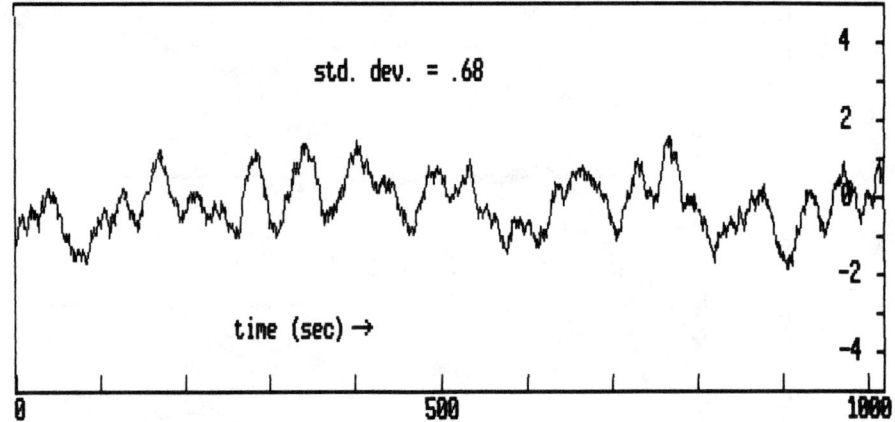

Figure 2.37   Smith predictor applied to a random-walk disturbance.

2.8 BOX–JENKINS CONTROL ALGORITHMS

An entirely different approach to the control of noisy processes is taken by Box and Jenkins (1970), where a process–noise model is developed that includes not only the deterministic parameters such as the process gain, time constant, and deadtime, but also the parameters that describe the stochastic disturbances to which the controlled variable is subjected. In Section 1.3, stochastic disturbances were described in terms of an autoregressive parameter α and a moving average parameter β, but these parameters have not been used in determining a control algorithm or in tuning a control algorithm. Although the Box–Jenkins approach is general, this section will deal only with the case where the process is first order with a deadtime (FOWDT) and with the case where the stochastic disturbance is a first-order autoregressive–moving average stochastic sequence, since this is a frequently occurring case and since it will serve the purpose of showing the features of the method.

For a FOWDT model, the Box–Jenkins approach would develop a model having the form of

$$C'_k = AC'_{k-1} + BM_{k-1-d}$$

$$C_k = C'_k + N_k$$

where, as was shown in Section 1.6,

$$A = \exp(-h/T_m)$$
$$T_m = \text{reference model process time constant}$$
$$B = g_m(1 - A)$$
$$g_m = \text{reference model process gain}$$
$$D_m = \text{reference model deadtime} = dh$$

The symbol N_k represents the stochastic disturbance, which, as was shown in Section 1.3, has the following form:

$$N_k = \alpha N_{k-1} + w_k + \beta w_{k-1}$$

where the parameters α and β describe the stochastic sequence and w_k represents white noise. For example, in this section the parameter α will be set to unity so that the resulting stochastic disturbance is an *integrated moving average*, which is a frequently occurring type of noise. Note that with the parameter α set to unity the stochastic disturbance is nonstationary.

Box and Jenkins show how to find the model-related parameters A, B, d, α, and β using a nonlinear least-squares technique significantly more complicated than the method presented in Section 1.10. Several software packages, such as MATLAB™, are capable of determining estimates of these parameters. In the following, the values of these parameters will be assumed known, and attention will be directed to the resulting control algorithms.

Box and Jenkins set up a predictor that, based on values of C_k and M_k, can predict the value of C_{k+1+d}, that is, the value of the controlled variable at the next time of control plus the deadtime. Having developed the predictor, Box and Jenkins specify that C_{k+1+d} be on target and then, effectively, invert the predictor and solve for the value of M_k that will drive C_k, which is known, to the target in $d + 1$ control intervals. The result, for the case of the FOWDT model, is

$$\Delta M_i = -(1 + \beta)(M_{i-1} - M_{i-1-d}) + IhE_i + P \, \Delta E_i \qquad (2.20)$$

where

$$E_i = R - C_i$$
$$I = (1 + \beta)/(hg_m)$$
$$P = (1 + \beta)A/B = (1 + \beta) \exp(-h/T)/\{g_m[1 - \exp(-h/T_m)]\}$$

The last two terms on the right side of Equation (2.20) are the integral and proportional terms from the conventional PI control algorithm, and the first term on the right side,

$$-(1 + \beta)(M_{i-1} - M_{i-1-d})$$

is the deadtime compensation term. For the case of no deadtime, that is, when $d = 0$, Equation (2.20) simplifies to an unmodified PI control algorithm. The expressions for the integral and proportional control gains are roughly similar to the tuning rules presented in Section 2.5. In Chapter Four, after the Z-transform tool has been introduced, it will be shown that the deadtime compensation term of Equation (2.20) appears in a Dahlin algorithm and in the Smith Predictor, even though both of these approaches ignore the structure of the disturbances.

In Chapter Five, where another look is given the Box–Jenkins algorithms, it will be shown that for certain combinations of nonstationary disturbances and process models this approach can cause excessive activity in the manipulated variable.

2.9 INTERCONNECTIONS AND MODIFICATIONS OF CONTROL LOOPS

So far there has been one controlled variable, one manipulated variable, and one control loop. Often the control problem is sufficiently complicated such that more than just one loop is needed.

2.9.1 Cascaded Master/Slave Combinations

Return for a moment to Figure 2.1, which shows a process for which a control loop was designed to keep the tank level on target by manipulating the valve in the inlet stream. For that loop to be effective, the pressure head of the liquid upstream of the inlet valve needs to be reasonably constant so that a given change in the inlet valve position always causes, within reason, the same change in the inlet flow rate. However, if the upstream pressure head were to vary widely, this would not be true, and this variation could have a strong impact on the ability of the level control loop to do its job.

To stabilize the inlet flow rate, a second loop can be installed that will adjust the valve in the inlet stream to keep the flow rate on target. This second loop would need a flowmeter that would supply its controlled variable. It would get its target or set point from a modified level control loop, which would now manipulate the target of the flow loop in order to keep the level on its target. This configuration is shown in Figure 2.38 and would have superior performance relative to the loop described in Sections 2.1 to 2.3, because variations in the upstream head would be dealt with immediately by the flow loop (labeled FC) and would have minimal impact on the tank level control loop (labeled LC). On the other hand, the control strategy consisting of just one level control loop could not deal with an upstream pressure head until it had shown up in a level variation.

In Figure 2.38 the level control loop is the master loop and the flow rate loop is the slave loop; that is, the master loop's manipulated variable is the set point of the slave loop. For a master/slave combination to be effective, the dynamics of the slave loop should be an order of magnitude quicker than that of the master. Thus, if the master loop had an effective time constant of 20 seconds, the slave loop's effective time constant should be approximately 2 seconds or shorter.

Cascade control is a widely used configuration, often with several levels of masters and slaves; that is, one master might supply a set point to its slave, which in turn supplies a set point to its slave, and so on. The reader should note that, when tuning a control system consisting of cascaded control loops, one should always start with the most internal slave loop and tune it while all its master loops are inactive. Once this most internal slave loop is tuned, attention should be directed to that slave's master loop. With that loop's slave active but with its masters inactive, that loop should then be tuned. This procedure should be followed, moving to higher-level masters one at a time.

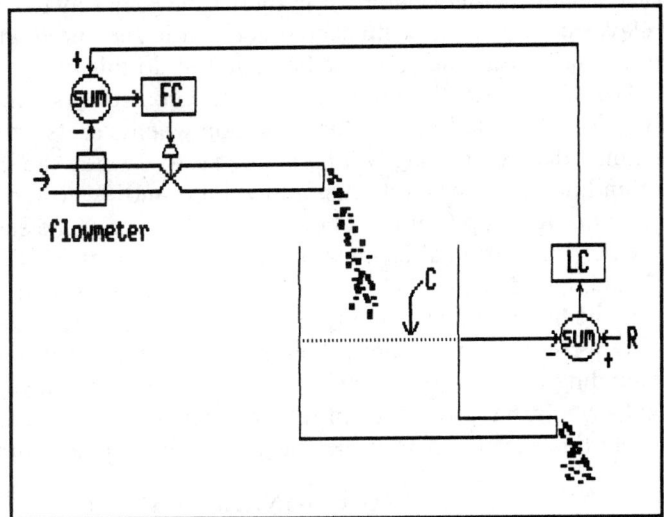

Figure 2.38   Master/slave cascade configuration.

2.9.2 Cascade Windup

Consider the case where a master loop has sent a set point, R_S, to its slave loop that is unobtainable because the manipulated variable of the slave loop, M_S, is at a limit. For the sake of discussion, assume that both loops are attempting to control processes that have positive process gains, that the master loop controlled variable, C_M, is below its set point, R_M, and that the master loop has sent a set point to the slave, R_S, that is higher than the present slave loop controlled variable, C_S. Finally, assume that the present slave loop manipulated variable, M_S, is on a maximum limit and that the master loop manipulated variable is below its maximum limit. That is,

$$\text{Master loop:} \quad C_M < R_M, \quad M_M < \text{max}$$

$$\text{Slave loop:} \quad C_S < R_S, \quad M_S = \text{max} \quad (R_S = M_M)$$

If the master loop controlled variable stays below its set point and if the master control algorithm has an integral component, the set point to the slave will continue to increase. In this situation, unless the master loop is aware that its slave loop's manipulated variable is on a maximum limit, it may continue to send out increasingly unobtainable set points to the slave loop, and the master loop will "wind up" its set point in a manner somewhat similar to the integral windup mentioned in Section 2.3.5.

This kind of windup is not usually addressed in most textbooks, but it can be important, especially in the distributed control systems that are purchasable as a package. Also, it can happen when single-loop controllers are combined in a cascade manner. In the latter case, there is no easy way to make sure the masters and slaves communicate with each other, but in the distributed systems it is possible to require that the software be coded to do this.

The solution to the problem is for the control system to set a flag, which will be cleverly called *FLAG*, for each loop whenever its manipulated variable is on a limit; that is, the flag will be set equal to 0 when the manipulated variable is within limits, -1 when it is on a low limit, and $+1$ when it is on a high limit. Then its master loop (if it has one) can use the sign of the slave loop's controller gain (either proportional or integral), which will be called G_a, to tell if the slave loop's manipulated variable is on a limit that warrants a change in procedure.

In the hypothetical example discussed above, if the slave loop's manipulated variable were on a lower limit, then there would be no problem with the master loop sending a higher set point to its slave, since the slave loop's manipulated variable would have room to move in order to keep its controlled variable on target. In this case the ΔM of the master loop would be positive, and the product

$$FLAG * \text{SIGN}\{\Delta M\} * \text{SIGN}\{G_a\}$$

would be negative and it would be acceptable for the master loop to send the new set point to the slave.

On the other hand, if the slave's manipulated variable were on an upper limit, *FLAG* would be positive and the above product would also be positive. In this case it would not be acceptable for the master loop to send the unmodified set point to the slave loop.

In general, when the product is less than or equal to zero, it is acceptable for the master loop to send the unmodified set point to the slave loop. Take the case where the master loop ΔM is negative and the slave's manipulated variable is on a lower limit. Here, FLAG is negative, the product is positive, and the master loop should not send the unmodified set point to its slave.

For the sake of completeness, consider the case where the slave loop is controlling a negative gain process, that is, where, except for disturbances, a positive change in the manipulated variable would cause a negative change in the controlled variable. If the master loop ΔM is negative and if the slave loop's manipulated variable is on a lower limit, the product will be

$$FLAG * \text{SIGN}\{\Delta M\} * \text{SIGN}\{G_a\} = (-1) * (-1) * (-1) = -1$$

so the master loop should send the unmodified set point to the slave loop. This makes sense: the master is lowering the slave's set point which, if the slave's controlled variable is near its old set point, will require the slave loop's manipulated variable to increase, because the slave loop's process has a negative gain. Since the slave loop's manipulated variable is on a lower limit, it has room to increase and attempt to satisfy the new set point.

In those cases where the product is positive, it is easiest to do nothing; that is, set the master loop's ΔM to zero. In some cases it is more effective to actually set ΔM equal to a small quantity according to

$$\Delta M = -\delta \ \text{SIGN}\{\Delta M\}$$

where δ has a small positive value. This has the effect of bringing the slave's manipulated variable just off its limit, thereby preventing even a small amount of cascade windup.

2.9.3 Parallel Loops

In the example process referred to in this section, there could be two manipulated variables. In addition to the valve position in the inlet stream, which has served as the manipulated variable in all the examples so far, there could also be a valve inserted in the outlet line. Many times a second manipulated variable is used to keep the first manipulated variable in the middle of its range so that it is in a good position to react to large disturbances. This is the idea behind the structure shown in Figure 2.39, where the LC loop manipulates the valve in the inlet stream to keep the level on target. There is a second loop (labeled VC) that uses the manipulated variable of the first loop as its controlled variable. This second loop manipulates the valve in the outlet stream in order to keep the manipulated variable of the first loop near its target. The second loop would not be tuned aggressively, since the goal is not to keep its controlled variable tightly near its target, but only to keep the manipulated variable of the first loop away from the nearly closed or nearly open position where it might not have room to move if a large disturbance occurred.

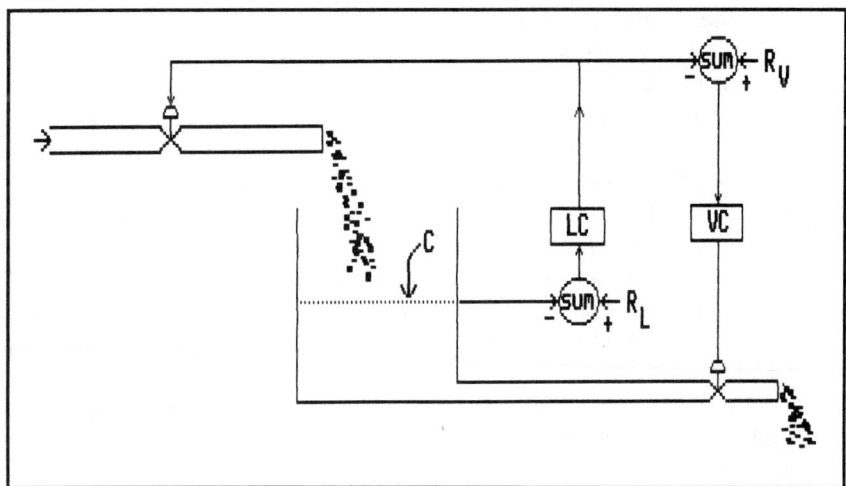

Figure 2.39   Parallel configuration.

2.10 FEEDBACK CONTROL AND STATISTICAL PROCESS CONTROL

Before attempting to compare statistical process control, or SPC, with the kind of feedback control that has been dealt with so far, some time has to be spent on a brief review of SPC. Before getting started with the review, a disclaimer must be made: some workers in the field of SPC would prefer that the following material be referred to as an illustration of the use of Shewhart control charts rather than SPC, since the latter often denotes a much wider concept of keeping a process under control. In any case, for the purposes of this section, SPC or Shewhart control-chart strategy, revolves around the WECO run rules, which say that the process is considered to be *out of control* when one or more of the following conditions is satisfied:

1. One sample of the controlled variable has deviated from the nominal by an amount greater than three standard deviations.
2. Two out of three samples have deviated from the nominal by two standard deviations.
3. Three out of four samples have deviated from the nominal by one standard deviation.
4. Eight samples in succession have occurred above or below the median line.

The derivation of these run rules is based on the assumption that the data sequence being tested is unautocorrelated or independent and that the probability of these conditions being satisfied if the data are truly independent is about 0.01. Thus these rules are statistical tests for nonindependence of data or for autocorrelation. For further discussion of the WECO run rules, the reader is referred to the text by the Western Electric Staff (1956).

Before proceeding, it should be pointed out that this is not the only approach to SPC. There are, for example, run rules for cumulative sums and for exponentially weighted moving averages of the controlled variable (Hunter, 1986). Although each has special features that make it attractive under certain conditions, the WECO run rules appear to be the most popular.

When a process is determined to be out of control by WECO rules, a search for the assignable cause is supposed to be initiated. Once the cause is determined, it is the obligation of the user to "make the process right." This may mean repairing a malfunctioning piece of equipment, correcting the chemical composition of an input stream, or adjusting a valve. However, there is nothing in the above strategy that tells the practitioner, if a valve is to be adjusted, how much that adjustment should be.

Figure 2.40 shows a simplified version of Figure 2.3 along with a block diagram of SPC (also severely simplified). Feedback control (or FBC) differs significantly from SPC at the process box. In both cases the output is the controlled

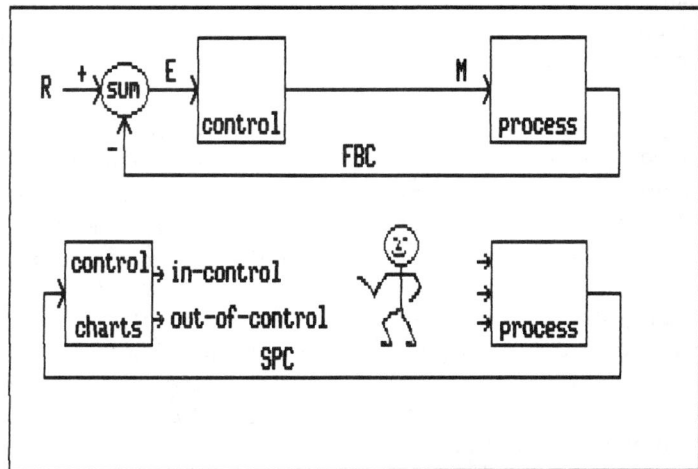

Figure 2.40 Feedback control (FBC) and SPC.

variable. However, in FBC there is *one* well-defined input, the manipulated variable, and its value is unequivocally dictated by the control algorithm box every *h* seconds. (It is important to note that *the implementation of FBC may be automatic or manual.*) In the control chart or SPC scheme, on the other hand, there is no clearly defined input to the process box since it is dependent on the assignable cause of the out of control condition.

The output of the box labeled "control charts" is binary. Based on the WECO run rules, the process is either *in control* or *out of control*. As was mentioned above, there is nothing in the control chart scheme that tells the user what to do with this status report. If the status is "out of control," the user is committed to look for the cause and "make the process right." After finding the assignable cause, the corrective action may be an adjustment of valve, or it may be the repair of a broken piece of equipment, or sometimes it may require several actions.

SPC probably is used incorrectly if the out-of-control corrective action is consistently the adjustment of one manipulated variable, say a valve, and if no attempt is made to find the assignable cause. If there is no cost associated with making the valve adjustment, this sort of corrective action is probably better done with a FBC strategy, where the control chart box would be replaced with a feedback control algorithm driving the adjustment of the manipulated variable either automatically or manually at a regular time interval.

Often defenders of making a control move only when the process is out of control maintain that more frequent moves would cause *overcontrol*. This concept of overcontrol (as yet undefined in this book and generally undefined in the statistical literature) is unfortunately the consequence of not understanding the dynamics of the process and the fundamentals of process control. Overcontrol is really a misnomer, since the erratic performance associated with overcontrol is usually the result of an overaggressive, poorly structured control strategy, rather than one that is carried out too frequently. A clearer definition of overcontrol will be given later in this section, and the concept will be revisited in Chapters Three and Five.

Should the SPC out-of-control output be used to trigger a change in the manipulated variable (instead of a search for an assignable cause), then, in effect, this use (or misuse) of SPC would be similar to using conventional feedback control with a deadband, which, as was pointed out in Section 2.6.2, can only lead to increased variance of the controlled variable about its target. There would also be an added deadtime that is detrimental to good control due to the time it takes for one of these rules to be satisfied. This latter time is often referred to as the *average run length,* or ARL. All run rules, whether they are based on individual samples, as these WECO rules are, or on exponentially weighted moving averages or cumulative sums have a nonzero ARL and therefore add an effective deadtime to the control strategy.

It should be pointed out again that the WECO rules say nothing about how much a manipulated variable should be changed, so if one persists in using SPC in place of FBC, one still has to concoct rules for specifying the change. Frequently, an integral-only control strategy is appended to the SPC apparatus, where the integral gain is chosen to be equal to the inverse of the process gain. In this case each adjustment to the manipulated variable should, assuming there are no other disturbances, bring the process back to set point.

2.10.1 Example with a Short Time Constant and No Deadtime

To illustrate the differences between FBC and a control chart based approach, two examples will be discussed. First, consider a process with no deadtime, a process gain of 2.0, a time constant of 0.5 second, and a control interval of 1.0 second (model 6). Because the control interval is larger than the time constant, integral-only control will be sufficient, and since the control interval is unity, the control algorithm will be

$$\Delta M_i = IhE_i = \frac{1}{G}(1)(E_i) = 0.5E_i$$

where the integral gain, I, is set equal to the inverse of the process gain, G; that is, every control move is designed to remove the error in time for the next control move (assuming that no more disturbances take place in the time between the last and the next control move).

Figure 2.41 shows how the process is controlled when it is subjected to white noise having a standard deviation of 0.707. In the first portion of the figure, there is no control and midway through this figure integral-only control is turned on. Note how the amplitude of the white noise appears to be amplified. Figure 2.42 shows the autocorrelation of 1024 points taken from a simulation of this process

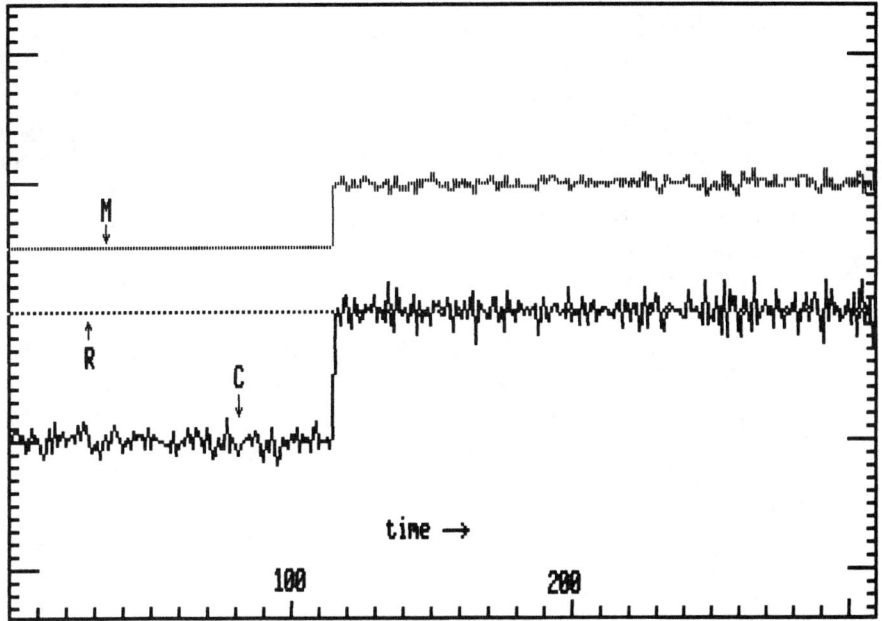

Figure 2.41   Integral-only control of model 6 (white noise disturbance).

while under integral-only control. Note how the autocorrelation at lag 1 is strongly negative, an indication of overcontrol. In other words, the aggressive nature of the control strategy tends to cause the control error at each instant of control to be the negative of the previous one. The autocorrelation is probably the best way to verify the presence of overcontrol, and the presence of a strong negative autocorrelation at lag 1 is a definitive characteristic of overcontrol.

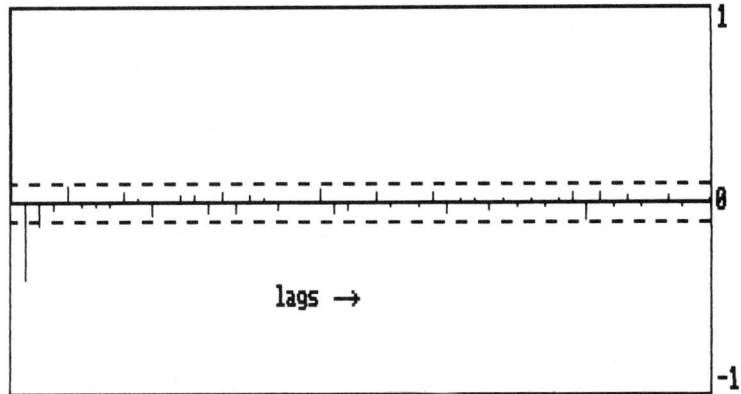

Figure 2.42   Autocorrelation of model 6 (integral-only, white noise disturbance).

Figure 2.43 shows how the manipulated variable changes its nature when, midway through the simulation, the disturbance changes from white noise to an autocorrelated stochastic sequence with an autoregressive coefficient of 0.99. Note that when the change in the nature of the disturbance is made the amplitude of the hash riding on the controlled variable appears to be slightly attenuated. The autocorrelation (Figure 2.44) of the controlled variable when subjected to this autoregressive type of stochastic disturbance shows that the integral-only control was able to transform the heavily autocorrelated disturbance into a nearly white noise sequence.

When the disturbance was white noise with no control, the standard deviation of the controlled variable was equal to that of the white noise, which in turn has a standard deviation of 0.707. With integral-only control, in the face of white noise disturbances, the standard deviation increased to approximately 0.93; that is, the effect of feedback control is to make things worse by a factor of 1.3. Since the process has a time constant that is small relative to the control interval, there is little filtering due to the process, and the activity of the manipulated variable is unattenuated.

When the disturbance becomes heavily autocorrelated (almost a random walk), the integral-only control strategy is able to lower the standard deviation to approximately 0.71. Using the formula derived in Section 1.3.2 [Equation (1.6)], the theoretical standard deviation of the uncontrolled autocorrelated disturbance can be calculated to be

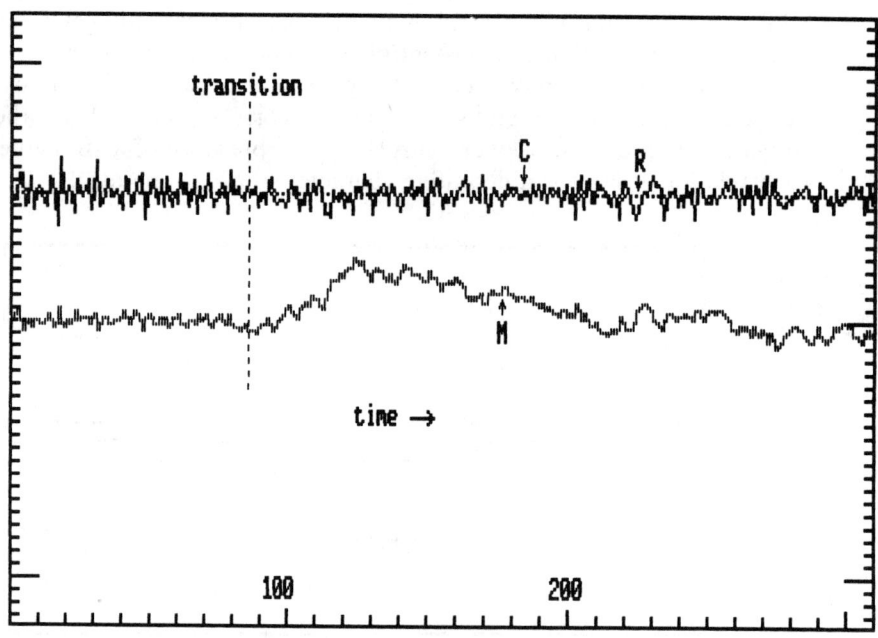

Figure 2.43 Transition from white noise to autocorrelated noise.

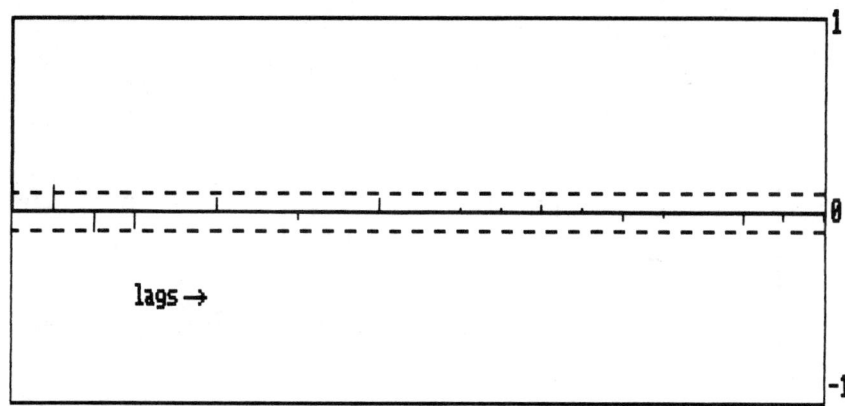

Figure 2.44 Autocorrelation of model 6 (colored noise disturbance, integral-only control).

$$0.7071 * \sqrt{\frac{1}{1 - (0.99)^2}} = 5.01$$

Therefore, the integral-only control reduces the standard deviation of the disturbance by a factor of approximately 7 to 1.

In Chapter Five this example process will be given another look, and it will be shown that for highly autocorrelated disturbances the addition of a proportional component further decreases the standard deviation.

To implement an SPC (or control chart based) type of control, the same integral-only control strategy mentioned above is applied whenever the WECO rules (using a standard deviation of 0.707) determine that the process is out of control. Figure 2.45 shows how the process is controlled when it is subjected to white noise. In the first portion of the figure there is no control, and midway through this figure the SPC-based control is turned on. Note how the amplitude of the white noise appears to be amplified, but not as much as it was when integral-only control was applied. Figure 2.46 shows the autocorrelation of 1024 points taken from a simulation of this process while under the SPC-based control. Note that, unlike the case when conventional integral-only control was used, the autocorrelation for the first lag is positive and slightly greater than the white noise limit.

Figure 2.47 shows how the manipulated variable changes its nature when midway through the simulation the disturbance changes from white noise into an autocorrelated stochastic sequence with an autoregressive coefficient of 0.99. Note that, when the change in the nature of the disturbance is made, the variation of the controlled variable appears to become autocorrelated and the amplitude appears to increase. The autocorrelation in Figure 2.48 indicates that there is a significant positive autocorrelation, showing that the control strategy was not able to remove the autoregressive nature of the disturbance.

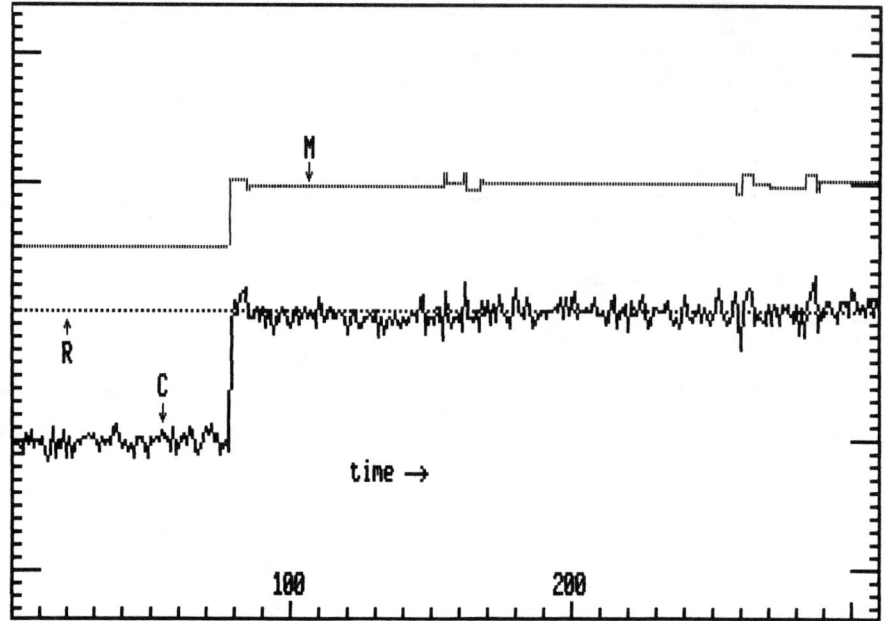

Figure 2.45 Control chart-based control of model 6.

When the disturbance was white noise, the standard deviation of the controlled variable under this type of SPC-based control was found to be approximately 0.85. Therefore, although the SPC type of control also made things worse when the disturbance was white noise, the penalty was less than with regular integral-only control. However, when the disturbance became heavily autocorrelated, the standard deviation was approximately 1.1, which is significantly worse

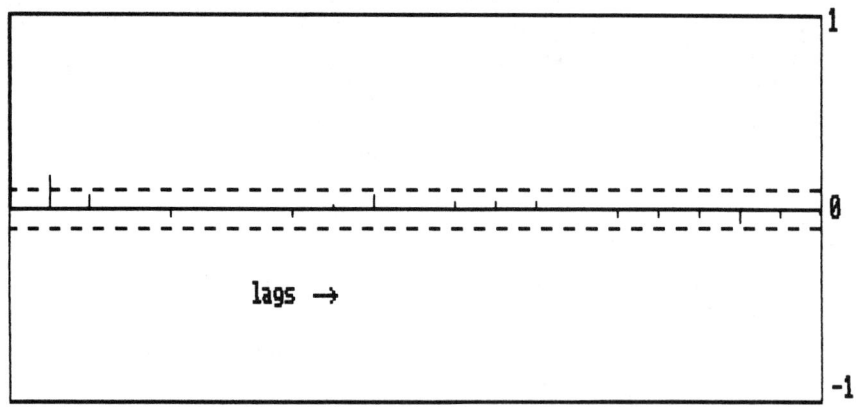

Figure 2.46 Autocorrelation of SPC-based control applied to model 6 subject to white noise.

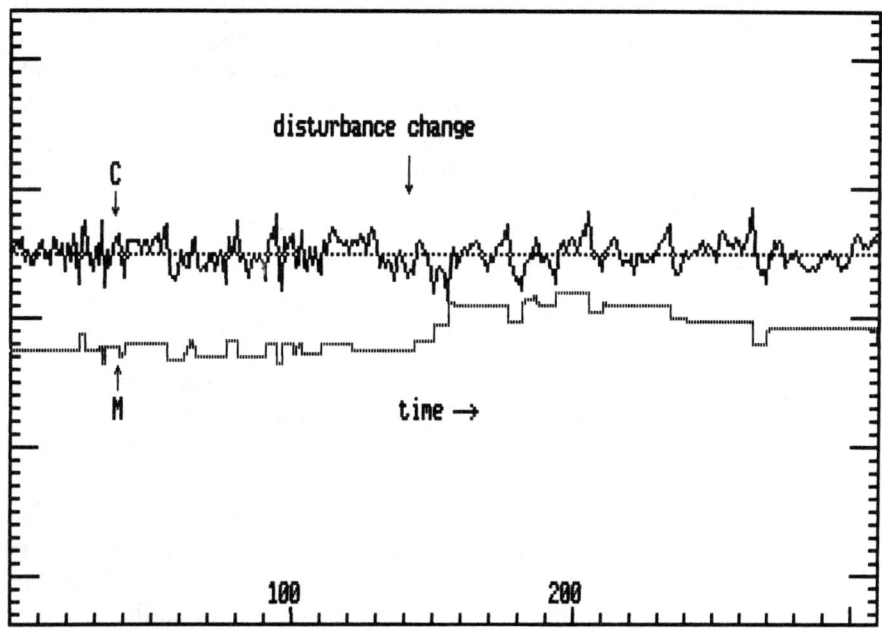

Figure 2.47   SPC-based control applied to model 6 (showing transition from white to colored noise).

than that for the regular integral-only control. Since most industrial disturbances are nonstationary, the advantages of making control moves at a regular time interval should be clear in this case where the time constant is less than the control interval.

The WECO run rules apply only to data that are assumed to be independent

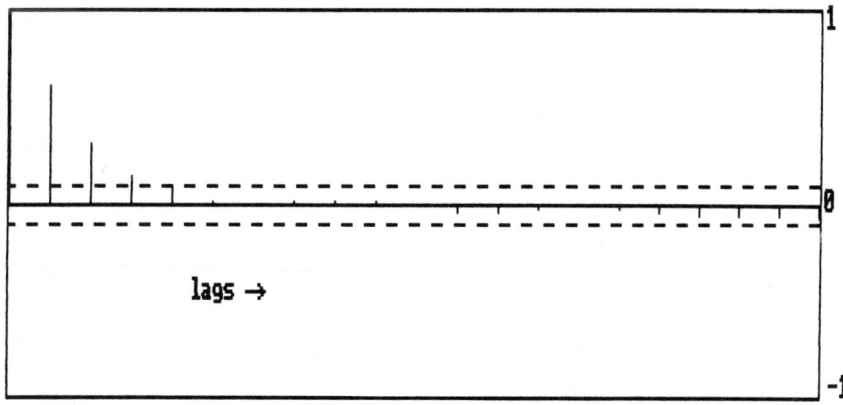

Figure 2.48   Autocorrelation of model 6 (SPC-based control subject to colored noise).

or unautocorrelated, and control moves are to be made only when the data become nonindependent or autocorrelated. Therefore, to apply the run rules to a situation where it is known that the data are autocorrelated is, strictly speaking, incorrect. There is no quarrel with this position; but, unfortunately, control algorithms based on run rules are often used when the disturbances are autocorrelated whether it is correct or not, and given this situation it is the object of this example to compare the methods.

In fairness to the integral-only control strategy, which amplified the white noise so severely, it should be pointed out that this situation could have been significantly improved had a less aggressive integral control gain been used. This will be demonstrated in Chapter Five using theoretical methods that do not require simulation.

2.10.2 Example with a Long Time Constant and a Deadtime

The second example (model 7) consists of a process with a time constant of 40 seconds, a deadtime of 10 seconds, a process gain of 2.0, and a control interval of 1.0 second. This is in direct contrast to the previous example since here the control interval is small relative to the time constant and the deadtime. For this reason, PI control instead of integral-only control will be chosen. The control gains of $P = 0.71$ and $I = 0.0175$ were obtained from the formulas of Section 2.5 using a desired time constant of 10 seconds. Without any control action, the standard deviation of the controlled variable is equal to that of the white noise, which is again 0.707. When PI control is activated in the face of the white noise disturbance, the standard deviation is, within statistical accuracy, unchanged. Unlike the first example process, which had the short time constant and amplified the activity of the manipulated variable, the large time constant of this process acts as a filter, and the activity of the manipulated variable is attenuated.

When the disturbance changes to an autocorrelated disturbance with an autoregressive coefficient of 0.99 (nearly a random walk) and when PI control is used, the standard deviation increases to approximately 3.0. In dealing with heavily autocorrelated sequences, one must run simulations over a large number of points in order to get a reasonable estimate of the standard deviation. For this example, repeated runs of 10,240 points each were made. In Chapter Five, the power spectral density will be used to show that the theoretical value of the standard deviation is 2.93. The controlled variable under these conditions is shown in Figure 2.49, and without calculating an autocorrelation the reader should be able to perceive that the controlled variable is still highly autocorrelated. (Note that for the 1024-point run shown in Figure 2.49 the estimated standard deviation is 2.64.) When there is no control, this nearly random walk disturbance would cause the controlled variable to have a theoretical standard deviation of 5.01. This result is similar to the previous case, which dealt with model 6.

To apply SPC to this case, the WECO rules are again based on a standard

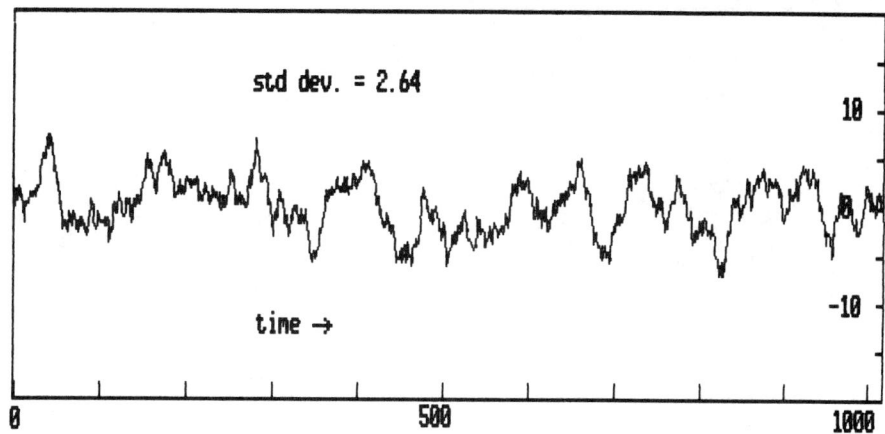

Figure 2.49   PI control of a long time constant process (colored noise disturbance).

deviation of 0.707, and integral-only control is applied whenever the rules indicate that the process is out of control. The integral gain here is the same as with the PI case: $I = 0.0175$. When the disturbance is white noise, the SPC/integral-only strategy, like the PI strategy, gives a standard deviation effectively unchanged from the standard deviation of white noise.

Figure 2.50 shows the performance of the controlled variable under SPC-based control when the disturbance is the same nearly random walk used above. The empirical standard deviation (based on repeated simulations of 10,240 points) is approximately 4.5. (Note that for the 1024-point run shown in Figure 2.50 the estimated standard deviation is 3.26.) As with the simulations mentioned above, a large number of points were required to get reasonable estimates of the standard deviation. Unlike conventional PI control, appending an integral-only control algorithm to the run rules generates a nonlinear system of equations that does not yield to analysis, so a value for the theoretical standard deviation cannot be derived.

The significantly worse performance of the SPC/integral-only control algorithm is a consequence of the large time constant and deadtime relative to the control interval. Integral-only rather than PI control was chosen since this type of strategy would usually be used in a manual mode, and it is relatively difficult to implement a PI algorithm manually, although Box and Jenkins (1970) do show a nomographic approach. When PI was coupled with the run rule-based approach, no appreciable difference in the estimated standard deviation was found.

If one is interested in detecting out-of-control conditions and then searching for an assignable cause, a control chart with the WECO rules is the tool to choose. On the other hand, if a well-defined pairing between a manipulated variable and a controlled variable has been made, and it is desired to keep the mean of the controlled variable on target in the face of nonstationary or at least highly au-

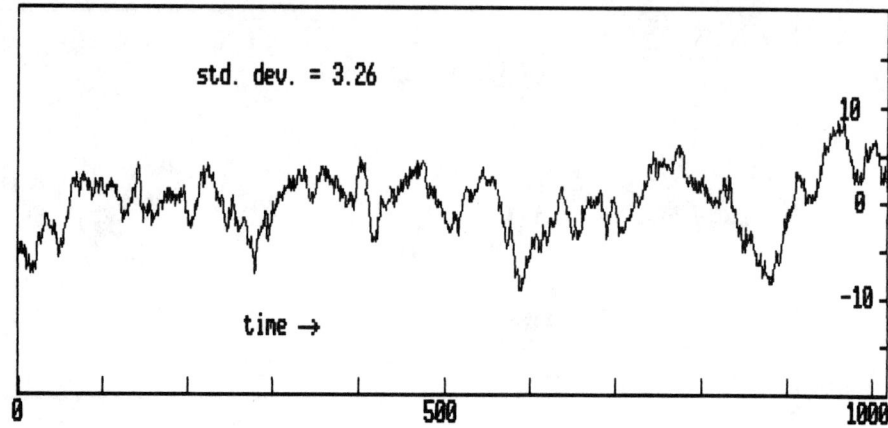

std. dev. = 3.26

time →

Figure 2.50 Control chart-based integral-only control of a long time constant process.

tocorrelated disturbances, the tool to choose is conventional PI or PID feedback control.

These two examples may shed some light on the idea held by many statisticians that one should not attempt conventional feedback control in the face of white noise. For short-time-constant processes, this may be true, since white noise is definitely amplified by the action of conventional feedback control when the control interval is unity and when the integral control gain is chosen to be the inverse of the process gain. Since statisticians have often dealt with short-time-constant processes, it is understandable why this folklore has persisted in statistical circles for so long.

On the other hand, for long-time-constant processes the amplification of white noise by the action of conventional feedback control is insignificant. Since control engineers have often dealt with these kinds of processes, it is understandable why they tended to dismiss the statistician's concern about applying conventional feedback control to processes subjected to white noise. It appears from the results of this section that under certain circumstances both groups were right.

As mentioned earlier, if the reader will look at the process to be controlled as a filter (refer to Section 1.9), the results can be explained succinctly. The short-time-constant process has little filtering ability, and the activity of the manipulated variable, as it attempts to control white noise, is passed through the system relatively unattenuated and is manifested as an out-of-control condition. On the other hand, because the long-time-constant process has significant filtering ability, the activity of the manipulated variable is heavily attenuated and causes relatively little problem with the controlled variable.

The dicussion in this section is not designed to suggest that FBC is superior to SPC, but only to point out that these are two different and powerful techniques designed for different problems. For additional comments on the problem discussed in this section, the reader is referred to MacGregor (1987).

2.11 MIXING MEASUREMENTS WITH MODELS: THE KALMAN FILTER

Whenever the controlled variable is corrupted with noise, there is a dilemma. No matter what kind of control algorithm is being used, a decision still has to be made as to what quantity will be chosen as the basis for making changes in the manipulated variable. Section 2.6.2 showed that applying a deadband to the controlled variable is not a particularly good idea even though it is used quite a bit. Filtering the controlled variable was shown in Section 2.6.3 to add lag and make the control problem more difficult. When there is a model of the process available, Section 2.7 showed that the Smith predictor can be used to lessen the deleterious effect of a large deadtime and that it appears to work reasonably well in the presence of noise. The Box–Jenkins control algorithms, which use models of the process as well as the noise, can be used effectively if the control interval is on the order of magnitude as the process time constant. Finally, in Section 2.10, techniques based on SPC/control chart run rules were tested and found to be satisfactory only in a few special cases. In short, none of these methods appeared to have any overwhelming advantage over the conventional PI control algorithm when the controlled variable is contaminated with noise.

In this section, yet another way of addressing a noisy controlled variable using the Kalman filter will be discussed. Although the Kalman filter was not introduced until the late 1950s (see Swerling, 1959, and Kalman, 1960), there is already a vast literature on the subject, and the material in this section will only skim the surface. The goal is to *uncover* rather than cover the Kalman filter and perhaps motivate the interested reader to delve into it more deeply. The Kalman filter is extensively used in space flight dynamics and ballistics, but it has not been widely used by control engineers and statisticians in the chemical process industries.

Consider a first-order process for which there is the following model:

$$C'_k = AC'_{k-1} + BM_{k-1} + w_k$$

where w_k is the *process noise* and represents the uncertainty associated with the model. This noise is white, has zero mean, and has a variance of V_w. The actual measured value C_k is related to the model value C'_k by

$$C_k = C'_k + v_k$$

where v_k represents the measurement noise, which is also white, has zero mean and has a variance of V_v.

The above model can be used to estimate the controlled variable at time k

based on information at time $k - 1$ by extrapolation. This estimate will be denoted by $C_k(-)$ and is calculated from

$$C_k(-) = AC_{k-1}(+) + BM_{k-1}$$

where $C_{k-1}(+)$ is the best estimate from time $k - 1$ and M_{k-1} is assumed to be known from the last control move. The symbol $(+)$ denotes the best estimate *after* the extrapolation and the measurement have been combined. The symbol $(-)$ denotes the extrapolated estimate *before* the measurement has been made.

The best estimate of the controlled variable at time k, denoted by $C_k(+)$, is a weighted sum of the extrapolated value $C_k(-)$ and the measured value C_k:

$$C_k(+) = K_k C_k + (1 - K_k)C_k(-)$$

The coefficient K_k is called the Kalman gain and expresses the level of confidence in the measured value. If the model is considered to fit the process nearly perfectly, as indicated by a small V_w, and if the measurement is relatively noisy, as indicated by a large V_v, then K_k would be expected to be nearly zero and

$$C_k(+) \approx C_k(-)$$

On the other hand, if the model is crude and the measurement variance is small, then K_k will be nearly unity and

$$C_k(+) \approx C_k$$

Before showing how the Kalman gain is calculated, two variances have to be introduced. The first is the variance associated with the extrapolation and is denoted as $P_k(-)$. The second is the variance associated with the effect of combining the measurement with the extrapolated estimate and is denoted as $P_k(+)$. As with $C_k(+)$ and $C_k(-)$, the symbol $(-)$ means before the measurement and the symbol $(+)$ means after the measurement. (The presence of the subscripts should prevent confusion of the P_k with the proportional gain P.)

As a consequence of the extrapolation, there is a change in the uncertainty or variance of the estimate, and this is expressed as

$$P_k(-) = A^2 P_{k-1}(+) + V_w$$

$P_{k-1}(+)$ is the variance based on the best estimate of C at time $k - 1$, and $P_k(-)$ is the variance associated with the estimate of C at time k based only on extrapolation. If A is nearly equal to unity and V_w is large, then the uncertainty increases. This makes sense because a large value of A means that C_k depends heavily on C_{k-1}, and a large value of V_w means that there is not much confidence in the model.

The variance or uncertainty of the estimate decreases as a consequence of the measurement and that decrease is given by

$$P_k(+) = (1 - K_k)P_k(-)$$

$P_k(+)$ is the variance associated with the best estimate of C at time k, which is based on both the extrapolation and the latest measurement at time k. The amount of the decrease depends on the Kalman gain. If the Kalman gain is nearly unity, as in the case of a nearly noiseless measurement, then the decrease is significant.

The Kalman gain is calculated from the variances according to

$$K_k = \frac{P_k(-)}{P_k(-) + V_v}$$

Note that because the variances are positive the Kalman gain will be positive and less than unity. These equations explain comments made earlier about the dependence of K_k on the variances asssociated with the extrapolation and the measurement.

To use the Kalman filter in a control mode, the following sequence is followed:

1. Calculate the best estimate of the controlled variable at time k by extrapolation:

$$C_k(-) = AC_{k-1}(+) + BM_{k-1}$$

The value of M_{k-1} will come from the control move made at time $k - 1$, and $C_{k-1}(+)$ will come from the last best estimate or from initial conditions if the calculations are just starting.

2. The change in the variance resulting from the extrapolation is calculated from

$$P_k(-) = A^2 P_{k-1}(+) + V_w$$

The value of $P_{k-1}(+)$ will come from the last best estimate or from an initial guess if the calculations are just starting.

3. Calculate the Kalman gain from

$$K_k = \frac{P_k(-)}{P_k(-) + V_v}$$

4. Combine the measured value with the extrapolated value:

$$C_k(+) = (1 - K_k)C_k(-) + K_k C_k$$

5. Calculate the decrease in the variance due to the measurement:

$$P_k(+) = (1 - K_k)P_k(-)$$

6. Determine the change in the manipulated variable by some control algorithm using $C_k(+)$.

7. Increment k and go back to step 1.

After running through several cycles of the above calculations, one finds that the variances and the Kalman gain reach steady values independent of the nature of C_k. Therefore, if the steady value of K_k is known, it probably makes more sense to avoid the calculations in steps 2, 3, and 5. Instead, one would treat K_k, $P_k(-)$, and $P_k(+)$ as constants and carry out steps 1, 4, and 6. This approach avoids having to estimate initial values of the variances and is more realistic for the control of processes that are not in a transient state such as start-up.

There are several comments to made. First, note that the above sequence would also work for the case where the model had a deadtime. Second, if the variance for the process noise, V_w, were zero, the P_k would gradually decrease toward zero, as would the Kalman gain. This means that ultimately no weight would be placed on the measured value of the controlled variable.

Third, the computation of the variances and the Kalman gain is independent of both the extrapolated values and the measured values of the controlled variable. The model could be turning out totally incorrect estimates of the controlled variable and there would be no effect on the Kalman gain. This is to be compared with the way the Smith predictor combines its two models with the measured value of the controlled variable.

Fourth, both the process noise and the measurement noise are white. For some of the example problems dealt with in Section 2.10, where the process was subject only to white noise, this method might be expected to perform impressively. However, in the kind of control problem dealt with most frequently in this book, noise is usually autocorrelated and sometimes nonstationary. The Kalman filter structure can be modified to deal with autocorrelated noise (see Bryson and Hendrikson, 1968).

Effective use of the Kalman filter depends on the user's ability to choose values for V_w and V_v, which in principle should be derivable from a priori knowledge of the process noise and measurement noise. However, in practice, estimation of these quantities is often difficult and it is sometimes referred to as "tuning the Kalman filter." Effective use also requires that the process model do a reasonably good job of representing the actual process.

This simple example has only skimmed the surface of the Kalman filter literature. The Kalman filter is most effective when the dimension of the system, that is, the number of controlled variables being analyzed simultaneously, is greater than 1. It can also be used when the number of measured variables is not equal to the number of controlled variables. For the interested reader, the book by Gelb (1974) is recommended.

2.12 SUMMARY

The basic goal of this chapter was to introduce the reader to the conventional PI control algorithm. Simple tuning rules were presented and illustrated via examples. The filtered derivative term was added to the PI algorithm and shown to be effective if one is willing to expend the extra effort of finding values of two additional tuning parameters.

As an aid in controlling processes with significant deadtimes, it was shown how the PI control algorithm could be embedded in a structure called the Smith predictor. As an example of a control algorithm that explicitly takes into account the stochastic as well as the deterministic aspects of the process, the Box–Jenkins approach was discussed. It was suggested that the Box–Jenkins method is best used when the process time constant is short compared to the control interval. In those cases where a single control loop is not sufficient, single loops were combined in cascade and parallel structures.

The concepts of feedback control in this chapter were compared with the ideas of statistical process control. Finally, the Kalman filter was presented as a way to combine noisy measurements with process models.

This chapter is certainly not meant to be all-inclusive. For example, only one-dimensional control structures have been dealt with. Multidimensional systems where there are n controlled variables and n manipulated variables are a subject far beyond the scope of this book. Also, there has been no mention of servomechanism systems, where the control algorithm is designed to force the controlled variable to follow a varying set point in the face of relatively insignificant disturbances. Instead, attention has been concentrated on control systems where the target is constant (although the target was occasionally given step changes to illustrate various points about the control algorithm) and the controlled variable is subject to autocorrelated stochastic disturbances.

2.13 DISCUSSION OF THE REFERENCES

Continuous time-domain control systems and servomechanism control theory, neither of which is covered in this book, are dealt with in detail in

SAUCEDO, R., and SCHIRING, E. E., *Introduction to Continuous and Digital Control Systems,* Macmillan, New York, 1968.

where, as an indication of how diverse the field of control theory is, the PID control algorithm is not mentioned once.

Another text that has much to offer, especially in the discrete time domain, but also makes no reference to the PID control algorithm, is

FRANKLIN, G. F., and POWELL, J. DAVID, *Digital Control of Dynamic Systems,* Addison-Wesley, Reading, Mass., 1980.

Other general texts on process control that cover all the mainstream topics are

STEPHANOPOULOS, G., *Chemical Process Control, An Introduction to Theory and Practice,* Prentice-Hall, Englewood Cliffs, N.J., 1984.

and

TAKAHASHI, Y., RABINS, M. J., and AUSLANDER, D. M., *Control and Dynamic Systems,* Addison-Wesley, Reading, Mass., 1970.

and
Harriott, P., *Process Control,* McGraw-Hill, New York, 1964.

A somewhat specialized text covering the regulation problem in a more mathematically oriented approach is
Gould, L. A., *Chemical Process Control: Theory and Applications,* Addison-Wesley, Reading, Mass., 1969.

A significantly more sophisticated approach to process control, including the multidimensional control problem, in the continuous time domain is given in
Ray, W. Harmon, *Advanced Process Control,* McGraw-Hill, New York, 1981.

An up-to-date summary of tuning methods, including those that are supposedly "self-tuning," is given in
Astrom, K. J., *Automatic Tuning of PID Controllers,* Instrument Society of America, Research Triangle Park, N.C., 1988.

Chapters 12 and 13 of the text by Box and Jenkins are devoted to their unique approach to feedback control, which is discussed in Section 2.8 and Chapter Five of this book:
Box, G. E. P., and Jenkins, G. M., *Time Series Analysis, Forecasting and Control,* Holden-Day, San Francisco, 1970.

The Smith predictor was first presented in
Smith, O. J. M., *Chemical Engineering Progress,* 53, 217 (1957).

An interesting and somewhat idiosyncratic text on the elementary aspects of automatic control, where the author claims to have invented a special case of the PI controller, is
Phelan, R. M., *Automatic Control Systems,* Cornell University Press, Ithaca, N.Y., 1977.

The standard reference text on statistical process control (SPC) is
Western Electric Handbook Committee, *Statistical Quality Control Handbook,* Western Electric, AT&T Technologies, Indianapolis, Ind., 1956.

An article reviewing Shewhart control charts and cumulative sum control charts and proposing a control chart strategy using exponentially weighted moving averages is
Hunter, J. S., "The Exponentially Weighted Moving Average," *J. Quality Technology,* vol. 18, no. 4, October 1986.
In this article, the author also proposes the exponentially weighted moving average as a basis for a feedback control algorithm. Before attempting to use this approach, the reader should keep in mind the comments made in Section 2.6.3 about introducing filters in feedback loops.

An interesting and popular article on the problem discussed in Section 2.10 is
MacGregor, J. F., "Interfaces between Process Control and On-line Statistical Process Control," *A.I.Ch.E. Computing and Systems Technology Division Communications,* vol. 10, no. 2, September, 1987, pp. 9–20. (This publication is also referred to as the *C.A.S.T. Newsletter of the A.I.Ch.E.*)

The original article by the fathers of the most widely known tuning method for the PID controller is

ZIEGLER, J. G., and NICHOLS, N. B., "Optimum Settings for Automatic Controllers," *Trans. ASME,* vol. 64, 1942, p. 759.

The first papers on the Kalman filter were

SWERLING, P., "First Order Error Propagation in a Stagewise Smoothing Procedure for Satellite Observations," *J. Astronautical Sciences,* vol. 6, 1959, pp. 46–52.

and

KALMAN, R. E., "A New Approach to Linear Filtering and Prediction Problems," *J. Basic Eng.,* March 1960, pp. 35–46.

One of the most readable books on the Kalman filter is

GELB, A., *Applied Optimal Estimation,* M.I.T. Press, Cambridge, Mass., 1974.

A modification to the Kalman filter for the case when the measurement noise is autoregressive is given in

BRYSON, A. E., and HENDRIKSON, L. J., "Estimation Using Sampled Data Containing Sequentially Correlated Noise," *J. Spacecraft and Rockets,* vol. 5, June 1968, pp. 662–665.

CHAPTER 3

EMPIRICAL SPECTRAL
ANALYSIS
OF NOISY PROCESSES

In the first chapter, several concepts of process dynamics were introduced using the first-order-with-deadtime (or FOWDT) model subject to stochastic disturbances. In the second chapter, the PID control algorithm and ways to tune it were discussed. For these topics it was quite satisfactory to stay in the discrete time domain. The Laplace and Z-transforms were avoided because, at that point, they simply would not have given the reader any added insight. However, in this chapter a transition from the discrete time domain to the discrete frequency domain is made in order not only to characterize the stochastic disturbances and analyze processes as candidates for new or improved control strategies, but also to search for dynamic problems not obvious to the naked eye.

The main tool will be the discrete-time Fourier series in the guise of the Fourier line spectrum, which is also often referred to as the power spectrum, although the power spectrum will be defined more carefully in Chapter Five. The approach will be to treat a sequence of sampled values of a process variable, usually the controlled variable, as a time series. A spectral analysis of this time series will consist of determining the harmonic content or, in other words, the elements of a time series will be treated as though they were samples of a musical sound, and an attempt will be made to find the strengths of the various harmonics that constitute it. While carrying out these frequency-domain analyses, continual reference will be made to the time-domain tool, the autocorrelation, which was discussed in some detail in Section 1.3.

The reasons for using spectral analysis are threefold. First is the ability to detect deterministic periodic components buried in the noisy signal that represents the process variable. The presence of such periodic components often is symptomatic of undetected problems in the process that have yet to manifest themselves as process downtime. Second is the ability to locate harmonic strength in the frequency spectrum. If there is harmonic strength at low frequencies, then often feedback control can remove that strength and produce better control. Conversely, if there is harmonic strength at high frequencies, then, although it probably cannot be removed by feedback control, it may be the consequence of a solvable process problem.

Third, the line spectrum can be used as a measuring stick for the goodness of control if the sampling interval has the same order of magnitude as the effective process time constant. When the line spectrum shows that the controlled variable is a white noise sequence with its mean on target, one can feel comfortable that even though the signal is noisy there is not much more that control can do to help the process. However, as will be shown, only with processes having time constants and deadtimes small relative to the control interval can a control strategy be expected to turn a colored (a highly autocorrelated) disturbance into a white noise sequence.

After a couple of introductory examples designed to whet the reader's appetite, the discrete-time Fourier series will be introduced. Unfortunately, there will be no way to avoid some of the associated mathematics since the discrete-time Fourier series is in fact a mathematical concept. However, an attempt will be made to keep it simple and cover just the essentials. For a more thorough discussion, the reader is referred to the text by Jenkins and Watts (1968) and the two texts by Oppenheim and Shafer (1975, 1989). Also, readers will be able to make much more of the material in this chapter if they have already read Section 1.3.

With the discrete-time Fourier series in hand, the line spectrum and some of the subtleties associated with the Nyquist frequency and aliasing will be discussed. The cumulative line spectrum will be shown to be useful in interpreting the degree of whiteness of a signal. As an important tool in interpreting data, the sum of the elements of the line spectrum will be shown to be equal to the variance. To condense the graphical information supplied by the line spectrum into one quantity, the color numbers will be introduced.

Many times, sharp discontinuities and nonstationarity in the time-domain data can make the interpretation of the line spectrum difficult; therefore, two sections will address these two topics. Two example processes will be used to illustrate how the line spectrum can be used to study processes under control: one with a short time constant and no deadtime and one with an appreciable deadtime and a long time constant. Some time will be spent analyzing a problem that was considered in Chapter Two: how to control a process subject to white noise.

The discrete-time Fourier series equations will be generalized to the exponential form of the discrete-time Fourier series pair and then, with this new notation in hand, frequency domain filtering will be discussed. With this tool, the time-domain data can be transformed into the frequency domain, where certain frequency bands can be removed. Then the altered frequency-domain data can be transformed back to the time domain for further analysis.

3.1 INTRODUCTION TO THE METHODS OF SPECTRAL ANALYSIS

3.1.1 Autoregressive Signal with a Hidden Periodic Component

Consider the time-domain plot of a stochastic sequence shown in Figure 3.1 and the histogram of that sequence shown in Figure 3.2. Here, 512 values of a signal sampled at 1-second intervals appear to the naked eye to be highly autocorrelated. The autocorrelation (discussed in Section 1.3) of the data shown in Figure 3.3 drops off slowly with increasing lags, indicating that it is probably an autoregressive stochastic sequence. At large lags the autocorrelations seem to have a periodic nature, but there is not much else to be found in the autocorrelation plot. Often with signals that have autocorrelations that drop off slowly, it pays to difference the data and repeat the autocorrelation. The result of such an effort is shown in Figure 3.4 (differencing of data will be mentioned later in Section 3.8). The peaks of the autocorrelation, after differencing, alternate with a period of four lags, suggesting strongly that the time-domain data have a deterministic periodic component with a period of 4 seconds.

The line spectrum shown in Figure 3.5 can give the same kind of insight into the nature of this stochastic sequence. Before proceeding, the line spectrum will be crudely defined as a plot of harmonic strength versus the harmonics of the

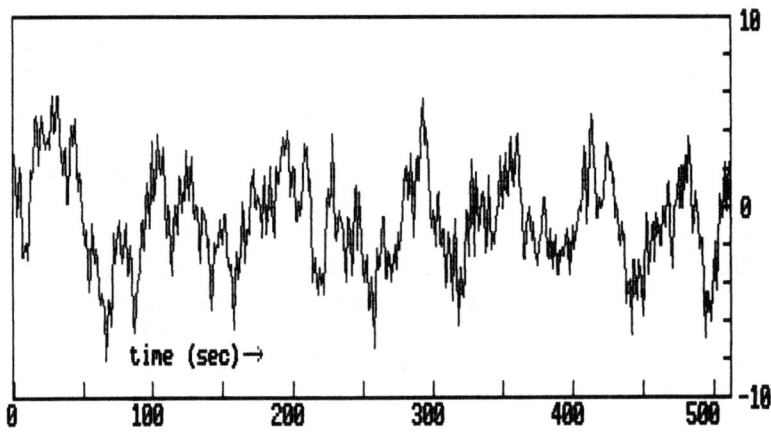

Figure 3.1 Time-domain plot of a complex signal.

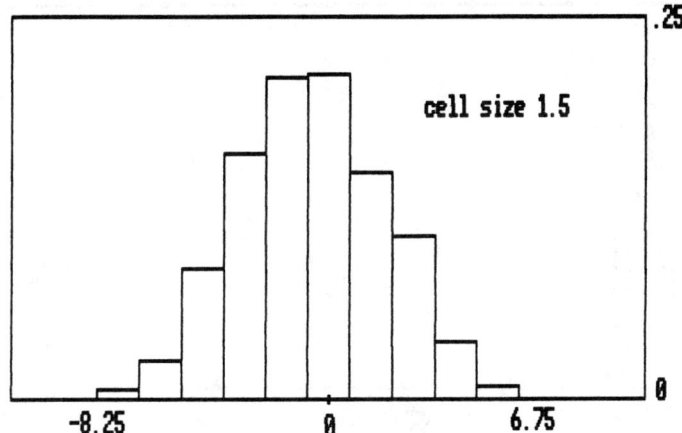

Figure 3.2   Histogram of a complex signal.

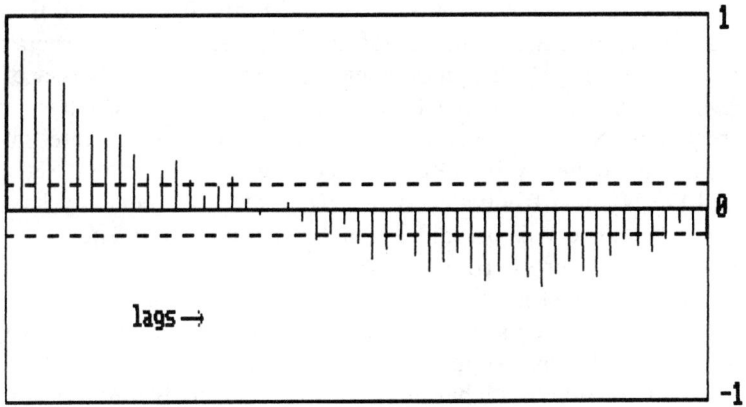

Figure 3.3   Autocorrelation of a complex signal.

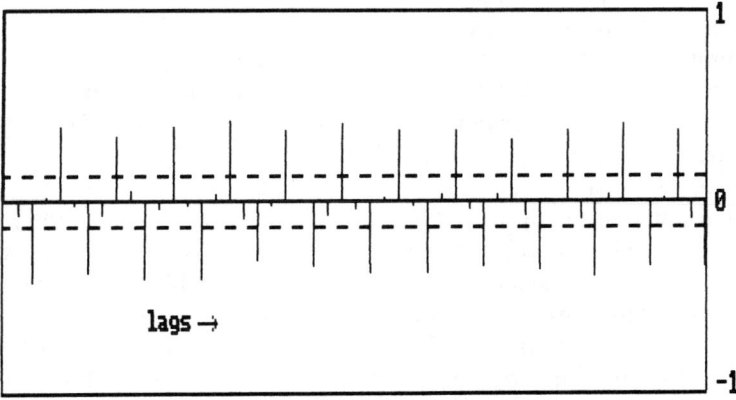

Figure 3.4   Autocorrelation of a complex signal after differencing.

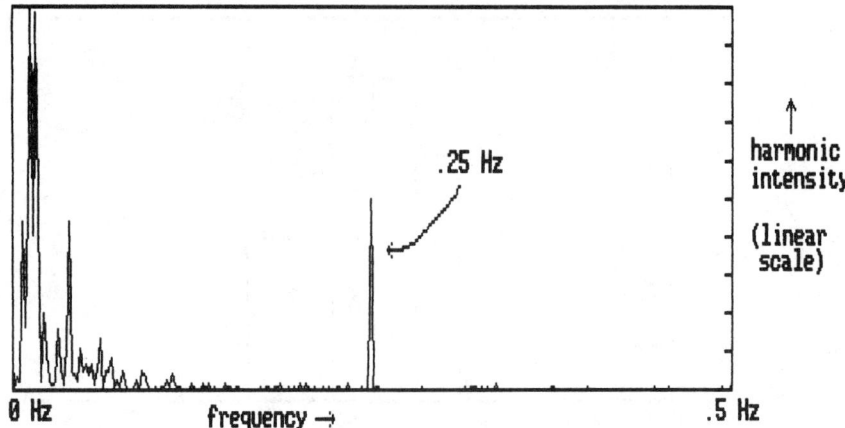

Figure 3.5   Line spectrum of a complex signal.

fundamental frequency. This last quantity, the fundamental frequency, has a pe-riod equal to length of the total data set. (A harmonic frequency is simply an integer multiple of the fundamental frequency.) The abcissa of a line spectrum ranges from zero frequency to the *Nyquist frequency*, which is half of the sampling frequency. Since this example sequence has a sampling interval of 1 second, the sampling frequency is 1.0 cycle/second or 1.0 Hz (where Hz stands for hertz), and the Nyquist frequency is 0.5 Hz, that is, a frequency having a period of 2 seconds. More will be said about the Nyquist frequency later. The frequency interval from zero hertz to the Nyquist frequency will be subsequently referred to as the Nyquist interval.

The ordinate of the line spectrum graph ranges linearly from zero harmonic strength to the maximum harmonic strength occurring over the Nyquist interval. The actual value of the harmonic strength is usually unimportant; it is the relative distribution over the Nyquist interval that counts. Often the harmonic strengths will be normalized by the maximum harmonic strength occurring in the Nyquist interval. Then 10.0 times the logarithm to the base 10 of this normalized harmonic strength will be plotted versus frequency, in which case the ordinate will have units of decibels (dB). This latter choice for the ordinate is sometimes helpful when dealing with data having high concentrations of harmonic strength at low frequencies. Further comments about these two types of plotting will be made in Section 3.12.2.

Right away the reader sees in Figure 3.5 that there is a concentration of harmonic strength at the very low frequencies and that the harmonic strength quickly tails off with increasing frequency. This is a characteristic of autoregres-sive stochastic sequences. However, in the middle of the line spectrum there is an isolated sharp peak at a frequency of 0.25 Hz, that is, a sinusoidal component having a period of 4 seconds. The presence of a deterministic sinusoidal com-ponent having a period of 4 seconds is not at all apparent from looking at the time domain trace in Figure 3.1.

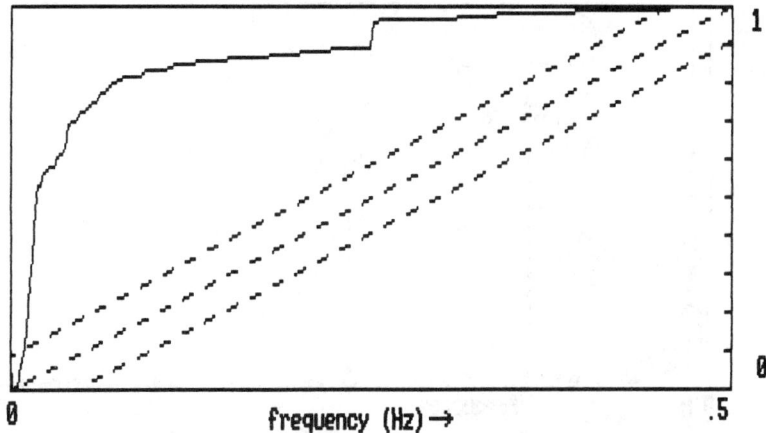

Figure 3.6 Cumulative line spectrum of a complex signal.

Figure 3.6 shows the cumulative line spectrum of the sequence plotted in Figure 3.1. The cumulative line spectrum is a normalized running sum of the line spectrum. In other words, the value of the cumulative line spectrum at frequency f is the sum of all the harmonic strengths at all the harmonics of the fundamental frequency less than f, normalized by the variance. As will be shown later, for a sequence that has had its mean subtracted from it, the sum of the line spectrum values from zero frequency to the Nyquist frequency is equal to the variance of the sequence about its mean. Therefore, if the cumulative line spectrum is plotted from zero frequency to the Nyquist frequency, the curve will always end up at unity on the y-axis at the Nyquist frequency. In Figure 3.6 the cumulative line spectrum rises smoothly but quickly due to the harmonic strength at the low frequencies and then takes an abrupt jump at 0.25 Hz. The dotted lines occurring in Figure 3.6 will be explained in Section 3.5.

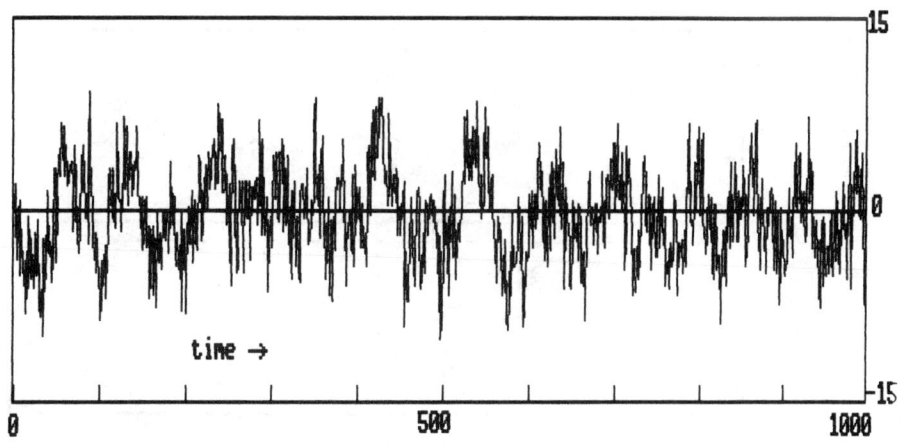

Figure 3.7 Time-domain plot of second example.

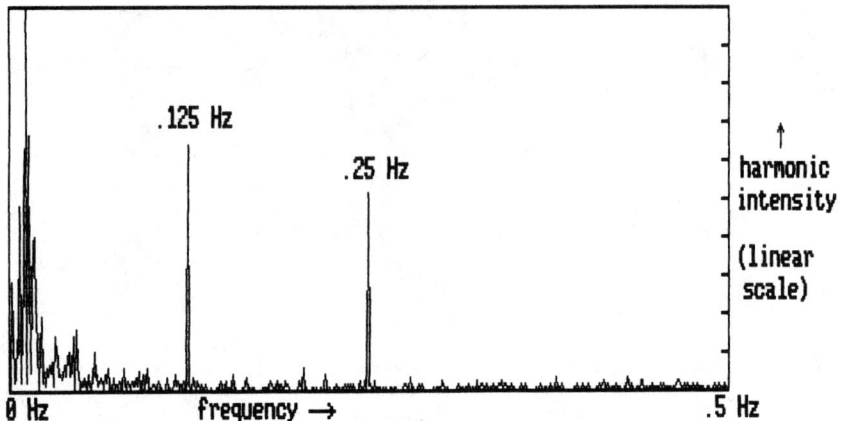

Figure 3.8 Line spectrum of second example.

This example should show that the autocorrelation, the line spectrum, and the cumulative line spectrum all, with close enough inspection, can tell the same things about a stochastic sequence. To further sharpen the distinction between what the line spectrum and autocorrelation can do, consider another example time series. Inspection of Figure 3.7, which shows the time-domain plot of this sequence, reveals little about its nature. However, the harmonic strength in the low frequencies in the line spectrum, shown in Figure 3.8, indicates that the sequence has an autocorrelated stochastic component. It also has *two* deterministic periodic components. The autocorrelation of this data set is similar to that of the first example shown in Figure 3.3. The autocorrelation of the differenced data is shown in Figure 3.9. Here, with much effort, one might be able to deduce the presence of two periodic components, but it is not anywhere as clear as in the first example. Thus, it appears that the line spectrum is more useful than the autocorrelation

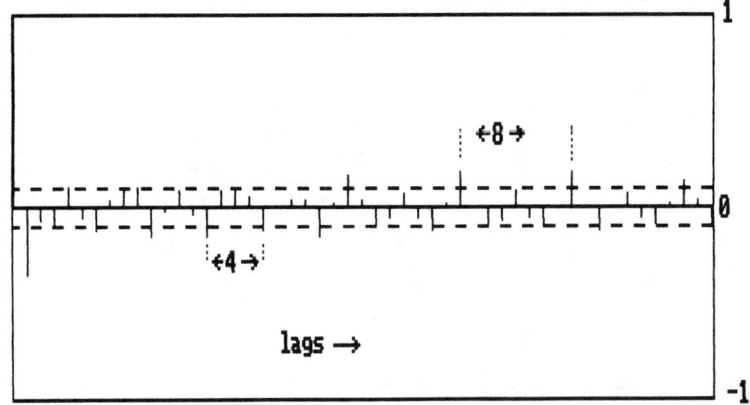

Figure 3.9 Autocorrelation of differenced data from second example.

when one is trying to determine the presence of periodic components in noisy data.

This ability of the line spectrum to reveal the presence of deterministic periodic signals in noisy data suggests it as a tool of the process analyst. For example, if the spectrum from the last example happened to pertain to data from an industrial process, the presence of the two hidden periodic components might be the result of malfunctioning equipment about to break down. Being able to detect this early warning signal and solve the associated problem before downtime occurred could mean significant savings in the cost of manufacturing.

3.1.2 Line Spectrum of White Noise

Figure 3.10 shows the time-domain plot of an unautocorrelated stochastic sequence (white noise), and Figure 3.11 shows the histogram of that sequence. Not only are there no autoregressive components present, but neither are there deterministic periodic components as there were in the previous example. Figure 3.12 shows the autocorrelation of the same sequence. Note that all the autocorrelation for nonzero lags lie between the dashed lines and there appears to be no pattern to them.

The line spectrum of the white noise sequence is shown in Figure 3.13. Although the variance of N samples of white noise is supposed to be distributed uniformly over all the harmonics of the fundamental frequency of $1/(Nh)$ Hz, Figure 3.13 suggests quite otherwise. Instead of being flat, which one would expect if all harmonics had equal power, the line spectrum of white noise is just as noisy as the original time-domain graph. The real question is as follows: is there any harmonic strength associated with all those peaks in the line spectrum?

The answer is no, and it is given by the cumulative line spectrum of white noise shown in Figure 3.14. Since the cumulative line spectrum is essentially the

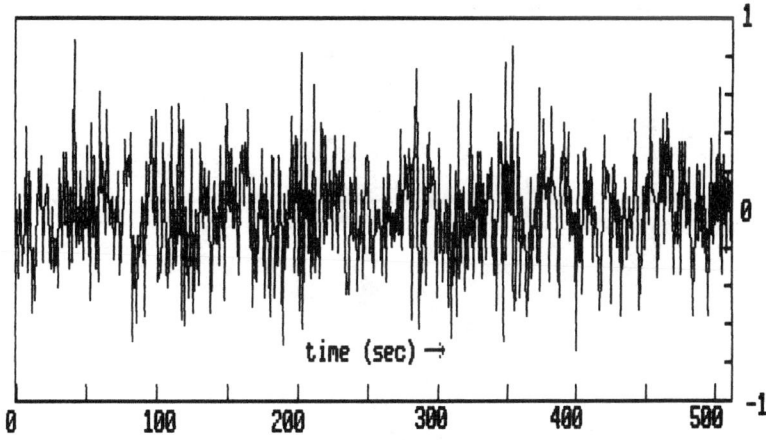

Figure 3.10   Time-domain plot of white noise.

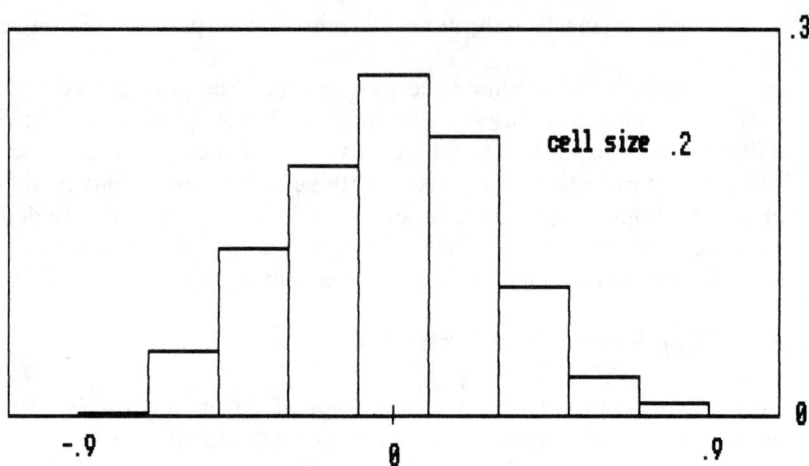

Figure 3.11 Histogram of white noise.

discrete indefinite integral of the line spectrum, the cumulative line spectrum of a sequence having a flat line spectrum should be a ramp from zero to unity, because the integral of a constant is a straight line with a slope equal to the constant. This is what Figure 3.14 shows. Here the dashed lines give boundaries that a white noise sequence must stay within. (These boundaries will be discussed in Section 3.5.) Thus the smoothing afforded by the summing action in the cumulative spectrum allows one to test for the "whiteness" of a sequence when the line spectrum by itself is relatively useless.

These two examples illustrate how the line spectrum and the cumulative line spectrum can be used to ferret out periodic components buried in a noisy signal

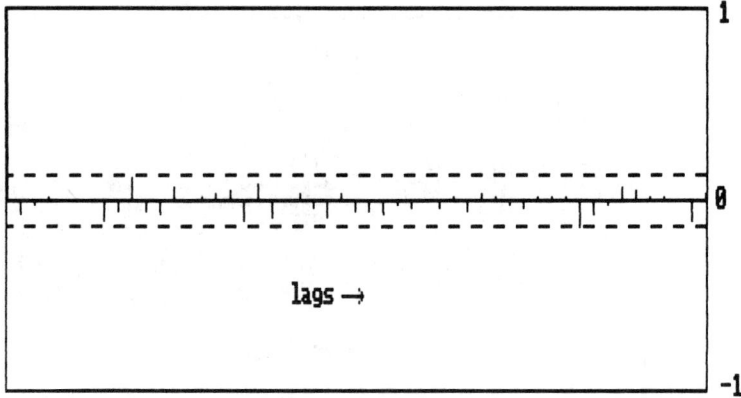

Figure 3.12 Autocorrelation of white noise.

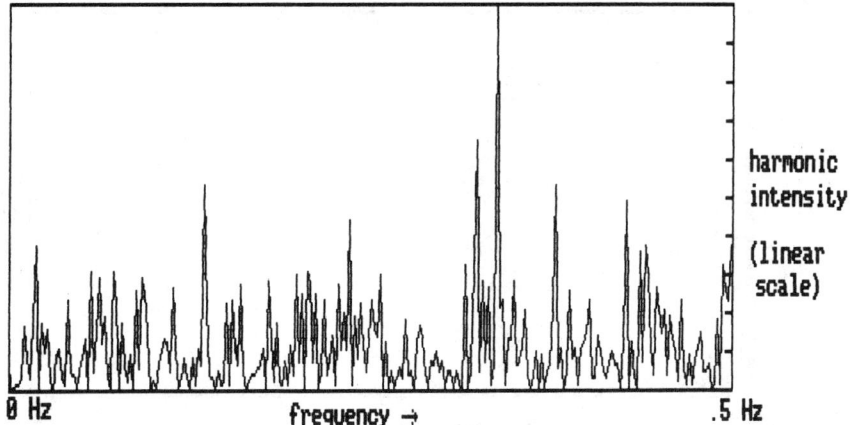

Figure 3.13   Line spectrum of white noise.

and also to test for the white noise. The term white noise comes from physics, where the visible spectrum is said to be white if all the frequencies in the spectrum are equally represented. Colors occur when various portions of the spectrum are more strongly represented. Therefore, in theory, discrete white noise would contain components from all harmonics of the fundamental frequency in the Nyquist interval at equal strength, while discrete *colored* noise would contain a nonuniform distribution of power over the same Nyquist interval. In practice, empirical white noise does not contain a perfectly even distribution of the variance over all the discrete frequencies in the Nyquist interval (as has been shown above), and this is why the cumulative line spectrum is so useful.

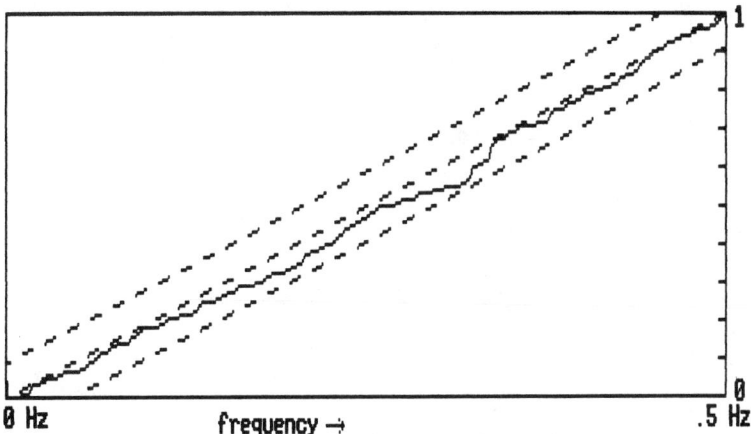

Figure 3.14   Cumulative line spectrum of white noise.

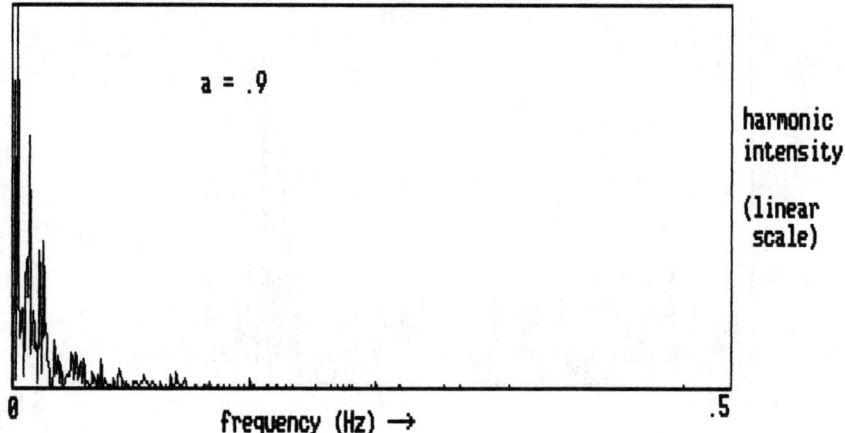

Figure 3.15 Line spectrum of an autoregressive stochastic sequence (coefficient = 0.9).

3.1.3 Line Spectrum of Some Common Stochastic Sequences

To complement Section 1.3, where the autocorrelations of some simple stochastic sequences were shown, the corresponding line spectra will be discussed here. First, consider autoregressive stochastic sequences. Figure 3.15 shows the line spectrum of 1024 samples of an autoregressive sequence with a coefficient of 0.9. Figure 3.16 shows the same thing when the coefficient is -0.9. Note that the harmonic power is concentrated at the lower frequencies in the former and at the higher frequencies in the latter.

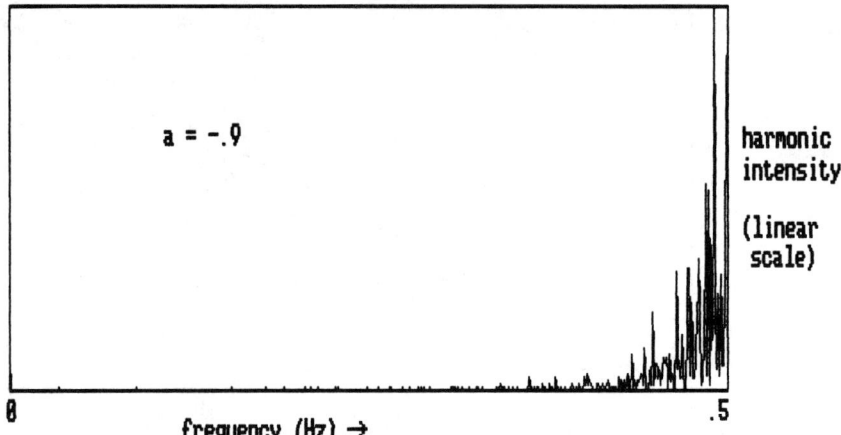

Figure 3.16 Line spectrum of an autoregressive stochastic sequence (coefficient = -0.9).

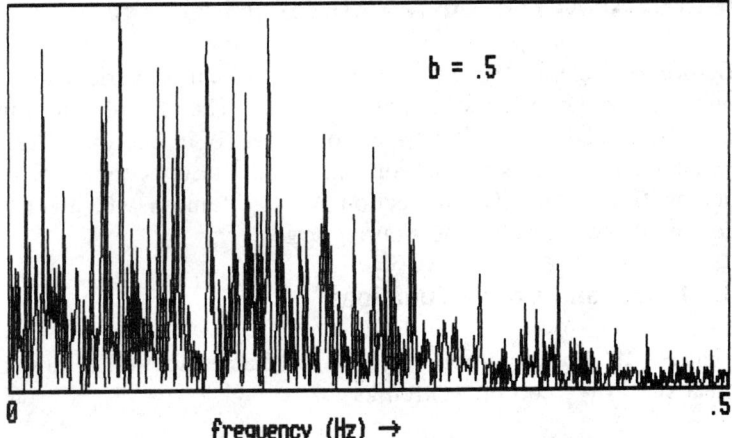

Figure 3.17   Line spectrum of a moving average stochastic sequence (coefficient = 0.5).

Figures 3.17 and 3.18 show the same thing for moving average stochastic sequences. The first figure of the pair has a moving average coefficient of 0.5 and the second figure has a moving average coefficient of -0.5. As with the autoregressive sequences, the harmonic power is concentrated at opposite ends of the Nyquist interval. Note that if, in Figure 3.18, the moving average coefficient had been -1.0 instead of -0.5, then the result would have been differenced white noise. More will be said about this in Section 3.8. Note also that if one is trying to distinguish between moving average and autoregressive stochastic sequences the autocorrelation, rather than the line spectrum, is the tool to use.

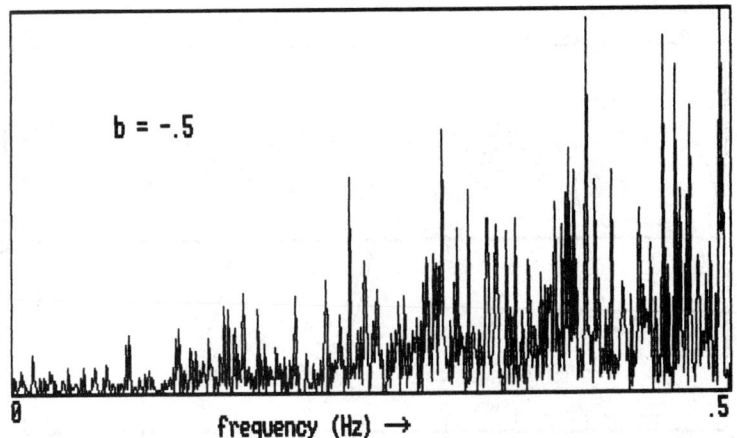

Figure 3.18   Line spectrum of a moving average stochastic sequence (coefficient = -0.5).

3.2 FOURIER ANALYSIS OF A TIME SERIES

Before getting into the actual details of the Fourier series, some elementary trig-onometry will be reviewed in Section 3.2.1. The idea of determining a polynomial least-squares data fit will be mentioned briefly in Section 3.2.2. This idea will be extended to cosines in Section 3.2.3 and the concept of orthogonality will be demonstrated. Finally, in Section 3.2.4, attention will given to the full-blown details of the discrete-time Fourier series.

3.2.1 Sine and Cosine Functions

Figure 3.19 shows a graph of a sine wave as a function of time t. The expression for a sine wave can be written as

$$y(t) = \sin\left(\frac{2\pi t}{H}\right)$$

where H is the period of the sine wave, that is, the length of time required for the sine function to repeat itself or the amount of time it takes the sine function to complete one cycle. The frequency of the sine function is given by

$$f = \frac{1}{H}$$

and the sine function could therefore also be written as

$$y(t) = \sin(2\pi f t)$$

The cosine function is shown in Figure 3.20 and is seen to have the same shape as the sine function except, that it takes on a minimum or a maximum where the

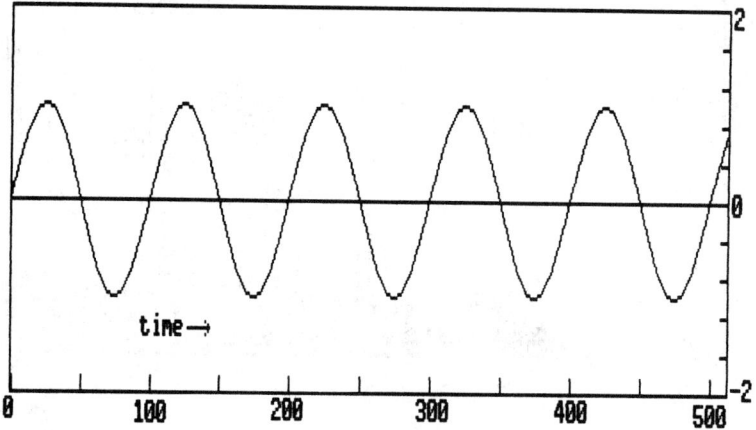

Figure 3.19   Time-domain plot of sin $(2\pi t/100)$.

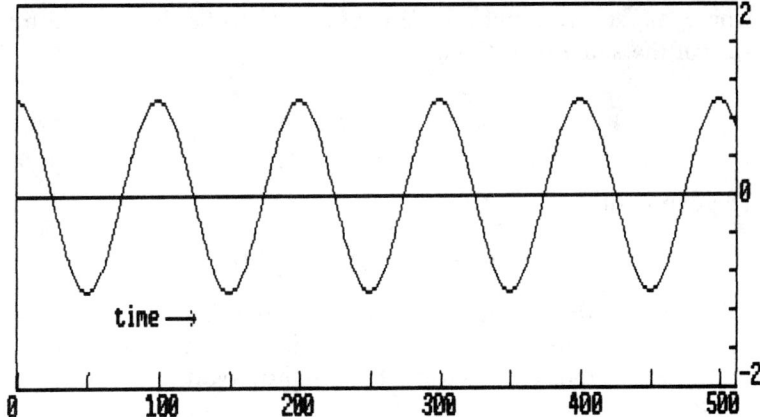

Figure 3.20   Time-domain plot of cos $(2\pi t/100)$.

sine function is crossing zero. The expression for a cosine function can be written as

$$y(t) = \cos\left(\frac{2\pi t}{H}\right)$$

The sine and cosine functions are said to out of phase by 90° or $\pi/2$ radians:

$$\sin(2\pi ft) = \cos\left(2\pi ft - \frac{\pi}{2}\right)$$

One could also say that the cosine leads the sine by 90°.

3.2.2 Approximating Sampled Data with a Polynomial in t

Using least squares to fit data with a polynomial will be discussed first. Then in the next section the same approach will be applied to fitting data with a sum of cosines.

Consider a sequence of sampled data values, $c_1, c_2, \ldots, c_N$, where t_i is the time that the ith sample is taken, where

$$t_i = t_{i-1} + h$$

and where h is the sampling interval, which is constant. The data might be approximated with a polynomial of, say, third order in t as follows:

$$c_i = A_0 + A_1 t_i + A_2(t_i)^2 + A_3(t_i)^3 + e_i \tag{3.1}$$

$$i = 1, 2, \ldots, N$$

where A_0, A_1, A_2, and A_3 are coefficients to be determined by minimizing the sum of the squares of the errors:

$$S = \sum_{i=1}^{N} e_i^2$$

The expression

$$c_i = A_0 + A_1 t_i + A_2 (t_i)^2 + A_3 (t_i)^3 \tag{3.2}$$

will be referred to as the *approximation equation*.

The theoretical procedure of "doing" least squares is the following. The sum S is minimized by setting its partial derivatives, with respect to the coefficients, equal to zero:

$$\partial S / \partial A_i = 0, \qquad i = 0, 1, 2, 3 \tag{3.3}$$

This condition generates four equations, sometimes called the *normal equations,* which can be solved for values of the coefficients.

The coefficients found by this approach define a *least-squares* approximation to the original data. If the order of the polynomial is less than $N - 1$, then the polynomial will come close to each sample, but in general will not pass through any of them exactly; that is, the e_i will not all be zero. If the order of the polynomial is equal to $N - 1$, then the polynomial will pass through each sample point, each e_i will be zero, and the resulting polynomial will be called an *interpolating polynomial*.

It turns out that the equations resulting from evaluating Equation (3.3) can also be generated by multiplying both sides of the approximation equation [Equation (3.2)] by powers of t_i and summing over the N points. For example, if Equation (3.2) is multiplied by the zeroth power, that is, unity, and the sum is taken over the N points, the result is

$$\sum_{i=1}^{N} c_i = A_0 N + A_1 \sum_{i=1}^{N} t_i + A_2 \sum_{i=1}^{N} (t_i)^2 + A_3 \sum_{i=1}^{N} (t_i)^3$$

Similarly, if Equation (3.2) is multiplied by the first power, that is, t_i, and the sum is taken over the N points, the result is

$$\sum_{i=1}^{N} t_i c_i = A_0 \sum_{i=1}^{N} t_i + A_1 \sum_{i=1}^{N} (t_i)^2 + A_2 \sum_{i=1}^{N} (t_i)^3 + A_3 \sum_{i=1}^{N} (t_i)^4$$

Doing the same thing for the second and third powers of t_i will generate two more equations that, when complemented by the first two equations, constitute a set of four simultaneous equations that can be solved for the four coefficients A_0, A_1, A_2, and A_3.

Note that in each of the normal equations all the coefficients appear linearly, but it is not possible, in general, to develop a closed-form solution for the A_i's unless the number of coefficients is two or less. Also note that the value of A_0 found for a second-order polynomial fit will not be the same as the value of A_0 found for a third-order fit. In other words, if one fits the data with a quadratic polynomial, getting values of A_0, A_1, A_2, and then refits the data with a cubic polynomial, getting values of A_0', A_1', A_2', A_3', one cannot expect that $A_0 = A_0'$, $A_1 = A_1'$, or $A_2 = A_2'$.

3.2.3 Approximating Sampled Data with a Sum of Cosines

Consider the N sampled data points $c_1, c_2, \ldots, c_N$, and fit them with the following expression:

$$c_i = A_0 + A_1 \cos(2\pi f_1 t_i) + A_2 \cos(2\pi f_2 t_i) + e_i \qquad (3.4)$$

This is similar to Equation (3.1) except that the cosines with different frequencies, f_1 and f_2, have replaced the different powers of t. Let f_1 be the *fundamental* frequency associated with the N samples:

$$f_1 = \frac{1}{Nh}$$

This frequency has a period equal the length of the sequence of sampled data, Nh. Next, let $f_2 = 2[1/(Nh)] = 2f_1$. In other words, the sampled data are being approximated with the sum of a constant, a term proportional to the cosine of the fundamental frequency, and a term proportional to the cosine of the second harmonic of the fundamental frequency. The constant represents the zeroth frequency component of the approximation function (which will be shown to be equal to the average of the N samples). So, the approximation consists of the zeroth, the fundamental, and the second harmonic frequencies associated with the N sample points.

As in Section 3.2.2, a least-squares solution to the problem is of interest. To develop equations that could be used to solve for the coefficients A_0, A_1, and A_2, an approach similar to the one used for the polynomial approximation will be tried. That is, the approximation equation will be multiplied by cosines with frequencies of the zeroth, the fundamental, and the second harmonic frequency, and sums over the data of the resulting equations will be formed. This procedure will generate three equations that can be solved for the three coefficients. These three equations could also be generated by setting the partial derivatives, with respect to the three coefficients, of the sum of the squares of the errors to zero.

To make the manipulations easier, assume that N is even and furthermore that $N = 4$. Next, assume that the sample interval h is unity and $t_0 = 0$. With this latter simplification, subscripts can be used to make notation more compact since

$$t_i = t_{i-1} + h = i$$

and

$$f_k = k \frac{1}{Nh} = \frac{k}{4}$$

Note that a function with three coefficients is being used to fit four data points, so the approximation function will not go through each point exactly.

To generate the first equation, multiply the approximation equation derived from Equation (3.4) by the zeroth frequency component, which is unity, and sum over the N (or four) points:

$$\sum_{i=1}^{N} c_i = NA_0 + A_1 \sum_{i=1}^{N} \cos(2\pi f_1 i) + A_2 \sum_{i=1}^{N} \cos(2\pi f_2 i) \tag{3.5}$$

It would be nice if the two sums in Equation (3.5) would drop out, allowing A_0 to be solved for directly. Amazingly enough, they do. Look at the first sum:

$$\sum \cos(2\pi f_1 i) = \cos\left(\frac{2\pi 1}{4}\right) + \cos\left(\frac{2\pi 2}{4}\right) + \cos\left(\frac{2\pi 3}{4}\right) + \cos\left(\frac{2\pi 4}{4}\right) = 0$$
$$= 0 \quad\quad + \quad (-1) \quad + \quad 0 \quad + \quad (1)$$

The reader can verify that the second sum also equals zero. Therefore, the coefficient of the zeroth frequency component of the approximation can be solved for directly:

$$A_0 = \frac{1}{N} \sum_{i=1}^{N} \frac{c_i}{N}$$

Not only is A_0 the average of the N samples, but it is important to note that the value of A_0 would not change if the approximation function were modified by adding more cosine terms having frequencies that were higher harmonics of f_1. This was not the case for the previous example that dealt with polynomials.

The reader should start to suspect that there is a generalization about to appear concerning sums of sinusoids defined on equally spaced sampled data.

To generate a second equation that could be used for the determination of the coefficients, multiply Equation (3.4) by the cosine of the fundamental frequency associated with the sampled data and sum over the N data points:

$$\sum_{i=1}^{N} c_i \cos(2\pi f_1 i) = A_0 \sum_{i=1}^{N} \cos(2\pi f_1 i) + A_1 \sum_{i=1}^{N} \cos(2\pi f_1 i) \cos(2\pi f_1 i)$$

$$+ A_2 \sum_{i=1}^{N} \cos(2\pi f_1 i) \cos(2\pi f_2 i)$$

The previous analysis showed that the sum in the first term on the right side is zero, so consider the second sum on the right side. Fortunately, it is simply the sum of the squares of the components of the first sum, so it equals 2.0.

Finally, look at the last sum:

$$\sum_{i=1}^{4} \cos\left(\frac{2\pi i}{4}\right) \cos\left(\frac{2\pi 2i}{4}\right) = \cos\left(\frac{\pi}{2}\right) \cos(\pi) + \cos(\pi) \cos(2\pi)$$

$$+ \cos\left(\frac{2\pi 3}{4}\right) \cos(3\pi) + \cos(2\pi) \cos(4\pi)$$

$$= 0 + (-1)$$

$$0 + 1 = 0$$

Again the amazing has happened, all the troublesome sums drop out, leaving one equation to be solved for A_1:

$$A_1 = \frac{1}{2} \sum_{i=1}^{4} c_i \cos\left(\frac{2\pi i}{4}\right)$$

Finally, the approximation equation could have been multiplied by the cosine of the second harmonic, yielding an equation for A_2. Also, note that, just as with A_0, the equation for A_1 or A_2 will not change if the approximation function is modified by adding a cosine of the third harmonic.

Comparing the results of the last two sections suggests that an approximation function consisting of cosines rather than polynomials has some nice properties. In the next section these properties will be expanded and generalized.

3.2.4 Discrete-time Fourier Series

Again assume that there are N equally spaced (in time) samples of a process variable, c_i, $i = 1, 2, \ldots, N$, where N is an even number. This time it is desired to approximate them by a sum of sines and cosines given as

$$c_i = A_0 + 2 \sum_{m=1}^{n-1} \{A_m \cos(2\pi m f_1 ih) + B_m \sin(2\pi m f_1 ih)\}$$

$$+ A_n \cos(2\pi n f_1 ih) + e_i \tag{3.6}$$

where $N = 2n$ and $f_1 = 1/Nh$. There are a total of N coefficients: $A_0, A_1, \ldots,$ $A_{n-1}, A_n, B_1, B_2, \ldots, B_{n-1}$, that is, $n + 1$ A_i's and $n - 1$ B_i's. Since there are N coefficients and N data points, the approximation function passes through each point exactly; that is, the e_i's are all zero and the approximation function is an interpolation function.

The first term on the right side, A_0, is the coefficient of the zeroth-order harmonic or, as electrical engineers say, the dc (direct current) component. The coefficients A_1 and B_1 are associated with the cosine and sine for the fundamental component having a period of Nh and a frequency of $1/(Nh)$. The coefficients A_i and B_i for $i = 2, 3, \ldots, n - 1$, are associated with the cosines and sines for the higher harmonics up to the $(n - 1)$th harmonic, which has a frequency of

$$f_{n-1} = \frac{n - 1}{Nh}$$

Finally, A_n is associated with the cosine of the highest harmonic present, that with a frequency of

$$f_n = \frac{n}{Nh} = \frac{1}{2h}$$

This will be called the *Nyquist frequency*. The sinusoid having this frequency can pass through zero at every data point or N times over the time domain of the samples. For there to be a sinusoid present in the approximation equation having a frequency higher than the Nyquist frequency would not make any sense, because it would pass through zero in between the data points, that is, more than N times over the time domain of the samples. There is no B_n because, if there were, it would be associated with a sine function that would be identically zero:

$$\sin(2\pi f_1 nh) = \sin\left(2\pi \frac{1}{Nh} nh\right) = \sin(\pi) = 0$$

To generate equations for N coefficients, the method of Section 3.2.3 will be used: the approximation equation will be multiplied by a sine or cosine of each harmonic that is present and a sum will be taken over the N data points. Start by multiplying by $\cos(2\pi f_1 hik)$ or $\cos(w_{ik})$, where $w_{ik} = 2\pi f_1 hik$, and summing over the data set:

$$\sum_{i=1}^{N} c_i \cos(w_{ik}) = A_0 \sum_{i=1}^{N} \cos(w_{ik}) + 2 \sum_{m=1}^{n-1} \left[A_m \sum_{i=1}^{N} \cos(w_{im}) \cos(w_{ik}) \right.$$

$$\left. + B_k \sum_{i=1}^{N} \sin(w_{im}) \cos(w_{ik}) \right] + A_n \sum_{i=1}^{N} \cos(w_{in}) \cos(w_{ik})$$

(3.7)

As the reader might have guessed from the results of Section 3.2.3, the sums of products of cosines and sines with different harmonics drop out of the right side of Equation (3.7), leaving just one sum on the right side:

$$\sum_{i=1}^{N} c_i \cos(w_{ik}) = 2A_k \sum_{i=1}^{N} \cos(w_{ik}) \cos(w_{ik})$$

where k is any of the indexes except the last one, n. This last nonzero sum can be shown to equal $N/2$, so an expression for the kth coefficient, where $k = 1, 2, \ldots, n - 1$, is given as

$$A_k = \frac{1}{N} \sum_{i=1}^{N} c_i \cos(w_{ik}) \tag{3.8}$$

For the case where $k = n$, the sum of the squared cosines is N instead of $N/2$, but, since the last term in the approximation function has a coefficient of unity instead of 2.0, the expression for A_n is the same as for other A_k. For the case where $k = 0$, Equation (3.8) is simply the sum of c_i over the N data points, yielding an expression identical to one derived in the previous section showing that A_0 is the average or the mean of the data.

If this procedure is repeated using the multiplying factor $\sin(w_{ij})$ instead of $\cos(w_{ij})$, then the following expression for B_k will result:

$$B_k = \frac{1}{N} \sum_{i=1}^{N} c_i \sin(w_{ik}), \qquad k = 1, 2, \ldots, n - 1 \tag{3.9}$$

Equations (3.8) and (3.9) effectively transform the original time-domain data, $c_k, k = 1, 2, \ldots, N$, into the parameters $A_k, k = 0, 1, \ldots, n$, and $B_k, k = 1, 2, \ldots, n - 1$, in the discrete frequency domain. So far N has been even. In the case where N is odd, that is, $N = 2n - 1$, the above equations for A_k and B_k stay the same except that the last term in the series containing A_n vanishes, because with $2n - 1$ data points it only takes $2n - 1$ coefficients, $A_0, A_1, \ldots, A_{n-1}$, $B_1, B_2, \ldots, B_{n-1}$, to fit the data points exactly. Note that the units of the coefficients A_k and B_k are the same as those of c_i.

The fact that the sum of the products of sines and cosines of certain harmonics equals zero has been repeatedly used. This is the property of *orthogonality* and plays an important role throughout applied mathematics. The interested reader can pursue the subject in the text by Hamming (1989).

Equations (3.6), (3.8), and (3.9) have been referred to as the discrete-time Fourier series in order to distinguish it from the continuous-time Fourier series, where an infinite number of sinusoids are used to approximate a known function defined on the continuous time domain.

3.3 FOURIER LINE SPECTRUM OF A TIME SERIES

If the A_k and B_k from Equation (3.6) are combined as follows,

$$R_k^2 = A_k^2 + B_k^2, \qquad k = 0, 1, 2, \ldots, n$$

and if the R_k^2 are plotted versus the frequencies f_k, where $f_k = k f_1$ and where $f_1 = 1/(Nh)$, then the Fourier line spectrum associated with the sampled data

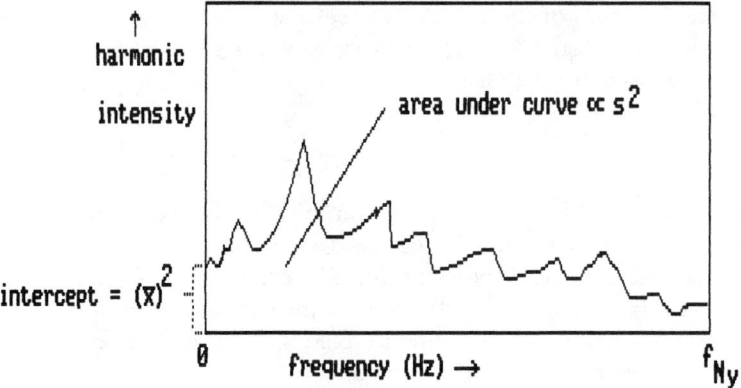

Figure 3.21 The line spectrum is a distribution of the variance.

c_i, $i = 1, 2, \ldots, N$, is the result. Note that, for $k = 0$, $R_0 = A_0$, which is the mean of the time series.

The quantity R_k^2 can be interpreted as the strength (or power) of the kth harmonic that constitutes the time-domain time series. If the units of c_i are volts, then the units of R_k^2 are (volts)^2. If all the elements of the time series are equal to a constant, unchanging with time, then all the R_k^2's, save the zeroth one, will be zero, since all the harmonic strength is in the zeroth harmonic, that is, the mean. If the time series is a sine wave or a cosine wave with a period of, say, Q seconds, where $Q = Nh/q$, q being an integer less than or equal to N, then all

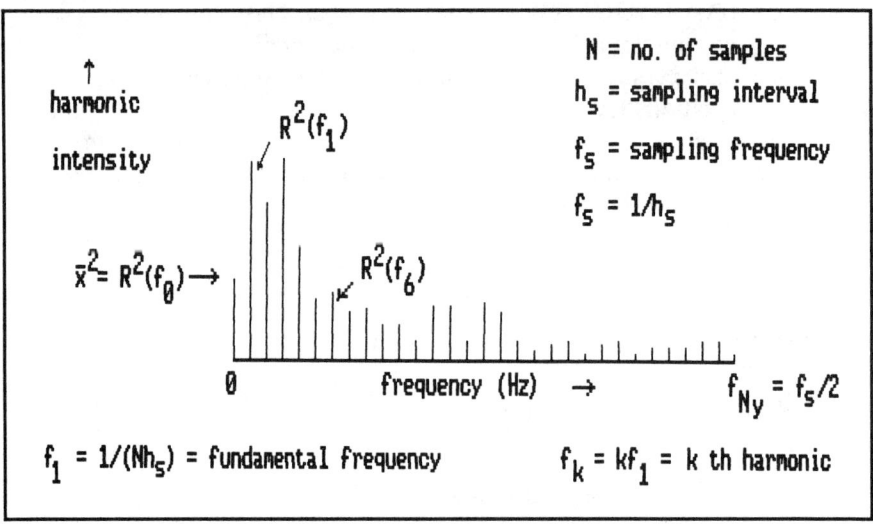

Figure 3.22 The line spectrum is discrete.

the R_k^2's save R_q^2 will be zero, since now all the harmonic strength is in the qth harmonic.

The time series plotted in Figure 3.1 contained a periodic component with a period of 4 seconds. Since the time series contained 512 samples with a sampling interval of 1 second, $N = 512$ and $h = 1$, so for the periodic component $Q = 4 = 512 * 1/q$ and $q = 128$. Thus the value of R_{128}^2 was significantly larger than its neighbors and appeared as a spike in the line spectrum shown in Figure 3.5.

Figures 3.21 and 3.22 summarize all the features of the line spectrum and can serve as a handy reference.

3.4 NYQUIST FREQUENCY AND ALIASING

If the sampling interval is h seconds, then the sampling frequency, f_s, is 1/h hertz. If the time series contains a periodic component with a frequency equal to the sampling frequency, its presence could not be detected because it would always be sampled at the same point in its cycle and it would appear as a constant (the same thing would happen if the periodic component had a frequency equal to an integer multiple of the sampling frequency). Likewise, the presence of a periodic component with a frequency higher than the sampling frequency could not be detected because it would be impossible to sample enough points on each cycle.

In fact, only when the periodic component has a frequency less than the *Nyquist frequency*, f_{Ny}, which is half the sampling frequency, is it possible to sample enough points on each cycle to confirm its presence. If the periodic component of interest has a frequency exactly equal to the Nyquist frequency, then it would be sampled twice per cycle and therefore there would be a chance albeit small, of detecting its presence by looking for a relatively large value of R_n^2.

If there is a periodic component in the process variable that has a frequency greater than the Nyquist frequency, it will show up as an *alias* at a frequency lower than the Nyquist frequency. The following equation gives the rule for determining the frequency of the alias, f_a, associated with a frequency f that is greater than the Nyquist frequency:

$$f_a = | f - mf_s | \tag{3.10}$$

where $m = 1, 2, 3, \ldots$.

Aliasing is one of the more subtle concepts in the field of spectral analysis of data. The presence of aliases can completely confuse the analysis of noisy data. To make aliasing a little more understandable, consider a clock face with a single hand that spins or rotates at a rate of f_{signal} hertz. Imagine that a snapshot of the clock face is taken every second; that is, $f_s = 1$ Hz and therefore $f_{Ny} = 0.5$ Hz. Figure 3.23 shows, from left to right, four snapshots or samples of the clock face spaced 1 second apart in time, where the single hand is rotating at $f_{signal} = 0.08333$ Hz $= \frac{1}{12}$ Hz $= 30°/\text{sec}$. Therefore, with this set of sampled data, it would be tempting to say that a periodic signal with a frequency of $\frac{1}{12}$ Hz had been identified. However, these four samples could just as well have come from a clock hand

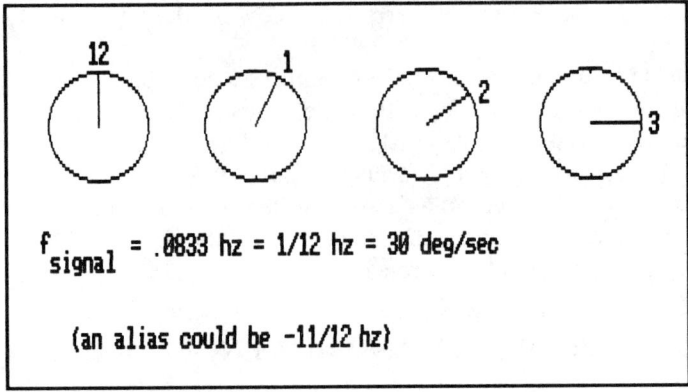

Figure 3.23   Snapshots for f_{signal} = 0.0833 Hz = 1/12 Hz = 30°/sec (an alias could be $-11/12$ Hz).

rotating counterclockwise at a rate of $-\frac{11}{12}$ Hz or $-330°$/sec. In other words, the $\frac{1}{12}$ Hz signal could be an alias of a signal having a frequency of $-\frac{11}{12}$ Hz.

Figure 3.24 shows snapshots of a hand rotating at f_{signal} = 0.5 Hz = 180°/sec, which is the Nyquist frequency. Here an alias could be -0.5 Hz. Figure 3.25 shows snapshots of a hand rotating at f_{signal} = 0.58333 Hz = $\frac{7}{12}$ Hz, which is slightly higher than the Nyquist frequency. This frequency would show up as an alias at $-\frac{5}{12}$ Hz. In fact, since only frequencies between 0 Hz and the Nyquist frequency of 0.5 Hz are considered, a frequency of $\frac{7}{12}$ Hz would not be seen. Instead, one would detect the alias of $-\frac{5}{12}$ Hz.

Figure 3.26 shows snapshots of a hand rotating at f_{signal} = 0.91666 Hz =

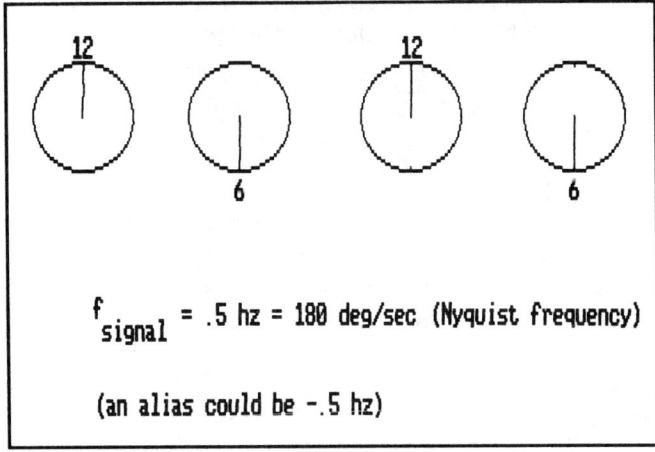

Figure 3.24   Snapshots for f_{signal} = 0.5 Hz = 180°/sec = f_{Ny} (an alias could be -0.5 Hz).

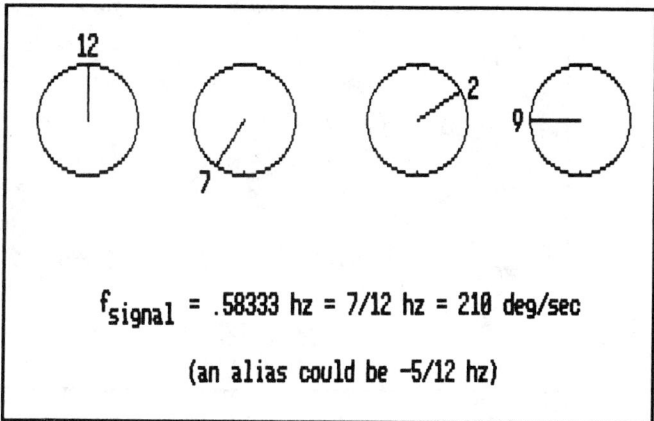

Figure 3.25 Snapshots for f_{signal} = 0.5833 Hz = 7/12 Hz = 210°/sec (an alias could be −5/12 Hz).

$\frac{11}{12}$ Hz = 330°/sec, which is also above the Nyquist frequency and would show up as a frequency of $-\frac{1}{12}$ Hz, which is the alias.

Finally, consider Figure 3.27, which shows a hand that is supposed to be rotating at f_{signal} = 1.08333 Hz = $\frac{13}{12}$ Hz, which is greater than the Nyquist frequency *and* the sampling frequency. The only signal detectable in this case would be the alias having f_{signal} = 0.08333 Hz, which is the signal that appeared in Figure 3.23. Note that in these five rotating clock hand examples the aliases have been arrived at by common sense, but they also satisfy Equation (3.10).

Consider now an example where the time series consists of white noise plus a periodic component with a frequency of 2.95 Hz sampled at 1 Hz. Since the period of the 2.95-Hz signal is 0.339 second and the sampling interval is 1.0 second, the true frequency of the signal cannot be detected. But it will show up as a

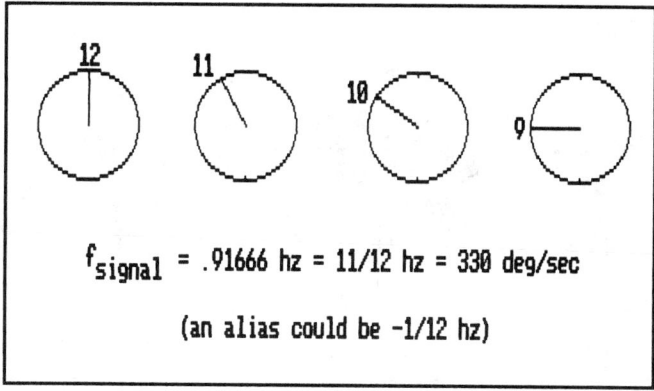

Figure 3.26 Snapshots for f_{signal} = 0.9166 Hz = 11/12 Hz = 330°/sec (an alias could be −1/12 Hz).

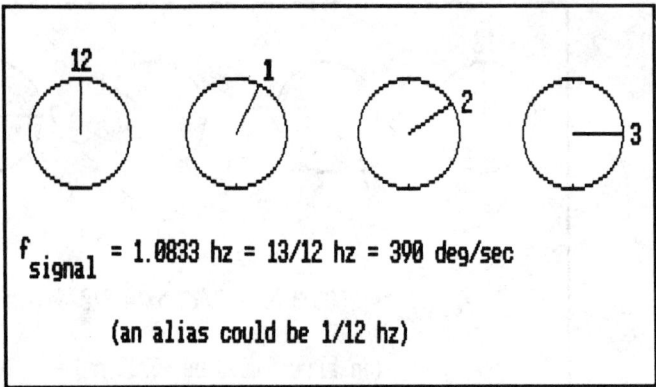

Figure 3.27   Snapshots for f_{signal} = 1.0833 Hz = 13/12 Hz = 390°/sec (an alias could be 1/12 Hz).

periodic component having a frequency of 0.05 Hz and a period of 20 seconds, which is the result of applying Equation (3.10):

$$0.05 = | 2.95 - 3 * 1 |$$

Note that applying Equation (3.10) with $m = 1$ and $m = 2$ would give f_a values that were greater than the Nyquist frequency. Figure 3.28 shows the example sequence in the time domain, while Figure 3.29 shows the line spectrum with the peak at 0.05 Hz. For purposes of comparison, Figure 3.30 shows the autocorrelation of the example sequence. Here the periodic nature of the time series is also evident.

When analyzing experimental data where the line spectrum shows a strong concentration of power at a particular frequency in the Nyquist interval, there is

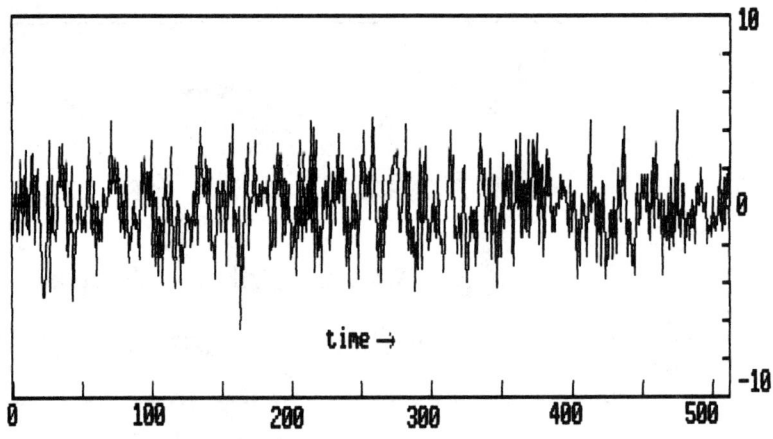

Figure 3.28   Time-domain plot of white noise plus a 2.95-Hz sinusoid.

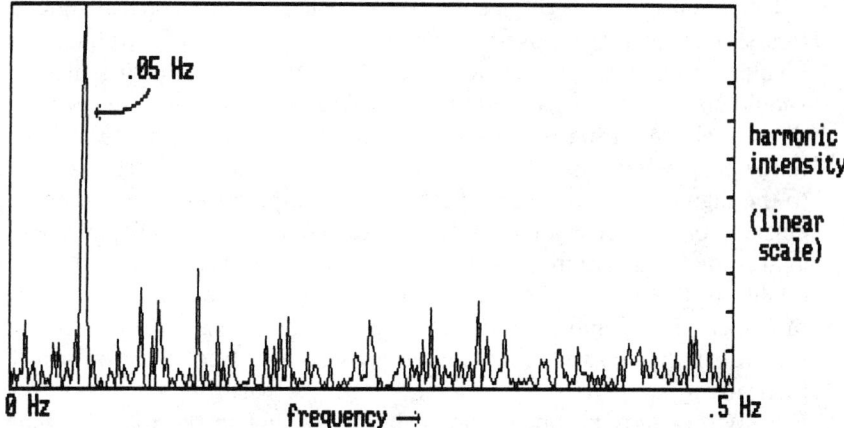

Figure 3.29   Line spectrum of white noise plus a 2.95-Hz sinusoid.

often the possibility that the frequency is an alias of a higher frequency. In this case, when it is feasible, one could resample the data at a higher frequency and recompute the line spectrum. If the frequency is not an alias, then the concentration of power will remain at the original frequency. On the other hand, if the frequency is an alias, then Equation (3.10) shows that it will move to some new place in the Nyquist interval.

Once it has been determined that higher frequencies have corrupted the data as aliases, it may also be feasible to resample at the original sampling interval, but with analog filters inserted at the input of whatever device is being used to sample the data. These analog filters would be designed to remove power at frequencies higher than the Nyquist frequency, thereby removing the threat of aliasing.

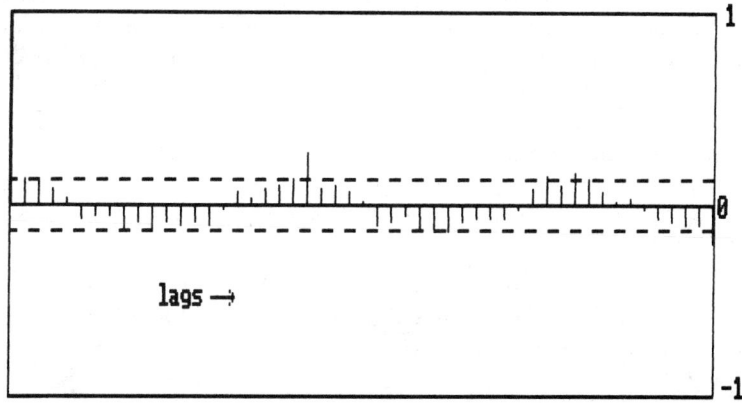

Figure 3.30   Autocorrelation of white noise plus a 2.95-Hz sinusoid.

Actually, whenever the analyst is dealing with electrical signals, analog filters should always be used to remove spectral power at frequencies above the Nyquist frequency. In addition to applying these *antialiasing* filters, the analyst should also check to see that proper wiring practice has been carried out, that is, that the signal wiring is properly shielded and that the shield has been connected to the proper terminals at the input of the device that converts the electrical signal to the digital signal used internally in the computations. Remember that the conclusions of a process analysis are only as good as the quality of the signals being analyzed. Furthermore, the quality of a control strategy is only as good as the quality of the input signals. A control strategy that executes at a control interval of h seconds will have a difficult time attenuating spectral power in the Nyquist interval if that power is an alias of power at a frequency outside the Nyquist interval.

Before moving on, it should be noted that in the digital signal processing field the Nyquist frequency is not necessarily half the sampling frequency. Instead, it often is described as the frequency above which the signal being analyzed has no power. Therefore, using this definition, one could say that if the sampling frequency were greater than twice the Nyquist frequency there would be no chance of aliasing. The definition of the Nyquist frequency as half the sampling frequency will be used in this book.

3.5 TOTAL VARIANCE AND THE CUMULATIVE LINE SPECTRUM

The cumulative line spectrum is simply a running sum of the line spectrum. It is useful for two reasons. First, since the line spectrum of white noise is itself a noisy signal and not easily interpretable, the smoothing effect of summing allows one to discern the original basic characteristic of theoretical white noise: it contains periodic components from all frequencies with equal harmonic intensity. Second, the cumulative line spectrum allows one to get a better feel for how much harmonic strength really lies under a peak occurring in a regular line spectrum. Used together, the line spectrum and the cumulative line spectrum can, along with the autocorrelation, give the analyst much insight into their process.

To develop the necessary equations, start with the basic definition of the Fourier series approximation function given in Section 3.2.4:

$$c_i = A_0 + 2 \sum_{k=1}^{n-1} [A_k \cos(w_{ik}) + B_k \sin(w_{ik})] + A_n \cos(w_{in}) \qquad (3.11)$$

Calculate c_i^2 by squaring both sides of Equation (3.11) and sum over the N data points. Use the amazing fact that the sums of products of cosines and sines of different harmonics equal zero, along with the definition of the harmonic strength, to get

$$\frac{1}{N} \sum_{i=1}^{N} (c_i)^2 = (R_0)^2 + 2 \sum_{k=1}^{n-1} (R_k)^2 + (R_n)^2 \qquad (3.12)$$

The left side of Equation (3.12) is simply the total variance of the time series. Therefore, by summing the harmonic strengths one can obtain the total harmonic strength, which is equal to the total variance. Many times it is enlightening to look at the line spectrum as giving the distribution of the variance over the frequencies between zero and the Nyquist frequency, f_{Ny}. This approach leads to an analysis-of-variance point of view. It also suggests that the area under a curve drawn through the components of the line spectrum would be proportional to the total variance.

One is usually more interested in the variance, s^2, of the time series about its mean and in the square root of this variance, s, which is called the standard deviation of the time series. (Note that s^2 is not the population variance that was discussed in Section 1.3.8.) An expression can be obtained for the variance about the mean by repeating the above derivation after replacing A_0 by the average of the time series, c, and subtracting the average from the time series. If this is done, one obtains

$$s^2 = \frac{1}{N} \sum_{i=1}^{N} (c_i - \bar{c})^2 = 2 \sum_{k=1}^{n-1} (R_k)^2 + (R_n)^2$$

To augment this derivation, consider the following exercise, where y_i is defined as

$$y_i = c_i - \bar{c}$$

That is, y_i is simply a zero-mean version of c_i. Squaring both sides of this equation, summing over N points, and dividing by N gives

$$\frac{1}{N} \sum_{i=1}^{N} y_i^2 = \frac{1}{N} \sum_{i=1}^{N} c_i^2 - 2\bar{c} \frac{1}{N} \sum_{i=1}^{N} c_i + \bar{c}^2 = s^2$$

$$= \frac{1}{N} \sum_{i=1}^{N} c_i^2 - \bar{c}^2$$

or

$$\frac{1}{N} \sum_{i=1}^{N} c_i^2 = s^2 + \bar{c}^2$$

Therefore, the total variance of the original c_i series is equal to the variance of the new zero-mean y_i series plus the square of the mean of the c_i series. The average of a series will frequently be removed before it is analyzed, so the reader should keep in mind that the total variance of the original series equals the variance of that series about its average plus the square of the average.

Define a new quantity, U_k, called the cumulative harmonic strength for the kth harmonic, as

$$U_0 = 0$$

$$U_k = 2R_k^2/s^2 + U_{k-1}, \qquad k = 1, 2, \ldots, n - 1$$

$$U_n = R_n^2/s^2 + U_{n-1}$$

A plot of the U_k versus f_k from $k = 0$ to n is called the cumulative line spectrum. From the discussion at the beginning of this section, the reader should see that U_n is unity. From the above derivation, the reader should see that, before the line spectrum calculations are carried out, it is a good idea to subtract the average from the time series.

If the time series is white noise, then all the harmonic strengths should be constant so that the U_k, when plotted against f_k, will give a straight line with a slope of $1/f_{Ny}$. As was shown in Section 3.1.2, the line spectrum of white noise is not constant. However, the cumulative line spectrum of white noise is quite close to being a straight line with the above-mentioned slope due to the smoothing effect of summing.

To test for white noise, dotted line bands about the theoretical straight line referred to above can be set up using the Kolmogoroff–Smirnov rules mentioned in detail in the text by Jenkins and Watts (1968). These lines are drawn at a distance of $K/\sqrt{[(N/2) - 1]}$ above and below the straight line and represent probability limits for white noise. For the case where the probability is 95%, which is used here, $K = 1.36$. Thus, if the cumulative spectrum falls within these lines, there is a 95% probability that the time series is white noise.

3.6 COLOR NUMBERS

Although a cumulative line spectrum is useful in determining whether a stochastic sequence is white noise, there is sometimes a need for a more compact measure of the degree to which the candidate sequence deviates from whiteness. The positive and negative color numbers, which are the normalized areas between the cumulative line spectrum and the diagonal line from zero cumulative power to unity cumulative power, are suggested here as possible measures. The positive color number, or PC#, is the normalized area between the cumulative line spectrum and the diagonal when the former is above the diagonal. The negative color number, or NC#, is the area between the cumulative line spectrum and the diagonal when the former is below the diagonal.

To derive the expressions for the color numbers, the dimensions of the cumulative line spectrum graph must be laid out. The y-axis on the cumulative line spectrum graph runs from zero cumulative power to unity cumulative power since the variance is used to normalize the cumulative sum of the power at each frequency. The x-axis runs from zero frequency to the Nyquist frequency, f_{Ny}, and, if there are N values of the time series, it is divided into $N/2$ discrete frequencies separated in frequency by $\Delta f = 1/(Nh)$, which is also the fundamental frequency f_1. The equation of the diagonal line connecting zero cumulative power and unity cumulative power on the cumulative line spectrum graph is given by

$$L_k = \frac{f_k}{f_{Ny}}$$

where $L_0 = 0$ and $L_{N/2} = 1$. If U_k (defined in the previous section) is the cumulative power at f_k, then the area between the cumulative spectrum and the diagonal line connecting zero power with unity, A_1, is given by

$$A_1 = f_1 \sum_{k=1}^{n} (U_k - L_k)$$

In calculating PC#, contributions are made to the sum only when U_k is greater than L_k. Conversely, when calculating NC#, contributions are made to the sum only when U_k is less than L_k. These areas can be made more useful for comparisons with other analyses based on different values of N by normalizing the areas with the total possible area for the case where the time series is totally colored. This area, A_2, is simply the product of the lengths of the x-axis and the y-axis divided by 2:

$$A_2 = \frac{1 * f_{Ny}}{2}$$

Therefore, the normalized areas, the color numbers, PC# and CN#, are given by

$$PC\# \quad \text{or} \quad NC\# = \frac{A_1}{A_2}$$

and the color numbers range from zero to unity.

Both the color numbers may be used as measures of how much color a time series contains. The positive color number can often be looked upon as a measure of the deviation from whiteness due to the presence of low-frequency power, while the negative color number can often be looked upon as a measure of the presence of high-frequency power. The use of these color numbers will be demonstrated in Section 3.9 when the effect of different combinations of control gains on the performance of test processes under control is investigated.

3.7 GENERATION OF HIGHER HARMONICS BY SHARP SHIFTS IN THE TIME DOMAIN

Sometimes the analyst comes across data containing repeated sharp shifts or discontinuities. Consider the time-domain plot of the time series in Figure 3.31. At first glance one notices that the data, a square wave, are periodic with a period of 40 seconds and a frequency of 0.025 Hz. Therefore, one might expect the line spectrum to show a spike at 0.025 Hz. Figure 3.32 shows that indeed it does, but it also shows that there are smaller spikes at higher frequencies, 0.075 Hz, 0.125 Hz, and 0.175 Hz, which are the third, fifth, and seventh harmonics of the frequency associated with the main spike. The presence of these odd harmonics can be explained if one remembers that the line spectrum is derived from a Fourier

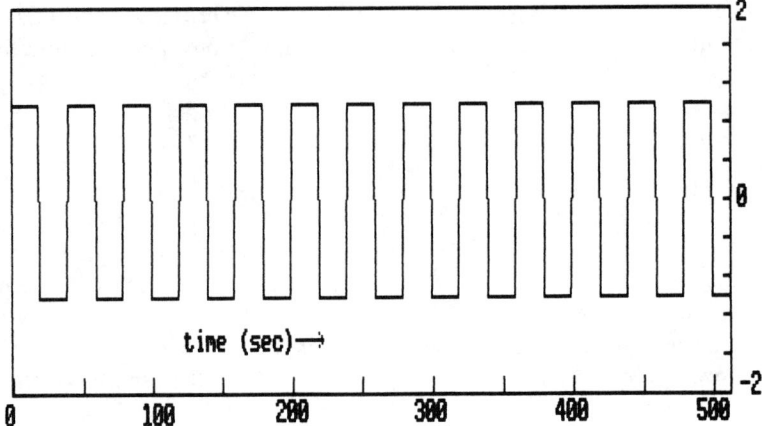

Figure 3.31   Signal with sharp shifts.

series that is trying to approximate the square wave by a sum of sinusoids. The sinusoid with a period of 40 seconds will crudely approximate the basic waveform, but the higher-frequency sinusoids are needed to handle the sharp edges of the square wave.

Figure 3.33 shows that the autocorrelation of a square wave with a period of 40 seconds is a triangular wave with a period of 40 lags. Finally, Figure 3.34 shows the line spectrum when the logarithm of the harmonic intensity multiplied by 10.0 is plotted versus the frequency. Here the ordinate has units of decibels (more about this in Section 3.12.2) and the higher harmonics are more evident.

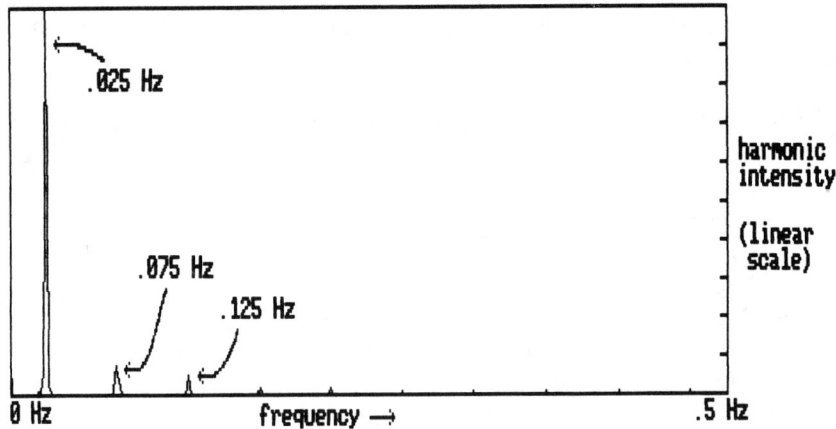

Figure 3.32   Line spectrum of a signal with sharp shifts.

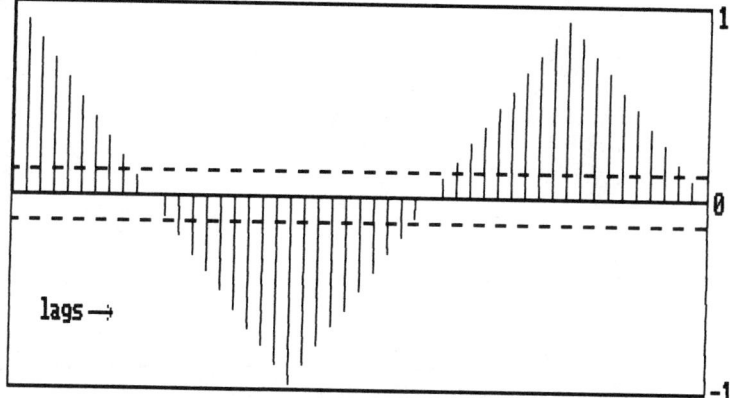

Figure 3.33 Autocorrelation of a signal with sharp shifts.

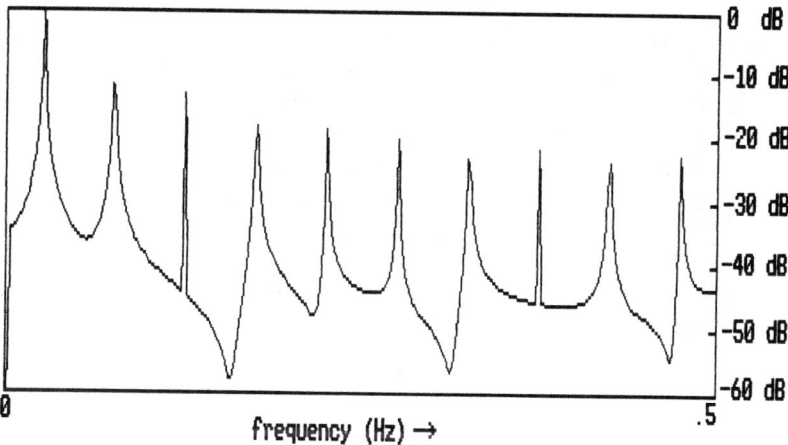

Figure 3.34 Line spectrum of a signal with sharp shifts (semilogarithmic scales).

3.8 ANALYSIS OF NONSTATIONARY TIME SERIES: DIFFERENCING

Consider the time-domain plot of the time series c_i, $i = 1, 2, \ldots, N$, shown in Figure 3.35. Note that this time series appears to be nonstationary. A line spectrum of such a time series would show huge peaks at a few frequencies near zero and little else. This is not surprising since the first step in computing a line spectrum is to fit a Fourier series to the time-domain data. Because of the lack of stationarity, the Fourier series coefficients of the low frequencies will dominate all the others.

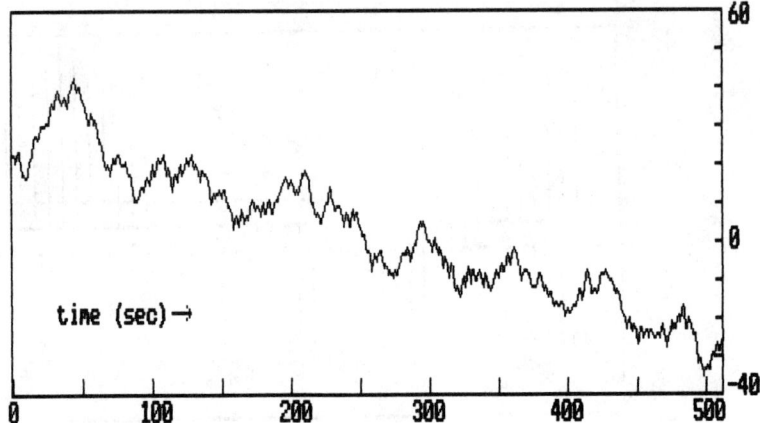

Figure 3.35 Time-domain plot of a nonstationary time series.

To cut down on the dominance of the Fourier coefficients associated with the zeroth frequency, the average can be subtracted from the time series. The line spectrum of the resulting series is shown in Figure 3.36. Still there is no insight gained due to the preponderance of the low-frequency harmonic strengths at frequencies just greater than zero. In any case, it is always good practice to subtract the mean from every sequence before the line spectrum is computed.

In Figure 3.37 the autocorrelation of this time series supports the contention that it is nonstationary.

The data are differenced such that a new series d_i, $i = 1, 2, \ldots, N$, is generated according to

$$d_1 = 0, \quad d_i = c_i - c_{i-1}, \quad i = 2, 3, \ldots, N$$

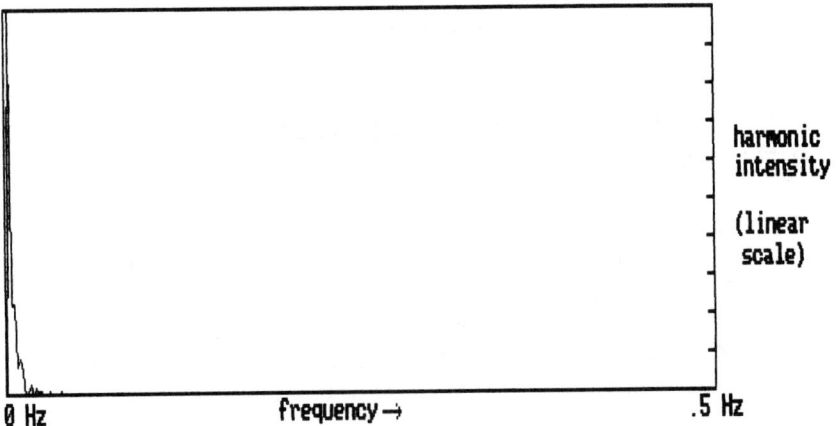

Figure 3.36 Line spectrum of a nonstationary time series.

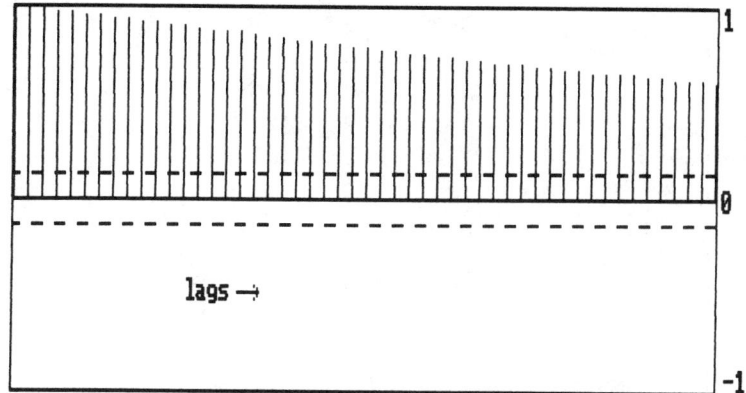

Figure 3.37   Autocorrelation of a nonstationary time series.

which has the time-domain behavior shown in Figure 3.38. The line spectrum of this new time series is shown in Figure 3.39, where it is seen that buried in the original time series was a sinusoid with a period of 4 seconds. The autocorrelation of this new time series, shown in Figure 3.40, indicates the presence of a sinusoid with the alternation of the peaks, each separated by 4 lags. So, by differencing the original time series it has become stationary, and a hidden periodic component has been discovered.

An alternative to plotting the line spectrum, as was done in Figure 3.36, is to plot 10.0 times the log of the normalized harmonic strengths. This gives Figure 3.41. The harmonic strengths have been normalized by the maximum value, which occurs at the fundamental frequency. Note that even without differencing the peak at 0.25 Hz shows up (although not quite as dramatically).

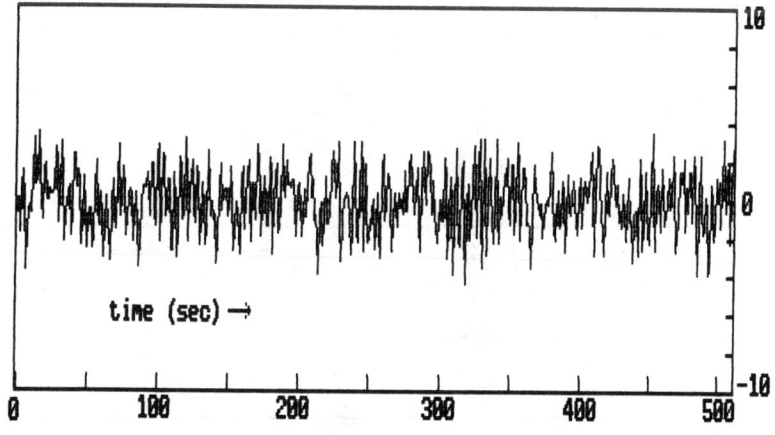

Figure 3.38   Time-domain plot of the differenced time series.

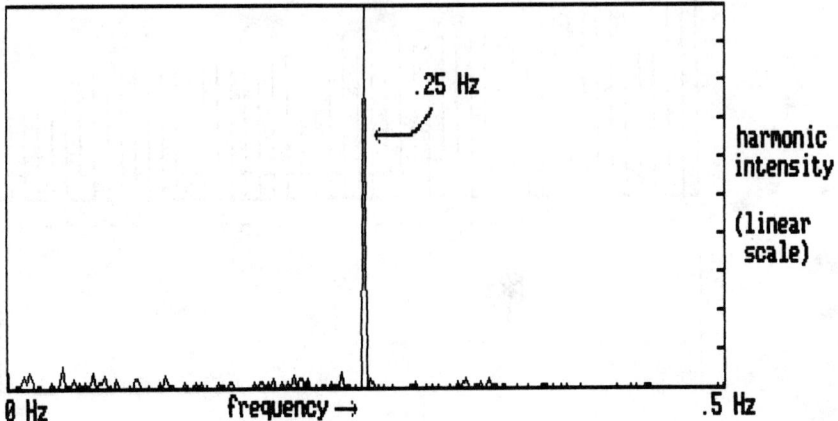

Figure 3.39   Line spectrum of differenced time series.

One must be careful in differencing time series because the process of differencing not only removes the power in the lowest frequencies, it also amplifies power at high frequencies linearly; that is, the higher the frequency in the time series being differenced, the higher the amplification of the power at that frequency. To illustrate this, white noise will be differenced. Figure 3.42 shows the line spectrum of the differenced white noise time series, and Figure 3.43 shows the cumulative line spectrum. Note that the harmonic strengths increase with increasing frequency. Therefore, in some cases differencing simply generates hash.

The autocorrelation of the differenced white noise is shown in Figure 3.44. Here the autocorrelation is significant only at one lag, the first one, and, to boot, it is negative. Upon reviewing Section 1.3, this will make sense because the differencing equation, when applied to white noise, is simply a special case of the first-order moving average stochastic equation with the coefficient set to -1.

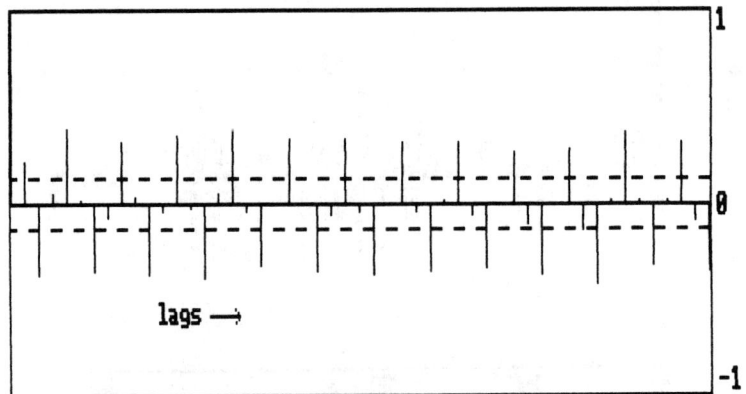

Figure 3.40   Autocorrelation of the differenced time series.

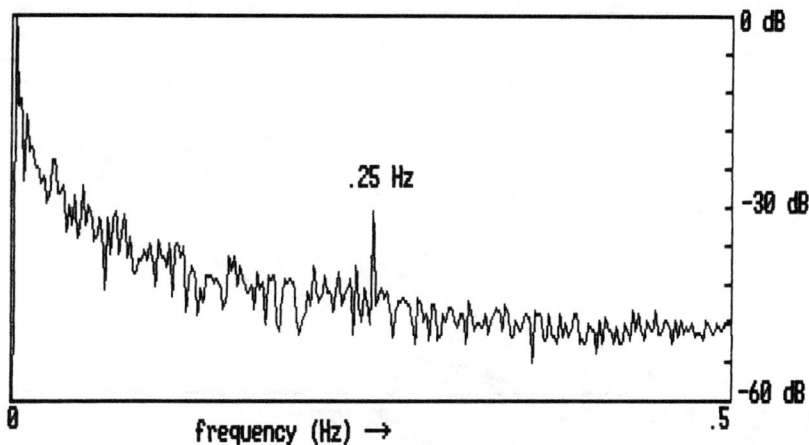

Figure 3.41 Line spectrum of a nonstationary time series (semilogarithmic scales).

The variance of the differenced data can be obtained by squaring both sides of the difference equation and then taking the expected value (or the average over many points) of the result:

$$d_i = c_i - c_{i-1}$$

$$d_i^2 = c_i^2 - 2c_i c_{i-1} + c_{i-1}^2$$

Since the input data are white noise, the expected value of the second term on the right side will be zero, leaving

$$V_d = 2V_c$$

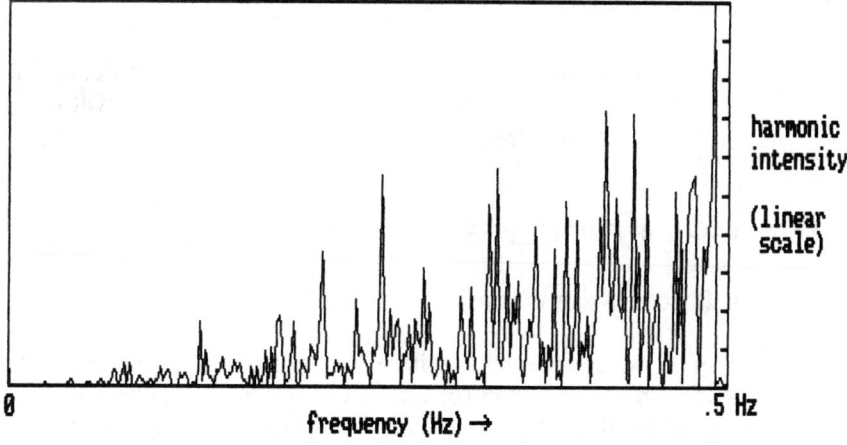

Figure 3.42 Line spectrum of differenced white noise.

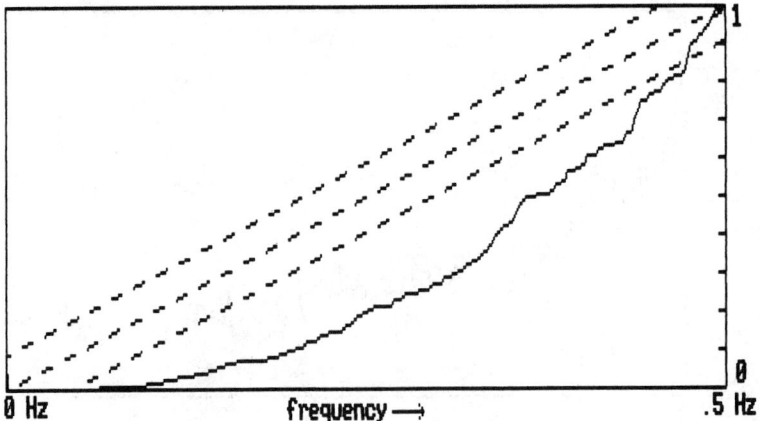

Figure 3.43 Cumulative line spectrum of differenced white noise.

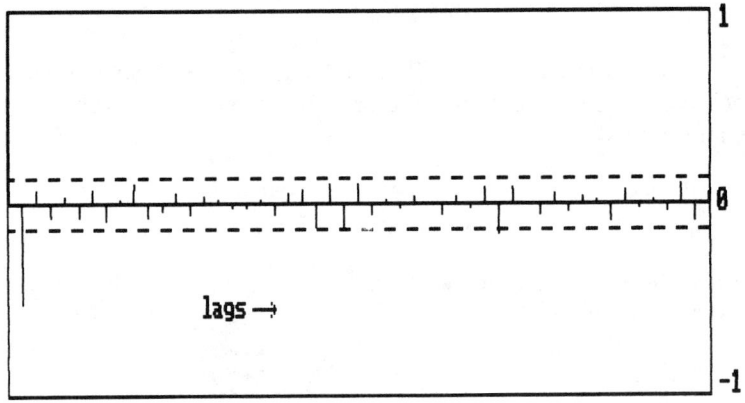

Figure 3.44 Autocorrelation of differenced white noise.

where V_c is the variance of the original sequence. Therefore, differencing white noise doubles its variance. In Chapter Four, differencing will be revisited, this time in the continuous frequency domain.

3.9 ANALYSIS OF PROCESSES UNDER CONTROL

3.9.1 Process with a Short Time Constant

Consider a process (model 8) with a time constant of 1.44 sec, a process gain of 2, no deadtime, and a control interval of 1 second, subject to two disturbance streams in parallel. The first stream X_k consists of a random walk:

$$X_k = X_{k-1} + w_k$$

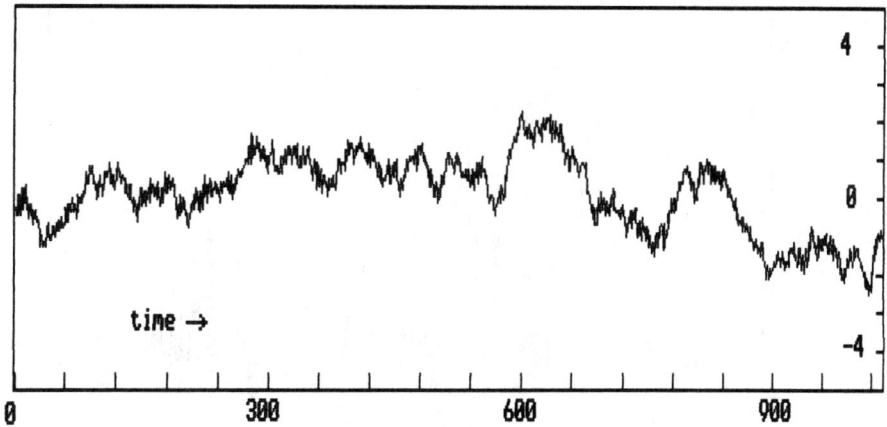

Figure 3.45   First test process with no control.

where the standard deviation of the white noise, w_k, is 0.118. The second stream consists of unautocorrelated white noise having a standard deviation of 0.144. This example process might be a model for many manufacturing processes where a "widget" is manufactured every second and it is desired to control its weight, which tends to drift off unless some action is taken. Assume that the weight is measured every time a widget is produced once per second and that there is a "knob" that can be adjusted to quickly affect the widget's weight. The white noise component in the disturbance may be due to sensor measurement error, it may be a consequence of the manufacturing process, or it might be the result of A/D round-off error. Figure 3.45 shows how the process meanders with no control.

Using the rough rules of thumb presented in Section 2.5, the following control gains can be calculated:

$$P = \frac{1}{G} = \frac{1}{2} = 0.5$$

$$I = \frac{1}{G * T} = \frac{1}{2 * 1.44} = 0.35$$

where

> G = process gain
> P = proportional control gain
> I = integral control gain
> T = process time constant

The tuning effort was begun with these control gains and then, by trial and error, control gains were obtained that were the best at making the controlled variable into a white noise process. Figures 3.46 and 3.47 show the line spectrum and cumulative line spectrum for 1024 points of the controlled variable, after the av-

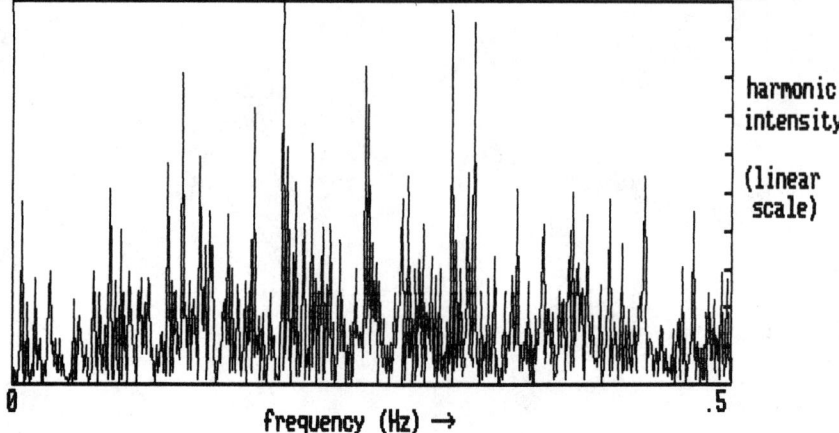

harmonic
intensity

(linear
scale)

frequency (Hz) →

Figure 3.46   Line spectrum of the first test process with best control gains.

erage was subtracted, for the case of $P = 0.2$ and $I = 0.4$. Note that, although the line spectrum is noisy, the cumulative line spectrum stays within the dotted lines, indicating that the controlled variable is probably a white noise sequence. The standard deviation about the target for this set of gains was 0.22.

Figure 3.48 shows the time-domain plot of the controlled variable with this set of gains. The mean is kept on target, but the controlled variable is still noisy. Since the line spectrum says that the noise is white, one assumes that the best gains have been determined. For reference purposes, note that the color numbers are as follows: PC# = 0.025, NC# = 0.025. Also, as an expansion of information

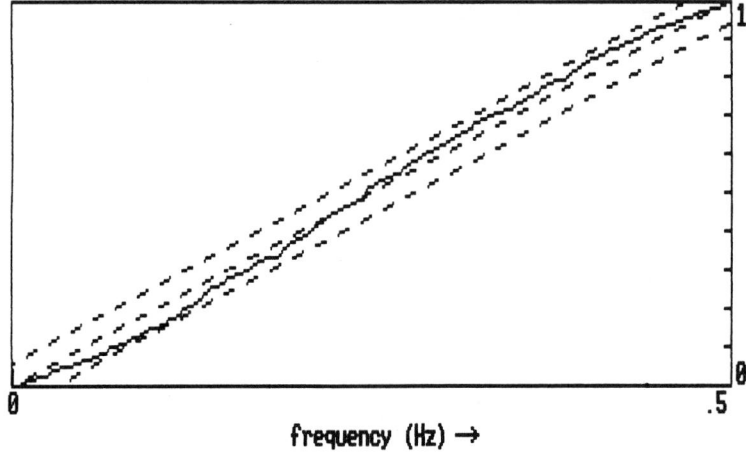

frequency (Hz) →

Figure 3.47   Cumulative line spectrum of the first test process with best control gains.

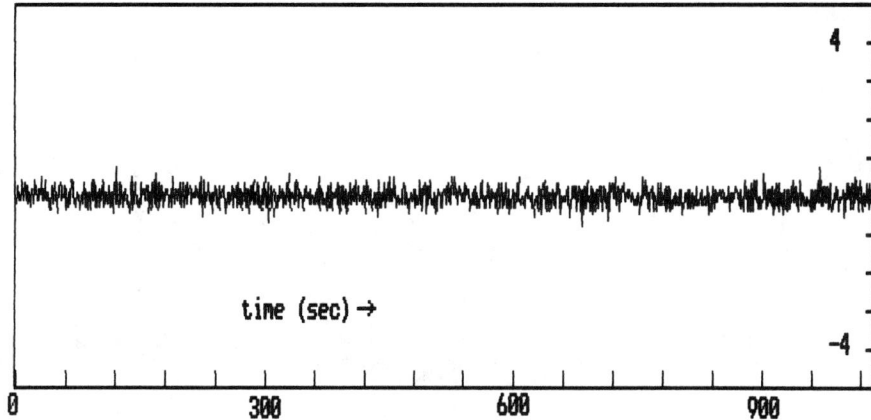

Figure 3.48   First test process under control with best gains.

contained in the standard deviation, the histogram of the controlled variable is shown in Figure 3.49.

It has been suggested that the goal is to develop a control strategy that can transform a controlled variable contaminated by autocorrelated noise into a white noise sequence. In many cases this may be impractical, since the effect of integral control is to remove any offset and as a result the line spectrum will show no strength at zero frequency. Therefore, the controlled variable sequence can never be truly "white" in the sense of having all frequencies equally represented.

When the proportional gain is tripled, the standard deviation increases insignificantly to 0.23, but the line spectrums in Figures 3.50 and 3.51 show a significant shift in the power to the higher frequencies. The positive color number

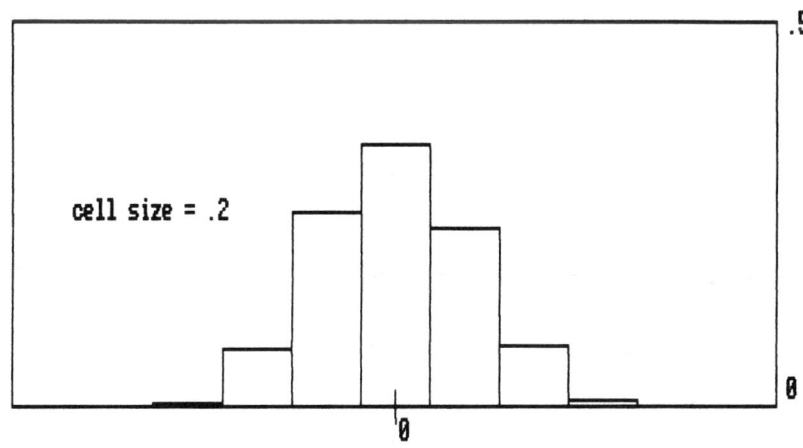

Figure 3.49   Histogram of the first test process with the best control gains.

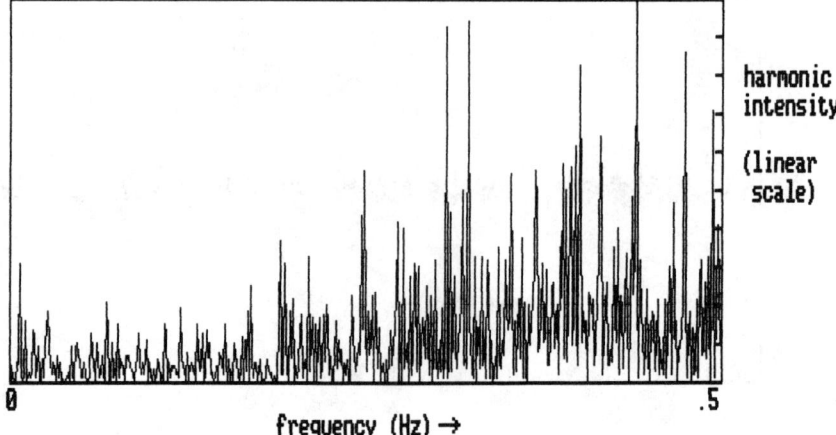

Figure 3.50   Line spectrum of the first test process with 3P.

is now zero, while the negative color number is 0.288, supporting the interpretation that the harmonic strength has shifted to the higher frequencies.

Figures 3.52 and 3.53 show how the low frequencies are attenuated relative to the middle and high frequencies when the integral gain is tripled. The standard deviation here is 0.288, the positive color number is zero, and the negative color number is 0.29. Therefore, although the power in the low frequencies has been attenuated by the increased integral gain, a penalty has been paid in that power in the higher frequencies has been significantly amplified.

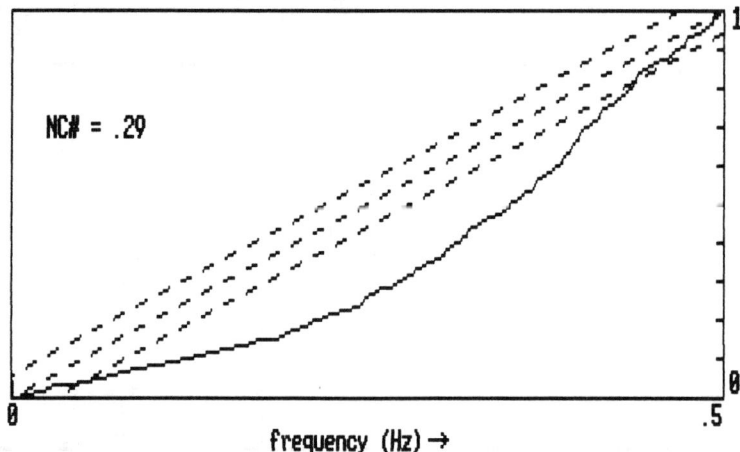

Figure 3.51   Cumulative line spectrum of first test process with 3P.

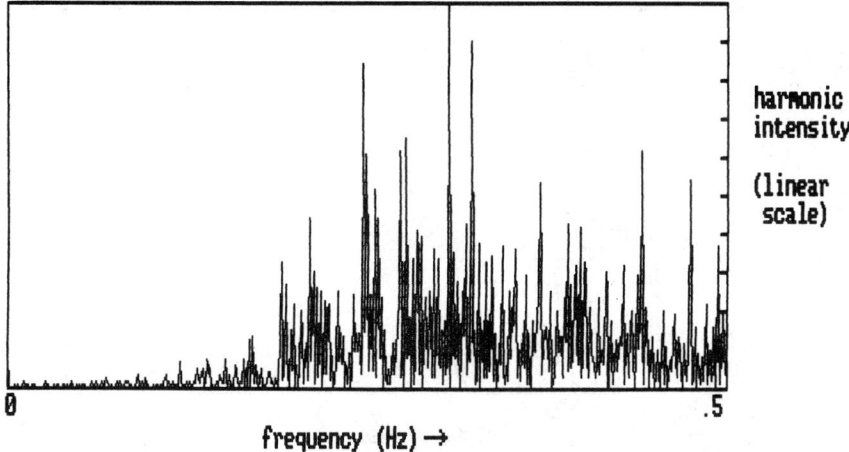

harmonic
intensity

(linear
scale)

frequency (Hz) →

Figure 3.52   Line spectrum of the first test process with 3I.

When the proportional gain is set to zero, the standard deviation remains unchanged at 0.23, but Figures 3.54 and 3.55 show that the harmonic strength has shifted to the middle to low frequencies. The positive color number is 0.15 and the negative color number is 0.012.

Figures 3.56 and 3.57 show how the low frequencies increase in harmonic strength when the best integral gain is divided by 3. The standard deviation here is 0.23, while the positive color number is 0.30 and the negative color number is zero.

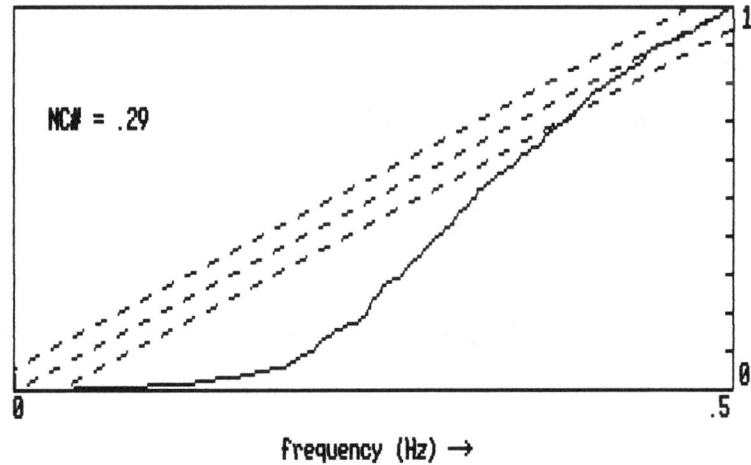

NC# = .29

frequency (Hz) →

Figure 3.53   Cumulative line spectrum of the first test process with 3I.

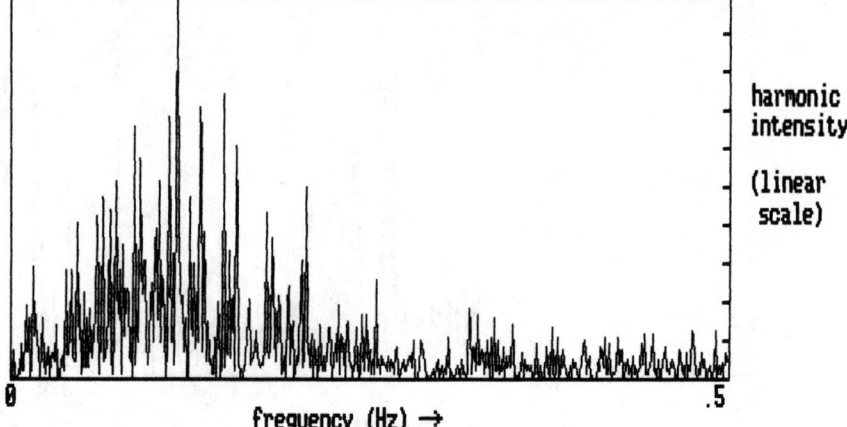

Figure 3.54 Line spectrum of the first test process with $P = 0$.

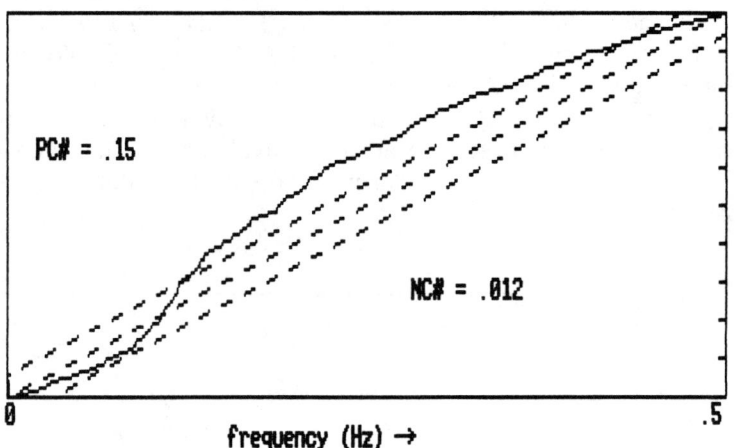

Figure 3.55 Cumulative line spectrum for the first test process with $P = 0$.

TABLE 3.1

P	I	s	PC#	NC#
0.2	0.4	0.22	0.025	0.025
0.6	0.4	0.23	0	0.288
0.2	1.2	0.29	0	0.290
0	0.4	0.23	0.15	0.012
0.2	0.133	0.23	0.297	0

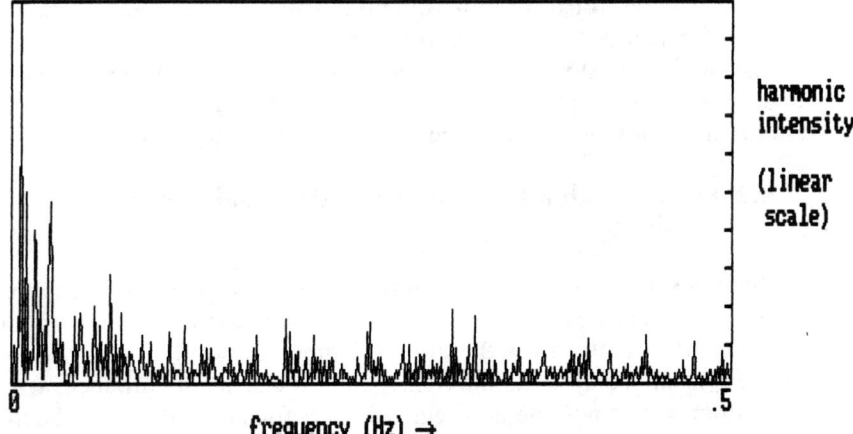

Figure 3.56   Line spectrum of the first test process with $I/3$.

The results of varying the control gains are summarized in Table 3.1, which shows that the standard deviation is relatively insensitive to changes in control gains but that the distribution of harmonic strength, as indicated by the color numbers, is strongly affected. Because of the short time constant and the absence of deadtime, the standard deviation shows less dependence on the proportional gain than on the integral gain. Notice how the presence of high-frequency harmonic strength in the case where the proportional gain is tripled supports the large negative color number and, conversely, how the presence of low-frequency harmonic strength in the case where the integral gain is divided by 3 supports the large positive color number.

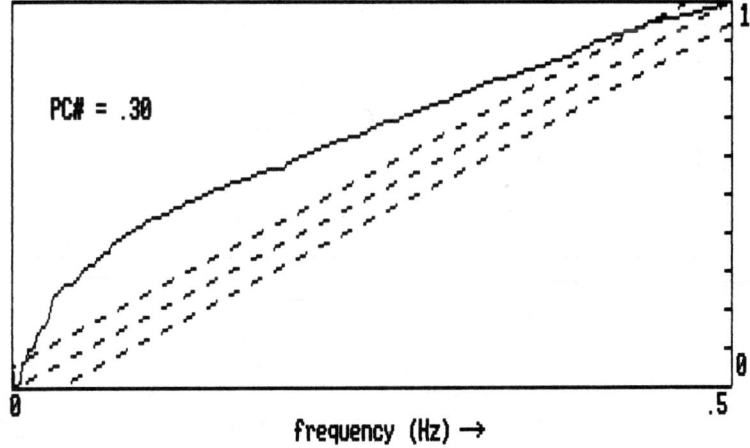

Figure 3.57   Cumulative line spectrum of the first test process with $I/3$.

In general, for processes with no deadtime and a time constant on the same order of magnitude as the control interval, an overly conservative integral gain is usually manifested in the presence of low-frequency power. Conversely, an overly aggressive proportional gain or integral gain usually causes harmonic strength to show up at the middle- to high-frequency range.

3.9.2 Process with a Large Time Constant and Other Challenges

In Sections 1.7.2 and 2.6.3, some time was spent with a process (model 4) that had an apparent deadtime (because of an inflection point) and a control interval significantly smaller than the effective time constant. In this section this same process is subjected to an autoregressive stochastic disturbance, which is nearly a random walk since the autoregressive coefficient is 0.99 (see Section 1.3 for a discussion of such a nearly nonstationary disturbance).

Figure 3.58 shows the step change response of this example process. It appears that the stochastic disturbances are going to influence the dynamics significantly. This example might be a model for an industrial process where a temperature is to be kept on target by manipulating the energy input by means of, say, an electrical heater. The inertia of the process would be modeled by the apparent deadtime and large time constant. This is to be compared to the example of Section 3.9.1, where there was no deadtime and the time constant was only slightly larger than the sampling/control interval.

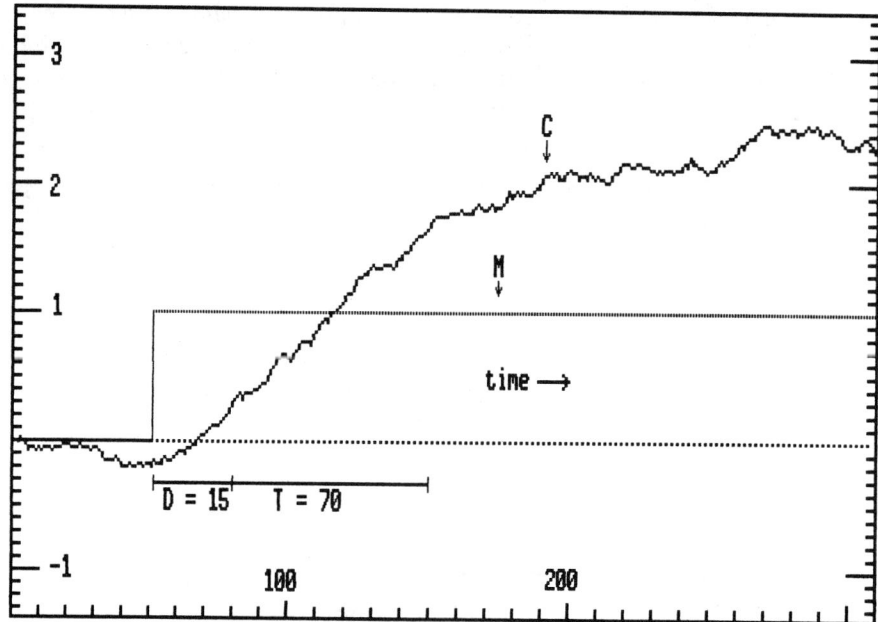

Figure 3.58 Step change response of the second test process.

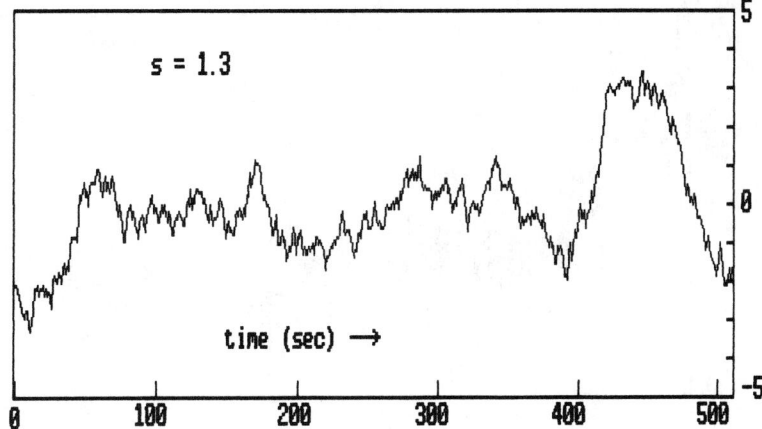

Figure 3.59   Time-domain plot of second test process sampled at 1.0 second.

From Figure 3.58 one might visually estimate, with some difficulty, the dead-time at 15 seconds, the time constant at 70 seconds, and the process gain at 2. The rough rules of thumb from Section 2.5 would then suggest starting gains of

$$P = \frac{1}{G} = 0.5$$

$$I = \frac{1}{G * T} = \frac{1}{2 * 70} = 0.007$$

Figure 3.59 shows the time-domain plot of the controlled variable using control gains of $P = 0.75$ and $I = 0.010$, which by trial and error were found to give the smallest standard deviation based on a sampling/control interval of 1 second. Note

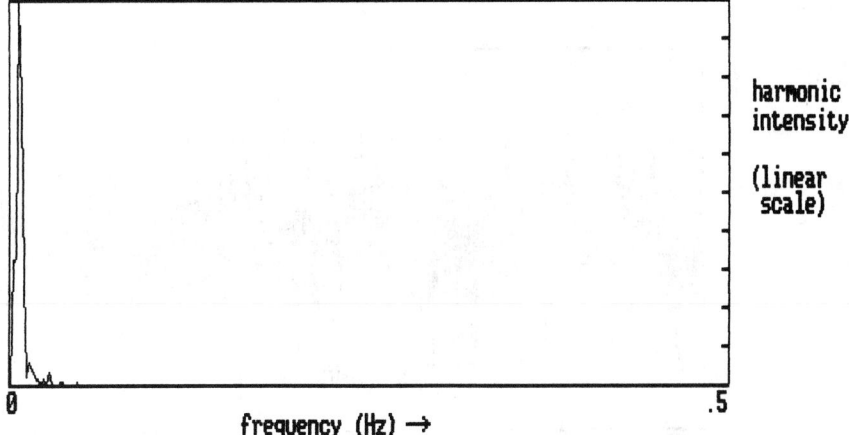

Figure 3.60   Line spectrum of second test process sampled at 1.0-second intervals.

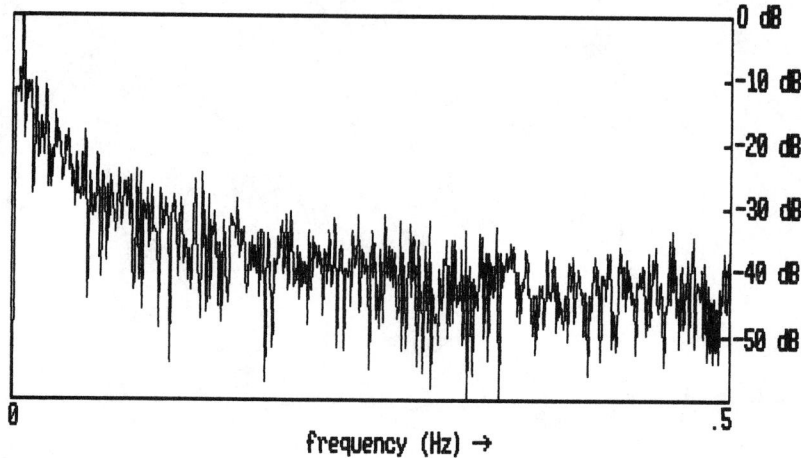

frequency (Hz) →

Figure 3.61 Line spectrum of second test process sampled at 1.0 second using semilogarithmic scales.

that these gains are slightly different from those used in Chapter Two, where the object was to obtain the best response to a set point change.

The line spectrum for this set of control gains and a sampling interval of 1 second, shown in Figure 3.60, indicates that the controlled variable is not even remotely white, but in fact shows strong harmonic strength at low frequencies. Figure 3.61 shows the line spectrum plotted on semilog scales.

This situation of not being able to make the controlled variable look white is a consequence of the short sampling interval relative to the process time constant and deadtime. A control strategy cannot be expected to be able to drive this process, with a time constant of approximately 70 seconds and a deadtime of approximately 15 seconds, back to zero error within the sampling interval used

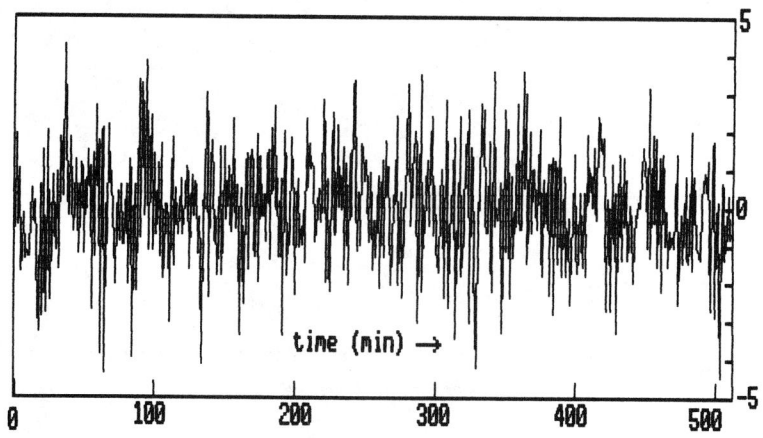

time (min) →

Figure 3.62 Time-domain plot of second process sampled at 1.0-minute intervals.

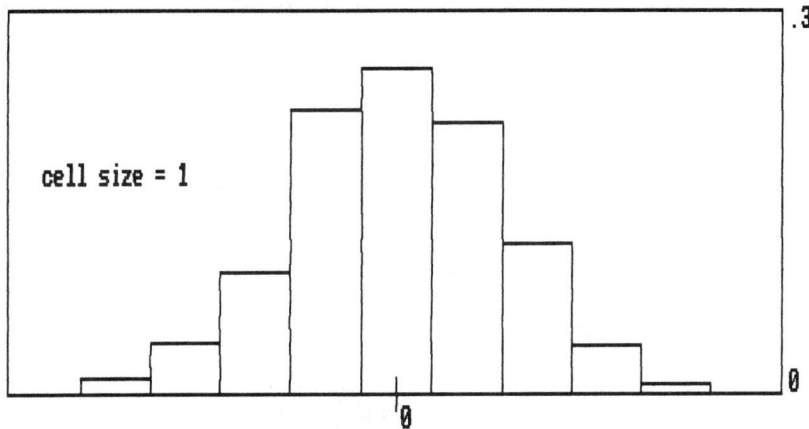

Figure 3.63 Histogram of second process sampled at 1.0-minute intervals.

to construct the line spectrum. This is in contrast to the example in the previous section, where the time constant was on the same order of magnitude as the control/sampling interval and there was no deadtime. There it was feasible to strive for a white noise error sequence.

Figure 3.62 shows the time-domain plot and Figure 3.63 shows the histogram of this system when sampled at a 60-second interval, but still controlled at a 1-second interval. The standard deviation is 1.50. Figures 3.64 and 3.65 show the line spectrum and cumulative line spectrum for this case. Note how both spectra confirm whiteness.

The proportional and integral control gains are now doubled. Figure 3.66 shows the time-domain plot of the controlled variable sampled at a 60-second interval. Note that the standard deviation of 1.61 is slightly greater than with the

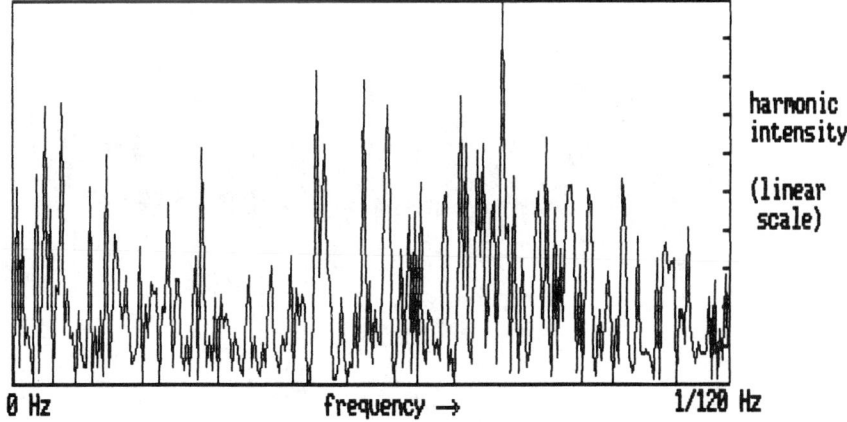

Figure 3.64 Line spectrum of second example sampled at 60 seconds.

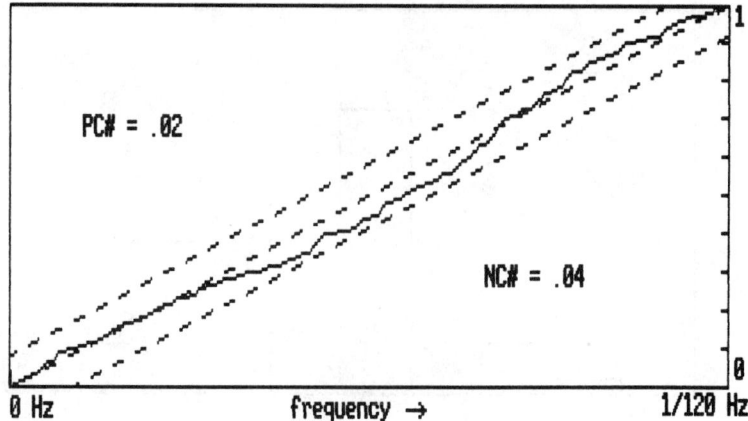

Figure 3.65   Cumulative line spectrum of second process sampled at 60 seconds.

previous set of gains. Figures 3.67 and 3.68 show the spectra, indicating that the high frequencies have been amplified.

When applying line spectrum or autocorrelation in the analysis of controlled processes, one must be careful when choosing the sampling interval. In the example of Section 3.9.1 the line spectrum sampling interval was the same as the control interval, and because of the short time constant, white noise could be used as a realistic goal. In the example of this section the long time constant and deadtime relative to the control interval required a larger sampling interval for the line spectrum.

In Chapter Two, PI plus a filtered derivative (or PIfD) was applied to the same process that is being studied in this section (model 4). By adding the derivative term, it was possible to increase the control gains from $P = 0.54$ and I

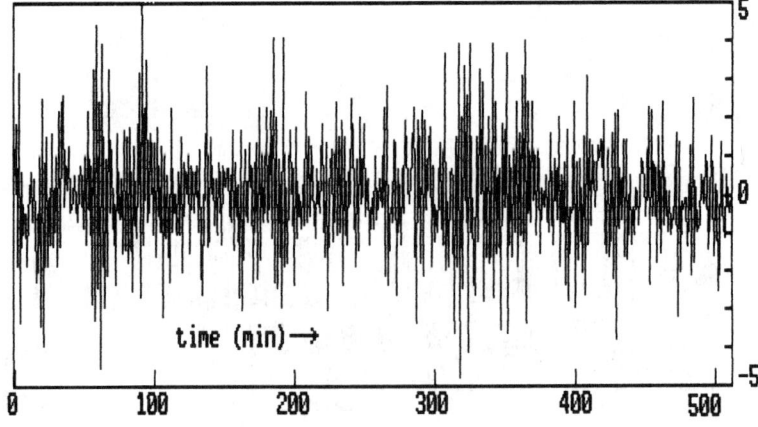

Figure 3.66   Time-domain plot of second process using doubled control gains.

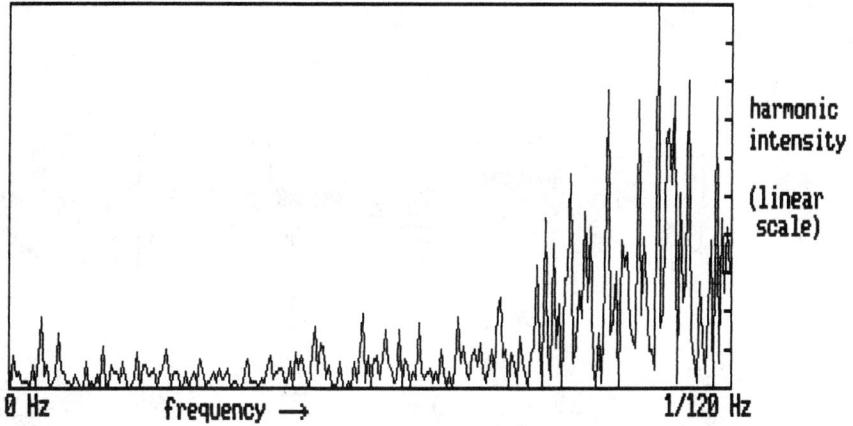

Figure 3.67 Line spectrum of second process with doubled control gains.

$= 0.009$ to $P = 1.1$ and $I = 0.012$ and $D = 15$, while the derivative was filtered with a coefficient of 0.1. The line spectrum and autocorrelation can be used to evaluate the choice of tuning constants.

Figure 3.69 shows the time-domain plot of the controlled variable sampled at 60-second intervals while being controlled at a 1-second interval. The standard deviation is 1.21, which is the lowest so far. Figure 3.70 shows the line spectrum, which suggests a variance distribution that could be close to white noise. The autocorrelation shown in Figure 3.71 also suggests that the controlled variable is reasonably white. This contention of whiteness is further strengthened by the cumulative line spectrum in Figure 3.72. Finally, comparing the histogram in

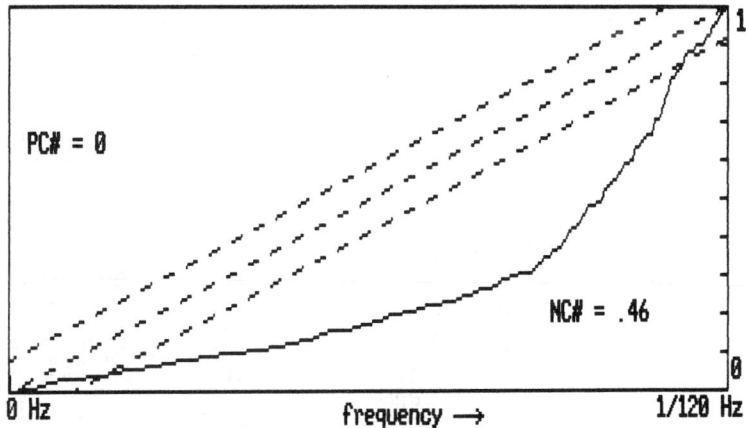

Figure 3.68 Cumulative line spectrum of second process with doubled control gains.

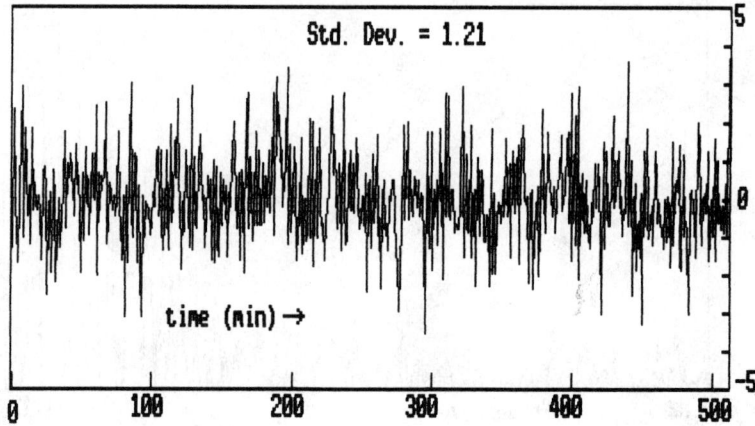

Figure 3.69 Time-domain plot of second example using PIfD.

Figure 3.73 with the one in Figure 3.63 suggests that PIfD gives a slightly tighter distribution.

In this and the previous subsection, conclusions have been based on one line spectrum, which in turn was based on but one realization of the stochastic disturbance. In the next subsection it will be seen that this approach must be used with care.

3.9.3 Control of Processes Subject to White Noise

In Section 2.10, conventional feedback control was compared to a control chart-based control strategy for two example processes (models 6 and 7). During this comparison, the question of controlling processes subject to white noise came up. To expand the comparison, the line spectra of the controlled variable for the two models will be displayed.

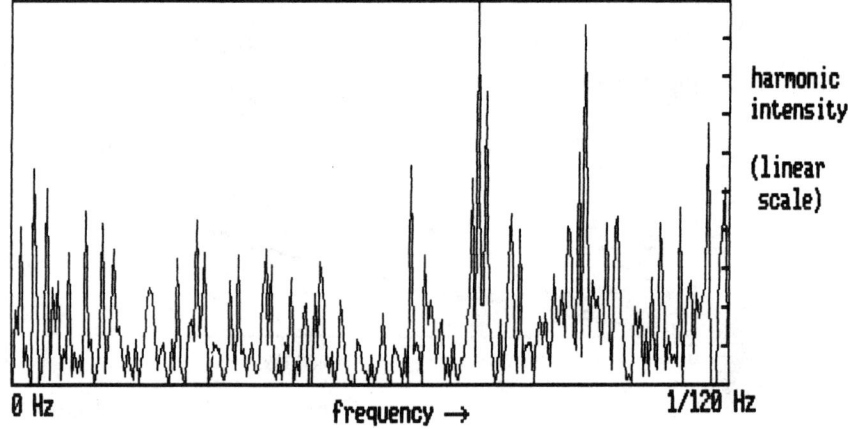

Figure 3.70 Line spectrum of second process under PIfD.

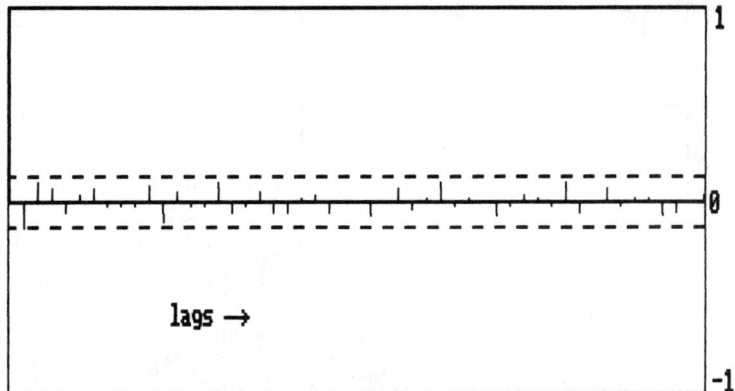

Figure 3.71   Autocorrelation of second process under PIfD.

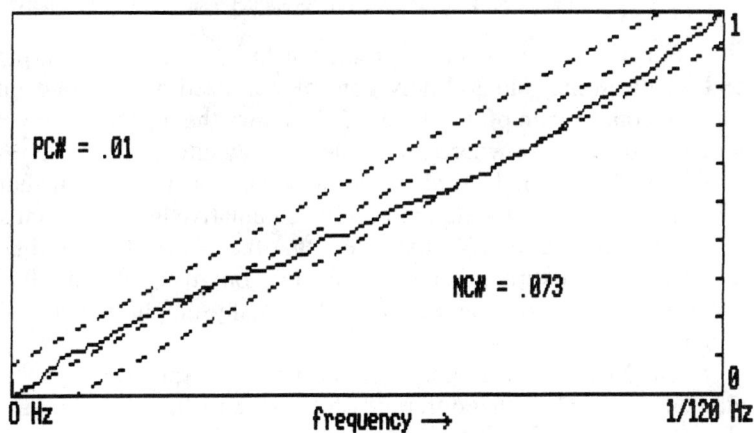

Figure 3.72   Cumulative line spectrum of second process under PIfD.

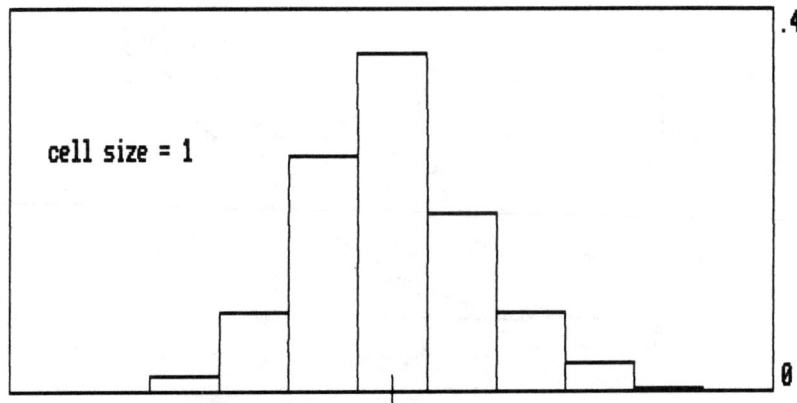

Figure 3.73   Histogram of second process under PIfD.

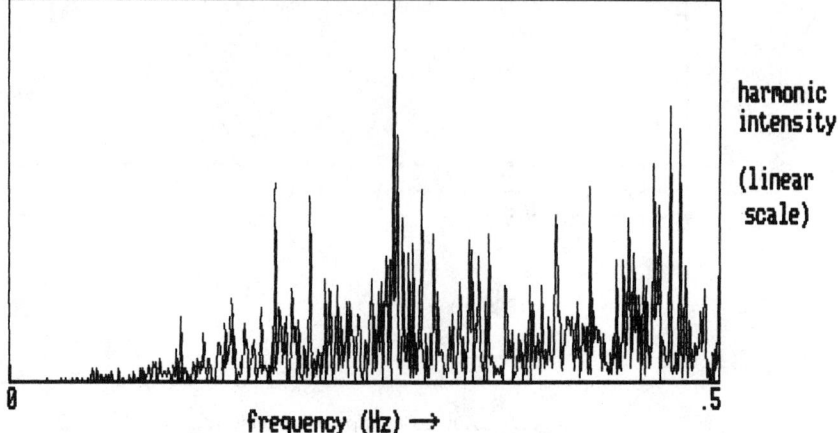

Figure 3.74 Line spectrum of model 6 subject to white noise.

For model 6, which has a time constant of 0.5 second, a process gain of 2, and no deadtime, integral-only control was used at 1-second intervals with an integral control gain of 0.5. Figure 3.74 shows the line spectrum of the controlled variable for the case where it is subjected to white noise. Note that the spectrum (based on 1024 samples), although noisy, is not uniform, in that the lower frequencies are relatively attenuated. The cumulative line spectrum shown in Figure 3.75 supports Figure 3.74 in showing that the effect of integral control is to turn a white noise disturbance into a nonwhite one by preferentially attenuating the lower-frequency components or preferentially amplifying the higher-frequency components.

To get a better estimate of the true line spectrum, line spectra from 15 simulations (each with a different white noise realization) were averaged, and the

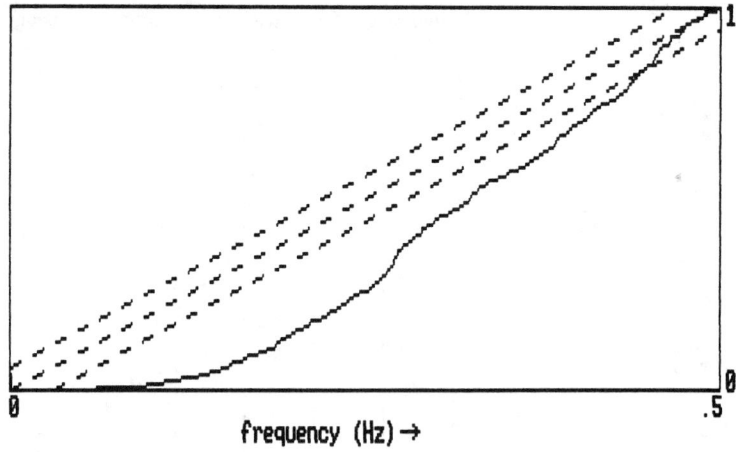

Figure 3.75 Cumulative line spectrum of model 6 subject to white noise.

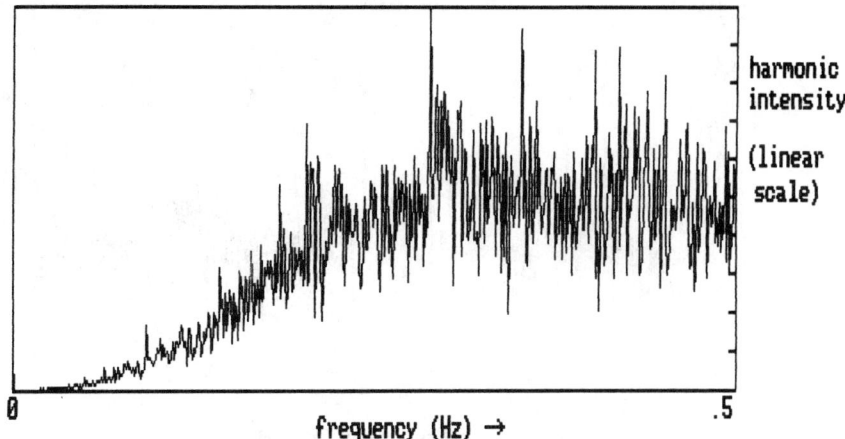

harmonic
intensity

(linear
scale)

frequency (Hz) →

Figure 3.76   Average of 15 line spectra for model 6 under integral-only control.

result is shown in Figure 3.76. In Section 5.4.2, this averaged line spectrum will be compared with the power spectral density derived from first principles.

In Chapter Two, this example controlled system was used to illustrate over-control. It is suggested that a better indication of overcontrol is given by the presence of a large negative value in the autocorrelation at lag 1 (see Figure 2.42), rather than the presence of higher harmonic strength at higher frequencies in the line spectrum.

For model 7, which has a time constant of 40 seconds, a process gain of 2.0, and a deadtime of 10 seconds, proportional–integral control was used in Chapter Two with a 1-second control interval and with a proportional control gain of 0.71 and an integral control gain of 0.0175. Figure 3.77 shows the line spectrum (based on 1024 samples) of the controlled variable for the case where it is subjected to

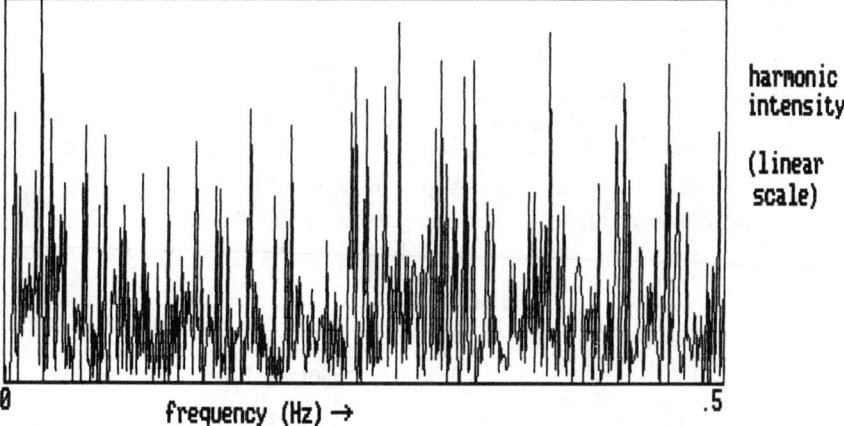

harmonic
intensity

(linear
scale)

frequency (Hz) →

Figure 3.77   Line spectrum of model 7 subject to white noise under PI control.

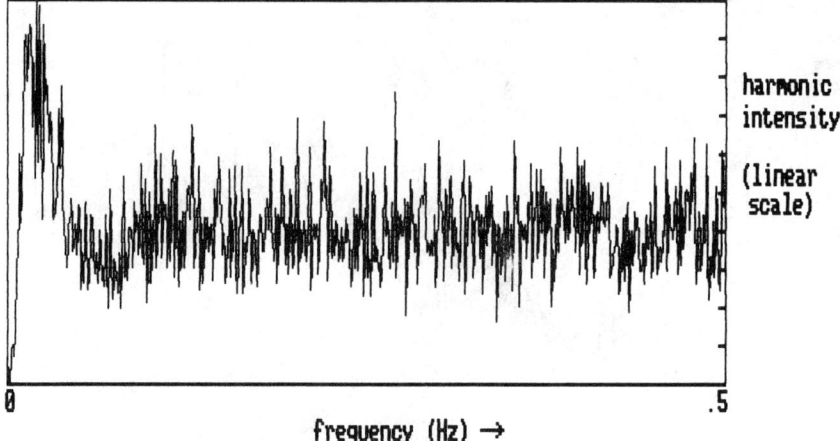

Figure 3.78   Average of 15 line spectra for model 7 under PI control.

white noise. Here only the lowest frequencies appear to be attenuated so that, in effect, all that is left is white noise. The cumulative line spectrum verifies this. As with the previous example, the line spectra from 15 different simulation runs (each with a different realization of the white noise disturbance) were averaged. The result is shown in Figure 3.78. Here there is the presence of a peak at low frequencies that did not show up in Figure 3.77. However, as will be seen in Section 5.5.2, even with this averaging, there are some other interesting features buried in the noise of the line spectrum.

3.10 EXPONENTIAL FORM OF THE DISCRETE-TIME FOURIER SERIES

The exponential form of the discrete-time Fourier series pair (also often referred to as the complex Fourier series) will be useful when frequency-domain filtering (Section 3.11) and error transmission curves (Chapter Five) are discussed. The following discussion requires only an elementary knowledge of complex variables, since no attempt will be made to go into any of the mathematical details. However, the mathematically uninclined reader will want to avoid getting bogged down in the algebra and should key on the highlights that deal with the compactness of the notation and the new insights into the symmetry and periodicity. For a fuller discussion, the reader is referred to the texts by Oppenheim and Schafer (1975, 1989).

3.10.1 Derivation of the Exponential Form of the Discrete-time Fourier Series Pair

The derivation begins with a review of the formulas for the discrete-time Fourier series discussed in Section 3.2. First, look at the discrete-time Fourier series interpolation function:

$$c_i = A_0 + 2 \sum_{k=1}^{n-1} [A_k \cos(w_{ik}) + B_k \sin(w_{ik})] + A_n \cos(w_{in})$$

$$i = 1, 2, \ldots, N, \qquad N = 2n, \qquad w_{ik} = 2\pi f_1 kih \tag{3.13}$$

In Section 3.2 it was shown that the coefficients could be determined from the following two expressions:

$$A_k = \frac{1}{N} \sum_{i=1}^{N} c_i \cos(w_{ik}), \qquad k = 0, 1, 2, \ldots, n$$

$$B_k = \frac{1}{N} \sum_{i=1}^{N} c_i \sin(w_{ik}), \qquad k = 1, 2, \ldots, n-1 \tag{3.14}$$

Thus there is a pairing between the N values of c_i in the discrete-time domain and the $2n$ values of A_k and B_k in the discrete-frequency domain. Equations (3.13) and (3.14) will be reformulated using Euler's formula, which relates the sine and cosine functions to an exponential with an imaginary argument:

$$\exp(j\theta) = \cos(\theta) + j\sin(\theta) \tag{3.15}$$
$$\exp(-j\theta) = \cos(\theta) - j\sin(\theta)$$

where the symbol j represents the imaginary operator or $\sqrt{(-1)}$ and the symbol θ represents an angle in radians. Having introduced this notation, take a look at the following expression for a set of N coefficients denoted as

$$F_m, \qquad m = -n, -(n-1), \ldots, -1, 0, 1, 2, \ldots, n-1,$$

which has a form somewhat similar to the original Fourier coefficients:

$$F_m = \frac{1}{N} \sum_{k=1}^{N} c_k \exp(-jw_{km}) \tag{3.16}$$

These new Fourier coefficients are called the *complex components of the discrete-time Fourier series* for the time-domain data, c_k, $k = 1, 2, \ldots, N$. The symbol $F_{\text{DTFS}}\{\ \}$ will be used to connect F_m and c_k; that is,

$$F_m = F_{\text{DTFS}}\{c_k\} \tag{3.17}$$

Applying Euler's formula to Equation (3.17) gives

$$F_m = \frac{1}{N} \sum_{k=1}^{N} c_k [\cos(w_{mk}) - j\sin(w_{mk})] \tag{3.18}$$

Replace the subscript m with $-m$ and write Equation (3.18) again:

$$F_{-m} = \frac{1}{N} \sum_{k=1}^{N} c_k [\cos(w_{mk}) + j\sin(w_{mk})] \tag{3.19}$$

where advantage has been taken of the trigonometric fact that

$$\cos(w_{-mk}) = \cos(w_{mk})$$

$$\sin(w_{-mk}) = -\sin(w_{mk})$$

Using Equations (3.14) and (3.19), F_m can be related to A_m and B_m as follows:

$$F_m + F_{-m} = 2 \frac{1}{N} \sum_{k=1}^{N} c_k \cos(w_{mk}) = 2A_m \tag{3.20}$$

$$F_m - F_{-m} = -j2 \frac{1}{N} \sum_{k=1}^{N} c_k \sin(w_{mk}) = -j2B_m \tag{3.21}$$

Another way of writing Equations (3.20) and (3.21) is

$$F_k = A_k - jB_k \quad \text{and} \quad F_{-k} = A_k + jB_k \tag{3.22}$$

or

$$\text{Re}\{F_k\} = A_k \quad \text{and} \quad \text{Im}\{F_k\} = B_k \tag{3.23}$$

where Re{ } means "take the real part of the argument" and Im{ } means "take the imaginary part of the argument."

Therefore, Equations (3.22) and (3.23) provide a means of determining the original Fourier coefficients, A_m and B_m, from the new set of complex discrete-time Fourier coefficients, F_m, and vice versa. As a result, just one expression relating the original time-domain data c_k, $k = 1, 2, \ldots, N$, to the new complex discrete-time Fourier coefficients need be written, instead of two expressions relating the c_k to A_m and B_m. The harmonic strengths R_m can be written in terms of the complex discrete-time Fourier coefficients as follows:

$$R_m = \sqrt{A_m^2 + B_m^2} = |F_m| = |F_{\text{DTFS}}\{c_k\}| \tag{3.24}$$

It follows from Equation (3.24) that

$$|F_k| = |F_{-k}|$$

That is, there is a symmetry in the harmonic strengths that will be useful later when frequency-domain filtering is discussed. Note that, for the zeroth frequency or the DC component, $k = 0$, $F_0 = A_0$, and that for the Nyquist frequency, $k = n$, $F_n = A_n$.

To complete the cycle, consider the inverse of Equation (3.16), which can be temporarily written as

$$F_m = \sum_{k=-n}^{n-1} F_k \exp(jw_{mk}), \quad m = 1, 2, \ldots, N \tag{3.25}$$

Note that (1) there is no $1/N$ factor in Equation (3.25) as there was in the defining expression [Equation (3.16)], that (2) the argument of the complex exponential is positive instead of negative, and that (3) the summation index k runs from $-n$ to $n - 1$. The above expression therefore transforms the N complex discrete-time Fourier coefficients into N quantities, F_m, which now must be interpreted.

Start by expanding Equation (3.25) into four terms:

$$F_m = F_{-n} \exp(jw_{-nm}) + \sum_{k=-(n-1)}^{1} F_k \exp(jw_{mk}) + F_0 + \sum_{k=1}^{n-1} F_k \exp(jw_{mk})$$

$$\tag{3.26}$$

using the fact that $\cos(0) = 1$ and $\sin(0) = 0$. Next, modify the summation indexes on the second term of Equation (3.26) and get

$$F_m = F_n \cos(w_{mn}) + \sum_{k=1}^{n-1} F_{-k} \exp(jw_{-km}) + F_0 + \sum_{k=1}^{n=1} F_k \exp(jw_{km}) \quad (3.27)$$

using the fact that $\sin(w_{nm}) = 0$. The second and fourth terms of Equation (3.27) are grouped into one sum, and

$$\exp(jw_{-km}) = \cos(w_{km}) - j\sin(w_{km})$$

$$\exp(jw_{km}) = \cos(w_{km}) + j\sin(w_{km})$$

are used to get

$$F_m = F_n \cos(w_{mn}) + \sum_{k=1}^{n-1} \{F_{-k}[\cos(w_{km}) - j\sin(w_{km})$$

$$+ F_k[\cos(w_{km}) + j\sin(w_{km})]\} + F_0$$

or

$$F_m = F_0 + \sum_{k=1}^{n-1} \{(F_{-k} + F_k)\cos(w_{km})$$

$$+ j(F_k - F_{-k})\sin(w_{km})\} + F_n \cos(w_{nm}) \quad (3.28)$$

Finally, Equation (3.28) is combined with

$$F_k + F_{-k} = 2A_k \quad \text{and} \quad F_k - F_{-k} = -j2B_k$$

to give

$$F_m = A_0 + 2\sum_{k=1}^{n-1} [A_k \cos(w_{km}) + B_k \sin(w_{km})]$$

$$+ A_n \cos(w_{nm}) = c_m \quad (3.29)$$

Thus, after all the algebraic dust has settled, Equation (3.29) shows that the inverse of the discrete-time Fourier series coefficients is simply the time-domain data, and the discrete-time Fourier series pair can be written as follows:

$$F_m = \frac{1}{N}\sum_{k=1}^{N} c_k \exp(-jw_{km}), \quad m = -n, -(n-1), \ldots, 0, \ldots, n-1 \quad (3.30a)$$

$$c_k = \sum_{m=-n}^{n-1} F_m \exp(jw_{km}), \quad k = 1, 2, \ldots, N \quad (3.30b)$$

where $w_{km} = 2\pi f_1 kmh = 2\pi[1/(Nh)]kmh = 2\pi km/N$. These expressions can also be written using the short hand symbology introduced in Equation (3.17):

$$F_m = F_{\text{DTFS}}\{c_k\}$$

$$c_k = F_{\text{DTFS}}^{-1}\{F_m\} \quad (3.31)$$

The above pair is equivalent to the equations that were derived in Section 3.2, but in some situations their compactness and symmetry make them preferable.

Equation (3.24) showed that the absolute value of the F_m gives the harmonic strengths R_m, which are symmetric about zero frequency. By using the above definition of F_m in terms of the complex exponential, it can be shown that F_m and R_m are periodic with period $2n$; that is,

$$F_{m+2n} = F_m$$
$$R_{m+2n} = R_m$$

(3.32)

Therefore, the harmonic strengths are both symmetric about the zero frequency and the Nyquist frequency.

The exponential notation can also be used to show that the time-domain data are periodic. Consider Equation (3.30b) with a slight change in the indexes:

$$c_{m+N} = \sum_{k=-n}^{n-1} F_k \exp\left[\frac{j2\pi k(m+N)}{N}\right]$$

$$= \sum_{k=-n}^{n-1} F_k \exp\left[\frac{j2\pi km}{N}\right] \exp(j2\pi k)$$

$$= c_m$$

which says that the discrete-time Fourier series assumes that the time-domain data are periodic with period Nh. Since only the first N values in the time domain are of concern, the fact that the next N values, which in effect are fictitious, are the repeats of the first N values is of little consequence. However, it is useful to keep this in mind when the data under analysis are significantly different in magnitude at the beginning of the data set than they are at the end of the data set. For example, if the time-domain data are a ramp, then c_1 is significantly different from c_N; but since c_{1+N} is assumed to be equal to c_1, there would be a sharp discontinuity at the end of the time-domain data set from the Fourier series point of view. For the Fourier series to fit this discontinuity, the amplitudes of the high-frequency components would have to be large, thereby causing the high-frequency harmonic intensity to be relatively high. In this case the analyst might want to consider using weights and padding to taper the starting and ending portions of the time-domain data (see Section 10.7 of Hamming, 1989, and Section 3.12.3 of this text for further details).

3.10.2 The Harmonic's Magnitude and Phase

In the previous section the following transform pair was derived:

$$F_m = \frac{1}{N} \sum_{k=1}^{N} c_k \exp(-jw_{km}), \qquad m = -n, -(n-1), \ldots, 0, \ldots, n-1$$

$$c_k = \sum_{m=-n}^{n-1} F_m \exp(jw_{km}), \qquad k = 1, 2, \ldots, N$$

Since the F_m components are complex, they can be written as

$$F_m = A_m - jB_m$$

This complex quantity can also be written in terms of its magnitude R_m and its phase, $\arctan(-B_m/A_m)$. Thus, for a time-domain sequence c_k, $k = 1, 2, \ldots,$ N, the magnitude and phase of each of the n harmonics that make up the associated discrete-time Fourier series can be determined. If, for example, the components F_m are all real, then each harmonic has zero phase. This viewpoint will come in handy in Section 3.11 when frequency-domain filtering is covered.

3.10.3 Alternative Notation and the Periodogram

The notation for the discrete-time Fourier series pair varies in the literature. The texts by Oppenheim and Schafer (1975, 1989), for example, place the $1/N$ factor in the expression for c_m and the summation indexes range from 0 to $N - 1$. Therefore, using their nomenclature, which, by the way, is widely accepted in the digital signal processing field, one would write

$$c_i = \frac{1}{N} \sum_{k=0}^{N-1} F'_k \exp\left[j\left(\frac{2\pi}{N}\right)ki\right], \qquad i = 0, 1, \ldots, N-1 \qquad (3.33)$$

$$F'_k = \sum_{i=0}^{N-1} c_i \exp\left[-j\left(\frac{2\pi}{N}\right)ki\right], \qquad k = 0, 1, \ldots, N-1 \qquad (3.34)$$

The Fourier line spectrum would then be written as

$$R'^2_k = \left|\sum_{i=0}^{N-1} c_i \exp\left[-j\left(\frac{2\pi}{N}\right)ki\right]\right|^2, \qquad k = 0, 1, \ldots, N-1 \qquad (3.35)$$

Primes have been used in these last three equations to denote that these quantities are different due to the different scaling factors. Oppenheim and Schafer refer to Equations (3.33) and (3.34) as the *discrete Fourier Transform* (DFT) pair, and this terminology is widely used in the digital signal processing literature. There are some advantages to the above form, but in this book the notation is closer to that in the text by Jenkins and Watts (1968).

Another notation employed by Marple (1987) in Chapter Two of his text presents the discrete-time Fourier series pair as

$$c_i = \frac{1}{Nh} \sum_{k=-n}^{n-1} F''_k \exp\left[j\left(\frac{2\pi}{N}\right)ki\right], \qquad i = 0, 1, \ldots, N-1$$

$$F''_k = h \sum_{i=0}^{N-1} c_i \exp\left[-j\left(\frac{2\pi}{N}\right)ki\right], \qquad k = -\frac{N}{2}, \ldots, \frac{N}{2} - 1$$

This form, if it were more popular, would be attractive because it is consistent with the definition that will be used in Chapter Five for the discrete-time Fourier transform of an infinite sequence.

Equation (3.35) can be generalized by taking a look at the argument of the exponential:

$$\theta = \frac{2\pi ki}{N} \qquad (3.36)$$

Since the fundamental frequency is $1/(Nh)$ and since the kth harmonic frequency, f_k, is $k/(Nh)$, Equation (3.36) can be written as

$$\theta = 2\pi f_k hi \qquad (3.37)$$

If the constraint that the frequency be a harmonic of the fundamental frequency is removed and a scaling factor of $1/N$ is added, Equation (3.35) becomes

$$\text{PER}(f) = \frac{1}{N} \left| \sum_{i=0}^{N-1} c_i \exp[-j(2\pi fhi)] \right|^2, \qquad k = 0, 1, \ldots, N-1$$

which is the *periodogram* of the c_i data. Unlike Equation (3.15), the periodogram can be computed for *any* frequency and therefore is a continuous function of the frequency. As will be pointed out in Section 3.12.1, the fast Fourier transform can be used to efficiently calculate the components of the discrete-time Fourier series pair but not the periodogram (unless the chosen frequencies happen to be the harmonics of the fundamental). Note that, when the periodogram frequencies *are* chosen to be the harmonics of the fundamental, the Fourier line spectrum and the periodogram, except for a $1/N$ factor, are the same.

3.11 FILTERING IN THE FREQUENCY AND TIME DOMAINS

There are many situations where there is an interest in removing the harmonic strength in certain bands of frequencies. In this section a method of "surgically" removing all the power from selected portions of the Nyquist interval will be discussed; it will be compared with two time-domain filters, one of which was discussed in Section 1.9.

3.11.1 Frequency-domain Filtering

The power in the frequency band of interest can be removed surgically by the following steps. First, transform the time-domain data c_k, $k = 1, \ldots, N$, into the frequency domain and obtain the complex Fourier coefficients using the first half of the discrete-time Fourier series pair derived in the previous section:

$$F_m = \frac{1}{N} \sum_{k=1}^{N} c_k \exp(-jw_{km}), \qquad m = -n, -(n-1), \ldots, 0, \ldots, n-1$$

where the F_m will in general be complex; that is, they will have real and imaginary components.

Second, a real-valued filter multiplier

$$Q_m, \qquad m = -n, \ldots, 0, \ldots, n - 1$$

is constructed such that it is zero at those harmonics where the power is to be removed and unity elsewhere. The line spectrum is symmetric about the Nyquist frequency and so is the filter multiplier; that is,

$$Q_m = Q_{-m} \quad \text{and} \quad Q_m = Q_{m+2n} \quad \text{and} \quad Q_{2n-m} = Q_{2n+m}$$

The filter multiplier is applied to the discrete-time Fourier series components to give modified components according to

$$F'_m = Q_m F_m, \qquad m = -n, -(n-1), \ldots, 0, \ldots, n - 1$$

where both F_m and F'_m have real and imaginary components. These modified discrete Fourier transform components represent the time-domain data in the frequency domain after the power has been removed from the selected harmonics. In view of the discussion in Section 3.10.2 about the magnitude and phase of each harmonic, surgical filtering consists of zeroing the magnitude and phase of selected harmonics.

Now all that remains is to transform the modified discrete-time Fourier series components back to the time domain using the second member of the discrete-time Fourier series pair derived in the previous section:

$$c'_m = \sum_{k=-n}^{n-1} F'_k \exp(jw_{km}), \qquad m = 1, 2, \ldots, N$$

The modified time-domain data c'_m, $m = 1, \ldots, N$, can now be analyzed from a variety of points of view depending on the problem. For example, the filtered data could be graphed in the time domain, and, if desired, the standard deviation could be computed. In passing, it should be mentioned that the standard deviation of the modified time-domain data can be calculated without transforming the modified Fourier components back to the time domain, since

$$R'_m = |F'_m|$$

and since the variance can be calculated from R'_m as shown in Section 3.5.

To illustrate frequency-domain or surgical filtering, consider the time series shown in Figure 3.79, which might represent the controlled variable of a process where there was still some opportunity to improve the control by removing the low-frequency power. Figure 3.80 shows the line spectrum plotted on semilogarithmic axes rather than linear axes, because the latter type of graph would emphasize the high-amplitude components near zero frequency and would wash out the low-amplitude components, which represent the remaining portion of the Nyquist interval. Using the semilogarithmic axes, the choice of the cut frequency is easier to make, at least in this case. Since every case will be different, the analyst should look at both the linear and the semilogarithmic presentations. In

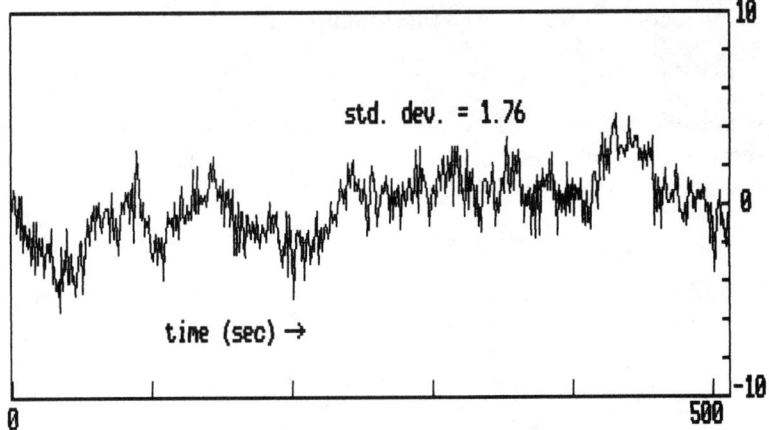

Figure 3.79   Time-domain plot of data before filtering.

the next section, more comments will be made about the semilogarithmic presentations.

Using frequency-domain filtering, the effect of some hypothetical (and also unrealistic) control strategy will be simulated whereby all the power at harmonics below 0.04883 Hz (or to the left of the dashed line in Figure 3.80) will be removed and all the higher harmonics will be passed. This high-pass filtering can be accomplished by defining a multiplier such that it is zero for harmonics less than 0.04883 Hz and unity for harmonics at and greater than 0.04883 Hz. The result of applying the multiplier to the complex Fourier components and transforming the result back to the time domain is shown in Figure 3.81. The high-pass filtering has reduced the standard deviation from 1.76 to 0.88. The cumulative line spec-

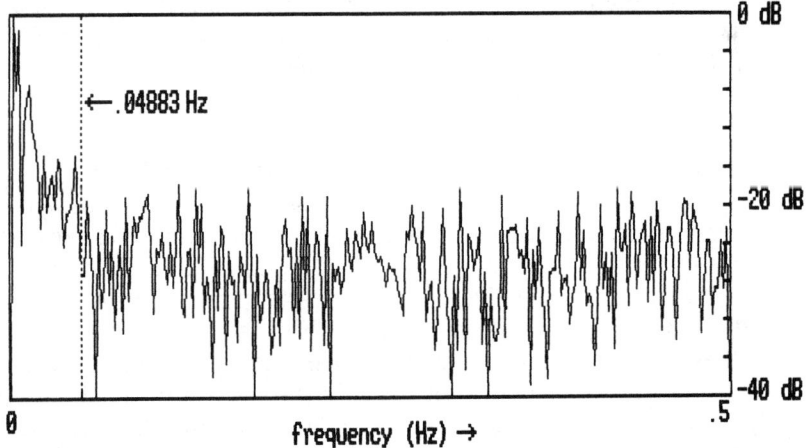

Figure 3.80   Line spectrum before filtering.

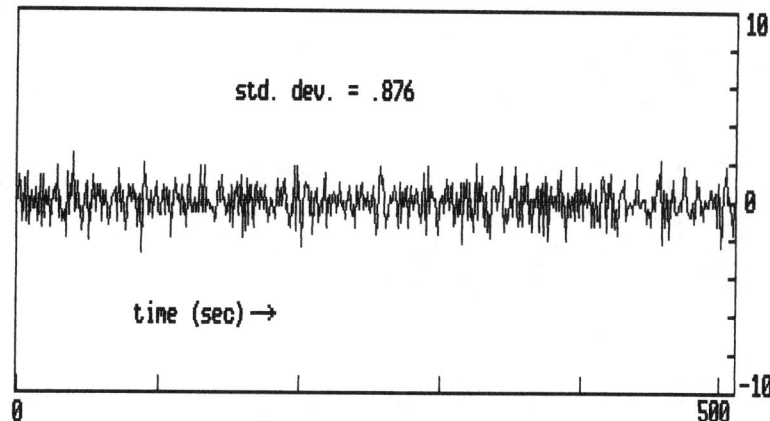

Figure 3.81   Time-domain plot after filtering.

trum of the surgically filtered data is shown in Figure 3.82, where it is seen that, as designed, below 0.04883 Hz the harmonic intensity is zero; but the linearity of the remaining portion of the curve indicates that the surgical filtering has removed the autocorrelated noise and left white noise.

The reader should note that this type of filtering can be designed to remove power from any portion of the Nyquist interval, so this method could be used to effect low-pass as well as high-pass filters. For example, consider the time-domain plot in Figure 3.83 and the associated line spectrum in Figure 3.84. All the harmonic power to the left of the dashed line in Figure 3.84 will be passed, and all the harmonic power to the right will be removed. The result in the time domain is shown in Figure 3.85.

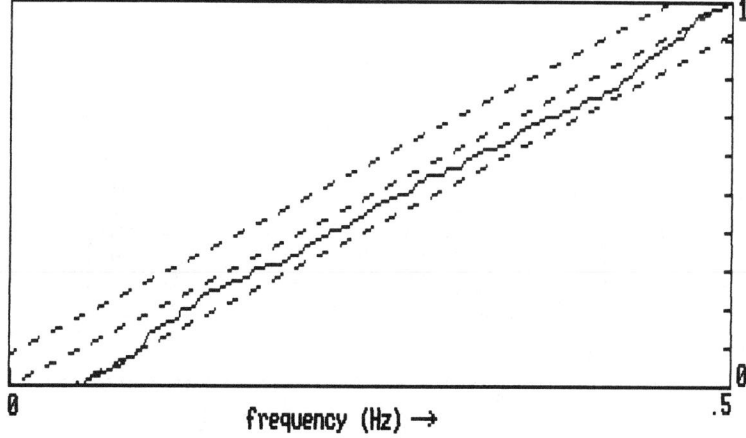

Figure 3.82   Cumulative line spectrum after filtering.

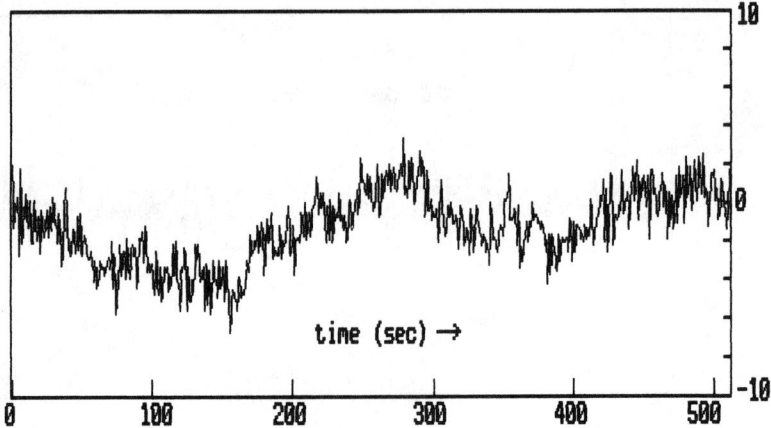

Figure 3.83 Time-domain data before filtering.

At this point the reader should be warned that this method of constructing a desired frequency response, although attractively simple and widely used, is not foolproof. It is beyond the scope of this book to demonstrate in detail the restrictions, but it stands to reason that, at best, this method is only as good as the relation of the line spectrum to the true spectrum of the original time-domain data. Since the line spectrum only provides estimates of the true spectrum at the harmonics of the fundamental frequency, there is a potential for error if there is significant variation in the spectral power at frequencies in between the harmonics.

In Section 3.12.4, it will be shown that padding the time domain data can be used to estimate the spectrum at frequencies in between the harmonics. Therefore, when the Q multiplier is applied to the components of the discrete-time

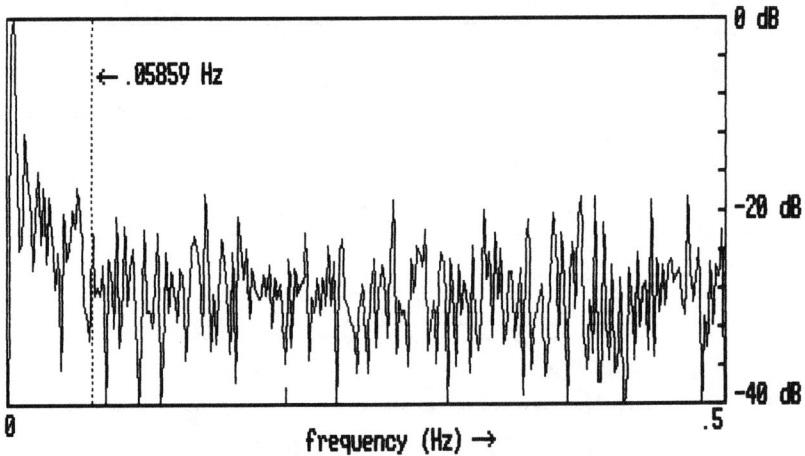

Figure 3.84 Line spectrum before filtering.

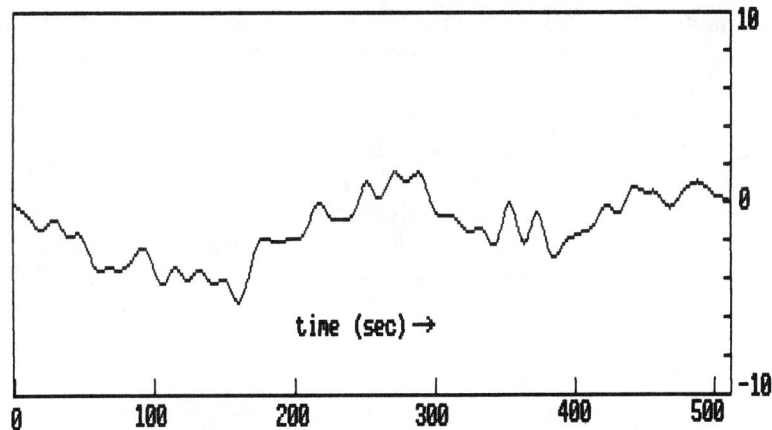

Figure 3.85   Time-domain plot after filtering.

Fourier transform, the discrepancy between the apparent frequency response, as defined by the Q multiplier, and the actual frequency response will be decreased if padding has been applied to the time-domain data. As will be pointed out in Section 3.12.4, windowing can augment the padding.

The reader is referred to Chapter Eight in the book by Karl (1989) for an interesting discussion of the pros and cons of this approach. More sophisticated methods of frequency-domain filtering are discussed in detail in Chapter 7 of Oppenheim and Schafer (1989).

3.11.2 Time-domain Filtering

In Chapter One, filtering in the time domain was mentioned in terms of the following first-order discrete filter:

$$Y_k = \alpha X_k + (1 - \alpha) Y_{k-1} \tag{3.38}$$

It was shown in Section 1.9 that application of such a filter could attenuate unautocorrelated noise and that the output of the filter could be characterized by a time constant. There are several differences between this filtering approach and that of surgical filtering.

First, this time-domain filter uses present and past data so it could be applied on line. (The frequency-domain filter approach must be done off line since it requires all the data to be present before filtering begins.) As a consequence of using past data, this time-domain filter will introduce lag that, if the filter were part of a control loop, could negatively affect the controllability. If this first-order time-domain filter is applied to data off line for analytical work, say in the case where the analyst simply wishes to smooth the data in order to get a better idea of the overall trend, the lag introduced by the above filter could muddy the interpretation. An alternative approach, using frequency-domain filtering, would be to apply a low-pass frequency-domain filter designed with a Q_m multiplier to be unity at the low harmonics, which contain the trend information, and to be zero at the high harmonics, which contain the noise that obscures the trend of the data.

Second, as will be seen in Chapter Four, the distribution of the attenuation across the Nyquist interval is gradual for the first-order time-domain filter instead of having a sharp discontinuity as in the case of the surgical filter.

Another approach for off-line analysis would be to apply time-domain central smoothing, which reduces the high-frequency noise but does not introduce lag. One of the simplest central smoothing formulas is

$$Y_k = \frac{X_{k+1} + X_k + X_{k-1}}{3} \tag{3.39}$$

which smoothes the kth data point by averaging it with its two neighbors. Since this is an off-line computation, the $(k + 1)$th data point is available to smooth the kth point, and since the computation is centered about the kth data point, no lag is introduced.

The variance attenuation for white noise can be determined by squaring both sides of Equation (3.39) and calculating the expected value. Because the noise is unautocorrelated, the cross terms cancel out, giving

$$V_y = \frac{V_x + V_x + V_x}{9} = \frac{V_x}{3}$$

so that the standard deviation of the smoothed signal is less than that of the original data by a factor of 1.732.

The distribution of the attenuation across the Nyquist interval for this kind of central smoothing, compared to the sharpness of the frequency-domain surgical filtering approach and the gradualness of the time-domain first-order discrete filter, is significantly different. Although the tools of Chapter Four are needed to pursue this question in detail, a few commonsense comments can be made here. Because of its structure, it is expected that Equation (3.39) would pass low frequencies and would attenuate high-frequency components, and it would also completely obliterate components having a period equal to three sample intervals, that is, a frequency of $1/(3h)$, as well as all the higher harmonics having frequencies of $2/(3h)$, $3/(3h)$, $4/(3h)$, and so on. Note that the Nyquist frequency is $1/(2Nh)$, so only the frequency at $1/(3Nh)$ would be obliterated. However, for central smoothing filters using more than three points, more than one of the obliterated frequencies will lie in the Nyquist interval. Since components with frequencies in between these obliterated frequencies would not be completely removed, it might appear as though they were being amplified. Consequently, spectral analysis of data after central smoothing could be misleading since it would show peaks in between the obliterated frequencies.

An interesting time-domain solution to the off-line smoothing problem that has been frequently used is the *double-sweep* filter. Here the raw time-domain data, X_k, $k = 1, \ldots, N$, are first filtered from left to right, or from $k = 1$ to $k = N$, according to

$$Y_1 = X_1, \qquad Y_k = \alpha X_k + (1 - \alpha) Y_{k-1}, \qquad k > 1$$

Next, the raw data are filtered right to left, or from $k = N$ to $k = 1$, according to

$$W_N = X_N, \qquad W_k = \alpha X_k + (1 - \alpha) W_{k+1}, \qquad k < N$$

The two filtered quantities are added to give the final filter output:

$$Z_k = \frac{Y_k + W_k}{2}$$

The sum of the two filters, Z_k, has the same phase as the original data while at the same time passing low frequencies and attenuating high frequencies. Alternatively, the output from the left-to-right filter could be fed as input to the right-to-left filter. In this case, the two first-order filters would be *cascaded*. After the Z-transform has been introduced, filtering and smoothing will be revisited in Section 4.4.2.

3.12 COMPUTATIONAL CONCERNS

3.12.1 Fast Fourier Transform

The equations defining the components of the line spectrum have been derived from the discrete-time Fourier series; however, these equations should not be used for computation. For example, if there are N data points, then there are $N/2 + 1$ values of the A_k's and $N/2 - 1$ values of the B_k's to be determined. If the equations presented in Section 3.2.4 are used, the number of calculations would be on the order of N^2. The number of calculations is approximately the same if the components of the exponential form of the discrete-time Fourier series are to be calculated by the equations presented in Section 3.10.1. A significantly more efficient method called the fast Fourier transform was developed by Cooley and Tukey (1965). To use this technique, the number of data points must be a power of 2.

There are many fast Fourier transform software packages available that can be used to calculate the Fourier coefficients and/or the components of the discrete-time Fourier series pair, so there is no point in taking up space here to review the theory of the fast Fourier transform. However, in case the reader wishes to start from scratch, Figure 3.86 shows a listing of a BASIC GOSUB routine that will perform a fast Fourier transform. Note that there are three options. If the variable IFT is 0, the routine calculates the discrete-time Fourier series coefficients for the input data residing in the array X1, where X1(1, I), I = 1, N, contain

```
'    ****************** FAST FOURIER TRANSFORM ************
'
'    This routine calculates the discrete Fourier series coefficients
'    for three cases:
'              1.   If IFT = 0, then it calculates the complex coefficients
'                   for time-domain data.
'              2.   If IFT = 1, then it calculates the time-domain data for
'                   the complex coefficients.
'              3.   If IFT = 2, then it calculates the components of the
'                   line spectrum.
'
'    This routine requires the real and imaginary components of the
'    input data to be in X1(1,I), X1(2,I), I=1,2^M, where X1(1,I) are
'    the real components and X1(2,I) are the imaginary components.
'    The record length (including zero padding if necessary) must be a
'    power (M) of 2.
'
'    This routine puts the calculated components back in X1(1,I),
'    X1(2,I), I=1,2^M.  When the output is the line spectrum, those
'    components are placed in X1(1,I), I = 1,2^M.  It also calculates
'    the average (XBAR) and the variance (VAR).
'
'    The subscripted variables used in this routine are shown in the
'    following dimension statements.  The maximum length is set to be
'    1024, but this can be adjusted to fit the needs of the user.  To
'    actually use this program, the user would probably want to
'    reconfigure it as a subroutine.

DIM X1(2,1024), U(2), W(2), D(2), T(2)

N = 2 ^ M          ' (the number of input data points)
PI = 3.14159265
FOR L = 1 TO M
     LE = 2 ^ (M + 1 - L)
     LE1 = LE / 2
     U(1) = 1
     U(2) = 0
     W(1) = COS(PI / LE1)
     IF IFT = 1 THEN W(2) = SIN(PI / LE1) ELSE W(2) = -SIN(PI / LE1)
     FOR J = 1 TO LE1
          FOR I = J TO N STEP LE
               IP = I + LE1
               T(1) = X1(1, I) + X1(1, IP)
               T(2) = X1(2, I) + X1(2, IP)
               D(1) = X1(1, I) - X1(1, IP)
               D(2) = X1(2, I) - X1(2, IP)
               X1(1, IP) = D(1) * U(1) - D(2) * U(2)
```

Figure 3.86 QUICKBASIC GOSUB for calculation of line spectrum.

```
                    X1(2, IP) = D(2) * U(1) + D(1) * U(2)
                    X1(1, I) = T(1)
                    X1(2, I) = T(2)
               NEXT I
               DUMMY = U(1) * W(1) - U(2) * W(2)
               U(2) = U(2) * W(1) + U(1) * W(2)
               U(1) = DUMMY
          NEXT J
NEXT L
NV2 = N / 2
NM1 = N - 1
J = 1
FOR I = 1 TO NM1
    IF I < J THEN
          T(1) = X1(1, J)
          T(2) = X1(2, J)
          X1(1, J) = X1(1, I)
          X1(2, J) = X1(2, I)
          X1(1, I) = T(1)
          X1(2, I) = T(2)
    END IF
    K = NV2
    WHILE K < J
          J = J - K
          K = K / 2
    WEND
    J = J + K
NEXT I
IF IFT = 1 THEN
     'return the inverse transform
     FOR I = 1 TO N
          X1(1, I) = X1(1, I) / CN
          X1(2, I) = X1(2, I) / CN
     NEXT I
ELSEIF IFT = 2 THEN
     'return the power spectrum
     CNS = CN * CN
     FOR I = 1 TO NV2 + 1
          X1(1, I) = (X1(1, I) ^ 2 + X1(2, I) ^ 2) / CNS
     NEXT I
     XBAR = SQR(X1(1, 1))
     VAR = 0
     FOR I = 2 TO NV2
          VAR = VAR + X1(1, I)
     NEXT I
     VAR = 2 * VAR + X1(1, NV2 + 1)
END IF
RETURN
```

Figure 3.86   (*continued*)

the real components and X1(2, I), I = 1, N, contain the imaginary components. If IFT = 1, then the routine calculates the inverse discrete-time Fourier coefficients. Finally, if IFT = 2, the routine calculates the components of the Fourier line spectrum from the real and imaginary components of the discrete-time Fourier series components. This routine is not guaranteed to be the fastest method of applying the fast Fourier transform, but it will produce correct numbers.

For readers wishing to look into alternative methods of spectral analysis, the review by Kay and Marple (1981), Chapter 11 in the first text by Oppenheim and Schafer (1975), and Chapter 9 in the second text by Oppenheim and Schafer (1989) are recommended. Also recommended are two texts by Kay (1988) and Marple (1987), which are expansions of their review.

3.12.2 Linear and Logarithmic Axes

Throughout this chapter, most of the x- and y-axes on the graphs of line spectra have been linear. The y-axis has been frequently scaled to range linearly from the minimum harmonic intensity to the maximum harmonic intensity without regard for the numerical values of the harmonic intensity. The idea of graphing the x-axis linearly from zero frequency to the Nyquist frequency is almost universally accepted, but many authors prefer, and in fact advocate, plotting the harmonic intensities on the y-axis in decibels (dB), that is, plotting 10.0 times the logarithm of the harmonic intensities. For example, instead of plotting R_k versus the fundamental frequency and its harmonics, one could plot

$$10 \log \left(\frac{R_k}{R_{\max}} \right)$$

where $R_{\max}$ is the maximum value of the R_k taken over the Nyquist interval. A typical range of the y-axis would be from -60 to 0 dB.

The logarithmic scale attenuates the low-frequency components of the line spectrum, so harmonic strength due to the presence of nonstationary or heavily autocorrelated stochastic disturbances will not overwhelm components located at higher frequencies. On the other hand, the presence of harmonic strength in the midrange of the Nyquist interval will not show up as strongly as in the linear plots. For example, Figure 3.8, when replotted as decibels versus frequency, is shown in Figure 3.87. Here the peaks do not stand out as much as in Figure 3.8. Years ago the analyst would have had to make a hard decision about how the spectra were to be graphed, but now with desktop computers and terminals everywhere the analyst should be able to graph the spectra both ways and choose which presentation affords the most insight.

As a second illustration of the effect of different graph scaling, consider Figure 3.88, where the sums of two stochastic sequences are plotted in the time domain. The first component is a stochastic sequence defined by

$$Y_k = 0.707 Y_{k-1} - 0.9025 Y_{k-2} + w_k$$

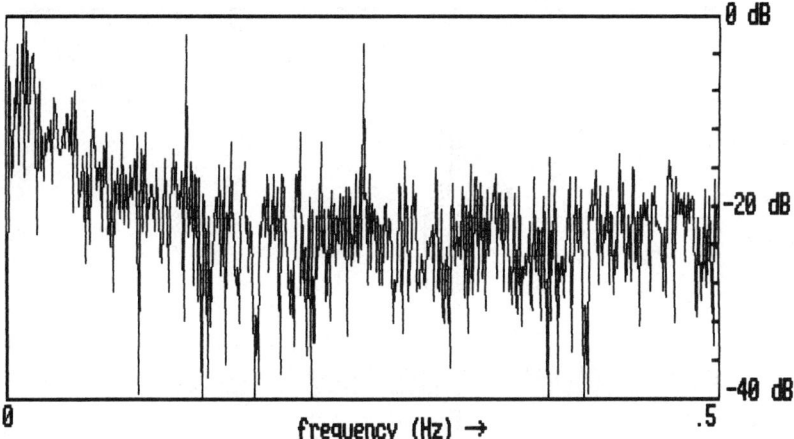

Figure 3.87   Semilogarithmic plot of Figure 3.8.

This sequence is different than those introduced in Chapter One in that it is second order instead of first order because Y_k depends on Y_{k-1} *and* Y_{k-2}. The second component of the stochastic sequence is simply white noise with an amplitude three times that of the white noise generating the first component. This second component could be looked at as background noise. The line spectrum, plotted on linear axes, is shown in Figure 3.89, where a peak occurs at approximately 0.2 Hz, but it is not a sharp isolated peak as in the case where a deterministic component is present. This type of line spectrum is sometimes seen when a control loop subject to autocorrelated stochastic disturbances is mistuned such that it oscillates, but the period of the oscillation is distributed over a small band of frequencies. In any case, Figure 3.90 shows the same line spectrum plotted on

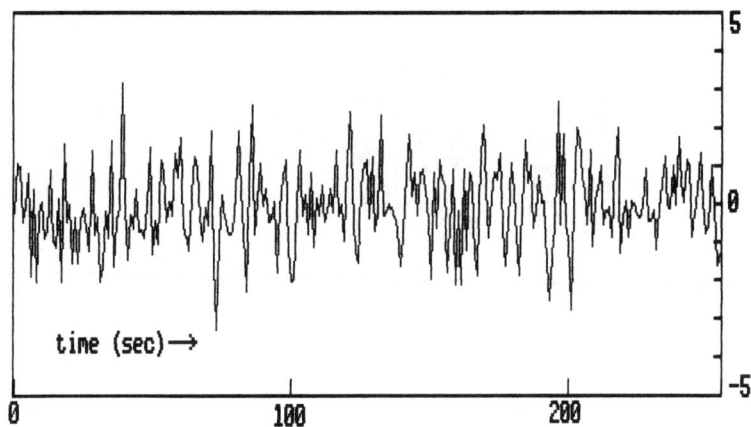

Figure 3.88   Time-domain plot of $Y(k) = 0.707Y(k - 1) - 0.9025Y(k - 2) + w(k)$.

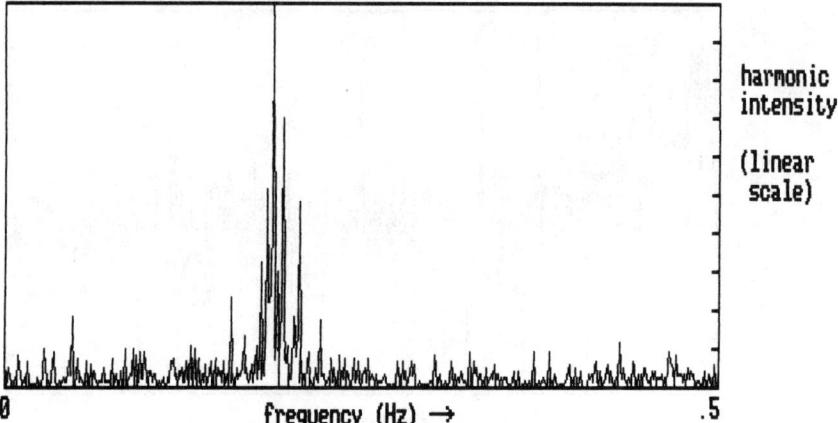

Figure 3.89 Line spectrum (linear scales).

semilogarithmic axes, and the reader should note that the peak is not as pronounced.

As a third example of how different graph scaling can show up different aspects of a spectrum, consider the case where there are 1024 samples, at one second per sample, of two sinusoids of equal amplitude in white noise. The frequency of the first sinusoid is 0.0625 Hz, which is the 64th harmonic of the fundamental frequency of $\frac{1}{1024}$ = 0.0009765625 Hz. The frequency of the second sinusoid is 0.08333333 Hz, which is *not* a harmonic of the fundamental frequency. Figures 3.91 and 3.92 show the line spectra plotted on linear and semilogarithmic scales. Note how the second sinusoid has a shorter and slightly more dispersed peak compared to the single spike for the first sinusoid, especially when semilogarithmic scales are used. Although both sinusoids have equal harmonic power,

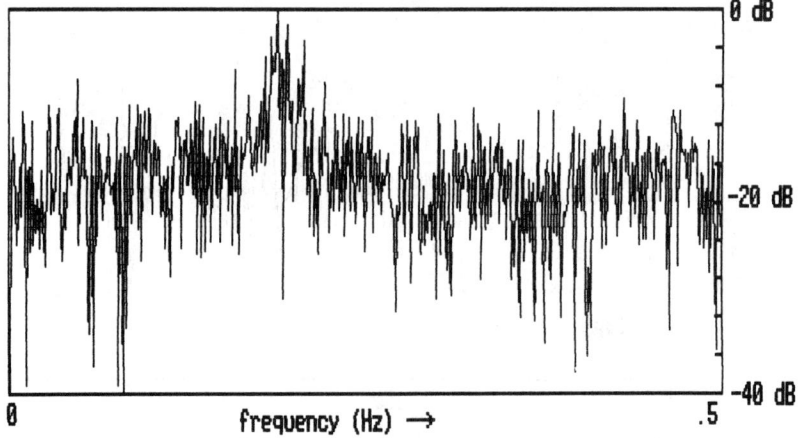

Figure 3.90 Line spectrum (semilogarithmic scales).

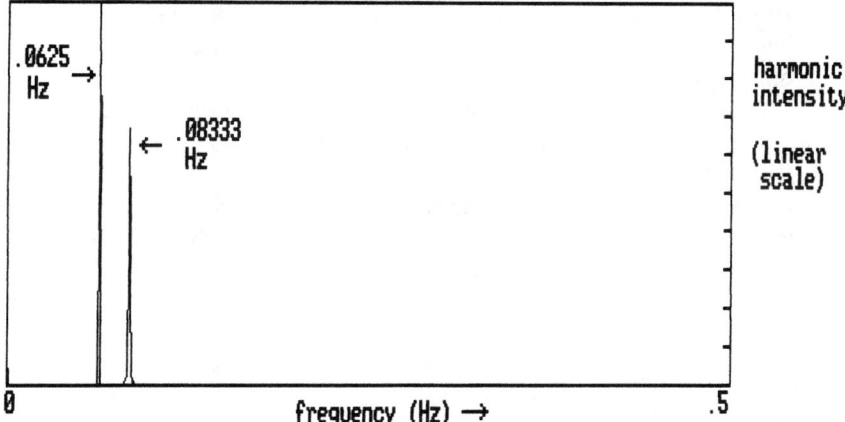

Figure 3.91 Line spectrum of two sinusoids in white noise (linear scale).

they appear different because one happens to fall exactly on a harmonic of the fundamental frequency of the data set.

3.12.3 Data Windows

In computing the Fourier line spectrum of a finite time-domain sequence, the distribution of harmonic power over the Nyquist interval is estimated. If the finite sequence is considered as a subsample of an infinite sequence, then in effect a time-domain window has been applied to the infinite sequence. The application of this time-domain window applies weights to the infinite sequence such that all points from 1 to N have weights of unity and all other points have weights of

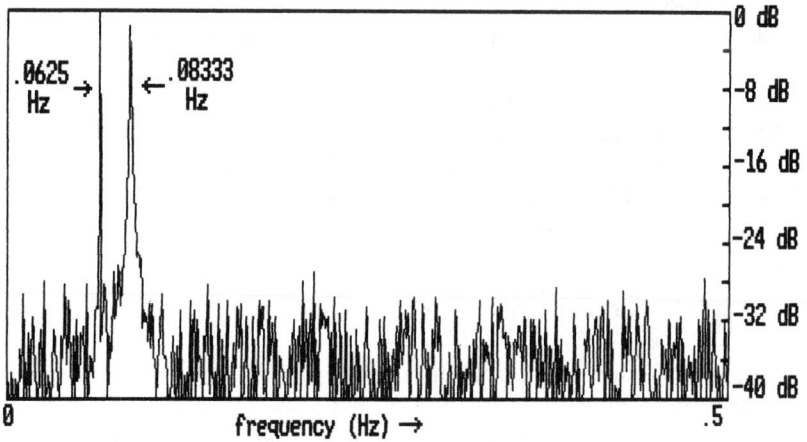

Figure 3.92 Line spectrum of two sinusoids in white noise (semilogarithmic scale).

zero. This type of time-domain weighting is called a rectangular window, and it has been used throughout the book. Because of the finite width of the window, only an estimate of the true spectrum can be obtained.

Using mathematics beyond the scope of this book, one could show that using a rectangular time-domain window introduces a problem that is described as *spectral leakage*. Crudely, this means that harmonic power at a particular frequency will leak into nearby frequencies, causing potentially misleading looking line spectra. This is most noticeable when there are two nearby sinusoids and one sinusoid has more power than its neighbor. In this case the presence of the weaker sinusoid may nearly be masked by its more powerful neighbor.

Nonrectangular types of windows that weight the time-domain data differently are sometimes recommended by workers in statistics and digital signal processing. These windows are designed to weight the time-domain data lightly at the beginning and end of the N-point sequence. For example, the Barlett or triangular window has weights, q_i, $i = 1, \ldots, N$, that satisfy

$$q_i = \frac{2}{N} i \qquad 1 \le i \le \frac{N}{2}$$

$$= 1 - \frac{2}{N} i, \qquad \frac{N}{2} \le i \le N$$

A more commonly used window is the Hamming window, where the weights are defined by

$$q_k = 0.54 + 0.46 \cos \left[\frac{2\pi(k - N/2)}{N} \right]$$

The idea behind weighting the time-domain data lightly at the ends of the sequence is derived from the fact that the Fourier series assumes that the time-domain data are periodic with period N. Thus, if the time-domain data are significantly different at the beginning compared to the end, a Fourier series analysis will show power in harmonics that might mislead the analyst. Applying windows decreases the impact of beginning–ending data differences. It also smooths the resulting spectrum, sometimes causing the spectral peaks to be less distinct.

For a more thorough discussion of the use of windows, the reader is referred to Chapter 11 in the second text by Oppenheim and Schafer (1989), Chapter 2 in the text by Jenkins and Watts (1968), and Section 5.3 in the text by Marple (1987).

3.12.4 Effect of Sample Size and Padding

Figure 3.93 shows the time-domain plot of 128 points of a white noise sequence plus a sinusoid having a frequency of 1/4.06 Hz. Figure 3.94 shows the line spectrum of this sequence. There is a peak at the frequency having a period of 4.0 seconds, but there are several other peaks that tend to cloud the issue. Because of the length of the data, the neighboring harmonics have periods of 3.878 and

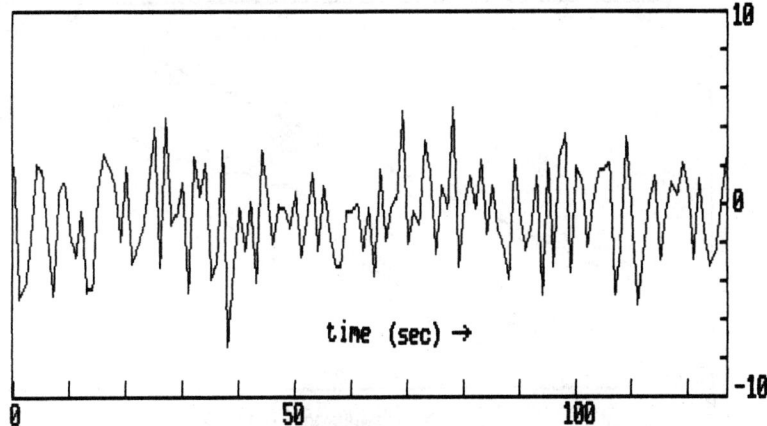

Figure 3.93   Time-domain plot of a 4.06-second period noisy sine wave.

4.129 seconds. Figure 3.95 shows the cumulative line spectrum of this sequence. Neither the line spectrum or the cumulative line spectrum give a clear indication of the presence of the sinusoid.

Figure 3.96 shows the line spectrum of 1024 points of the same sequence. Here the sampling interval is the same, but the data length is eight times longer. A clear peak appears with a period of 4.06349 seconds. The neighboring harmonics have periods of 4.04743 and 4.07968 seconds. Figure 3.97 shows the cumulative line spectrum of this sequence. Although there is sharp shift in the cumulative line spectrum at a frequency having a period of 4.06349 seconds, the Kolmogorov–Smirnov limits are not exceeded, suggesting that these limits must be used with care.

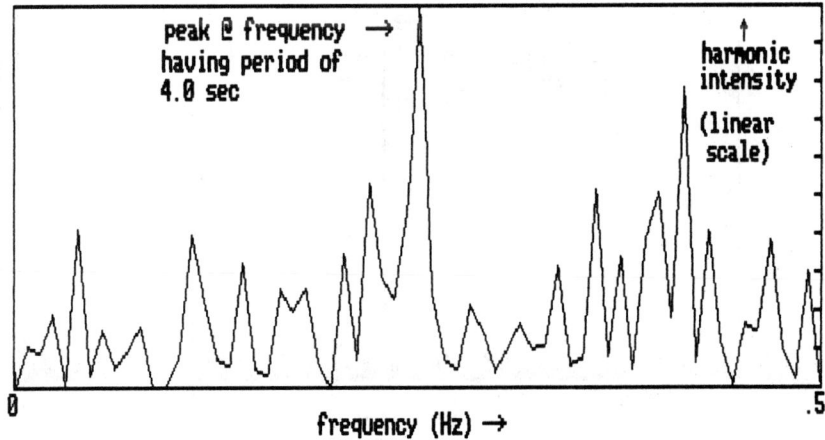

Figure 3.94   Line spectrum of 128 points of a noisy sine wave.

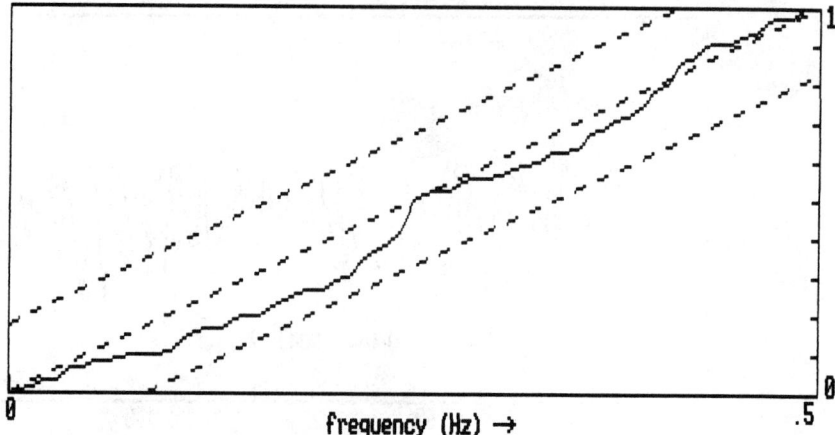

Figure 3.95 Cumulative line spectrum of 128 points of a noisy sine wave.

Sometimes a technique called *padding* is used to increase the apparent resolution in the discrete frequency domain. If N points are to be analyzed, then the fundamental frequency is $f_1 = 1/(Nh)$ hertz. This means that values of the line spectrum can be determined at the harmonics of this fundamental frequency, or at $f_k = k/(Nh)$, $k = 1, \ldots, N/2$, where each frequency is separated by $1/(Nh)$ hertz. If the N data points are padded with zeros such that there are $2N$ time-domain points with the second N points in the time-domain being zeros, then the fundamental frequency is now $1/(2Nh)$. This means that values of the line spectrum can be determined at the harmonics of this new fundamental frequency, which are now separated by $1/(2Nh)$ hertz. This suggests that the position of peaks in the line spectrum can be located twice as precisely. Note, however, that padding does not increase the ability to resolve *two* closely spaced peaks resulting from

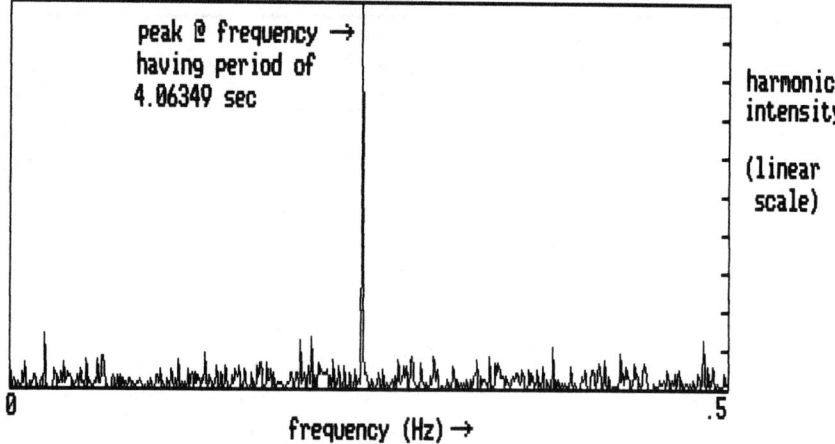

Figure 3.96 Line spectrum of 1024 points of a noisy sine wave.

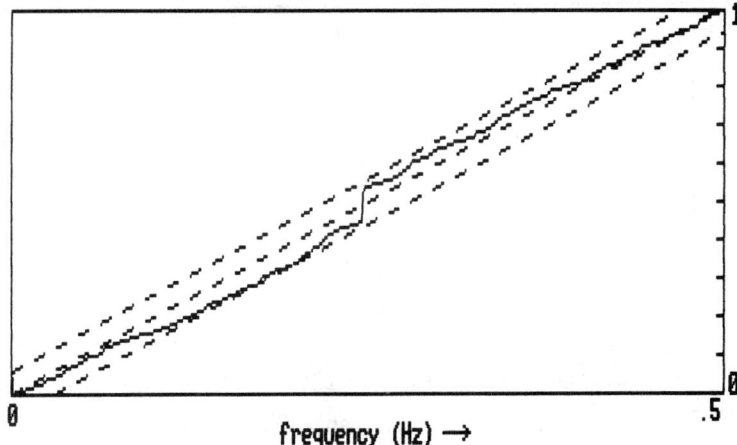

Figure 3.97 Cumulative line spectrum of 1024 points of a noisy sine wave.

two sinusoids of similar frequencies buried in noise. Instead, it allows for interpolation between the harmonics associated with the original data record. As a
result, the spectrum sometimes appears smoother, and the location of the peaks
associated with sinusoids having significantly different frequencies can be determined more accurately.

For example, consider 64 samples of a noiseless sine wave having a period
equal to ten samples and a frequency of 0.1 Hz, assuming that the sampling interval
is 1 second. The fundamental frequency for this data set is $\frac{1}{64} = 0.015625$ Hz.
The line spectrum for this data set shows a peak at 0.09375 Hz, which is closer
to 0.1 Hz than the next harmonic, which is at 0.109375 Hz. After padding with
64 zeros, the line spectrum shows a peak at 0.1015625 Hz. After padding with
192 zeros, there is still a peak at 0.1015625 Hz. After padding with 448 zeros, the
peak resides at 0.0996093 Hz. Finally, after padding with 960 zeros, the peak still
resides at 0.0996093 Hz. The fundamental frequency for this last padded data set
is $\frac{1}{1024} = 0.0009766$ Hz, so the next closest harmonic to 0.0996093 Hz is 0.1005859
Hz, which is farther away from the true frequency.

Padding of data should be done with care since the addition of a string of
zeros at the end of a data record may cause a discontinuity after the Nth point
which might generate misleading peaks in the line spectrum of the padded sequence.

Return to the example of 129 samples of a noisy sine wave plotted in Figure
3.93 and pad that data with 896 zeros to make a total of 1024 points. The line
spectrum of this padded sequence is shown in Figure 3.98. Here a peak occurs
with a period of 4.04743 seconds. Neighboring harmonics are located at frequencies having periods of 4.03150 and 4.06359 seconds. This last harmonic is nearer
to the actual frequency of the sine wave buried in the noise.

To illustrate the effect of padding and windowing, consider their application
to the frequency-domain filtering technique of Section 3.11. Figure 3.99 shows a

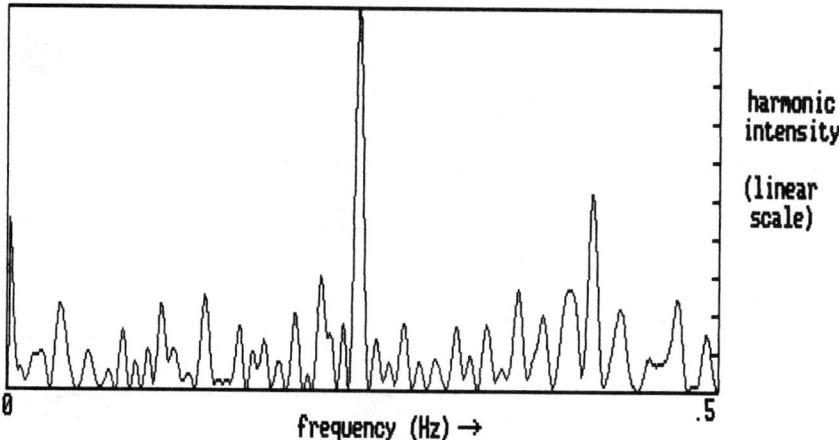

Figure 3.98 Line spectrum of 128 points of a noisy sine wave padded to 1024 points.

time-domain plot of a noisy sinusoid having a period equal to 22.75 seconds, which corresponds to a frequency halfway between two harmonics for this data set of 512 points. Figure 3.100 shows the Fourier line spectrum of these data. Note that the peaks occur at frequencies corresponding to periods of 23.2727 and 22.26087 seconds. These data will be padded with 512 zeros, but first the tapering window shown in Figure 3.101 will be applied to the original 512 points. This window uses the cosine function to smoothly taper the data from zero to full value in the first 50 samples. The first 50 components of the window are given by

$$w_k = \frac{1 - \cos[\pi(k - 1)/(n - 1)]}{2}$$

where n is the number of samples to be windowed (50 in this case).

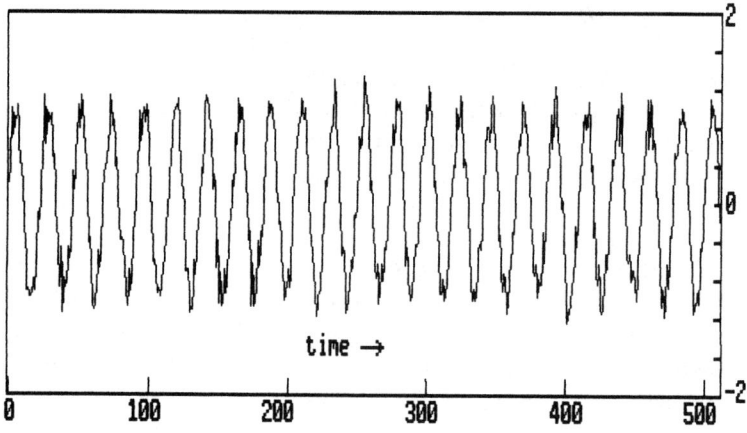

Figure 3.99 Time-domain data before filtering.

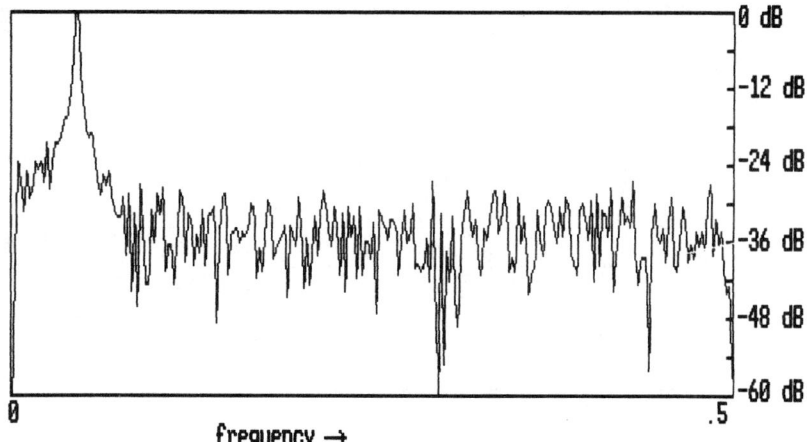

Figure 3.100   Line spectrum of data before padding and windowing.

A mirror image of this tapered cosine function is applied to the last 50 samples. As a result, the transition between the original time-domain data and the trailing zeros has been smoothed. Since the Fourier analysis assumes that the time-domain data are periodic, they can now be considered to have leading zeros as well. Therefore, the tapering window is applied to both ends of the data.

Figure 3.102 shows the line spectrum of the windowed padded data. Note that the spectrum is more symmetrical about the peak, which occurs at a frequency corresponding to a period of 22.75 seconds. The frequency-domain filter multiplier is designed to be zero between the frequencies of 0.04297 and 0.04492 Hz and unity elsewhere. These two frequencies correspond to periods of 23.2727 and 22.26087 seconds. Figure 3.103 shows the result of transforming these data back to the time domain. Note that the first and last 50 points are not shown since they felt the impact of the tapering window.

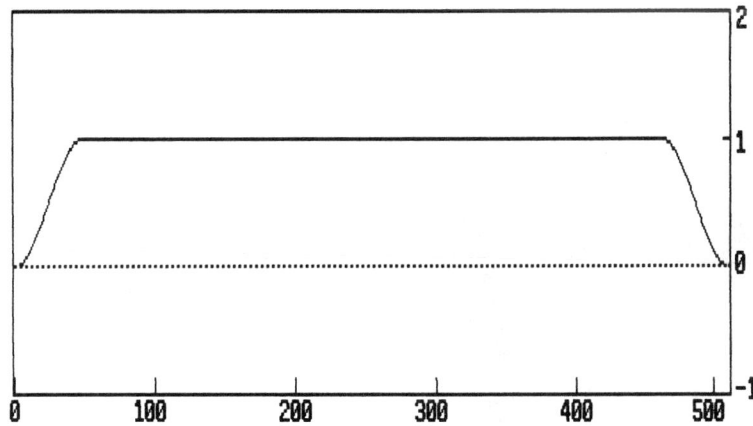

Figure 3.101   Tapering window used for frequency-domain filtering.

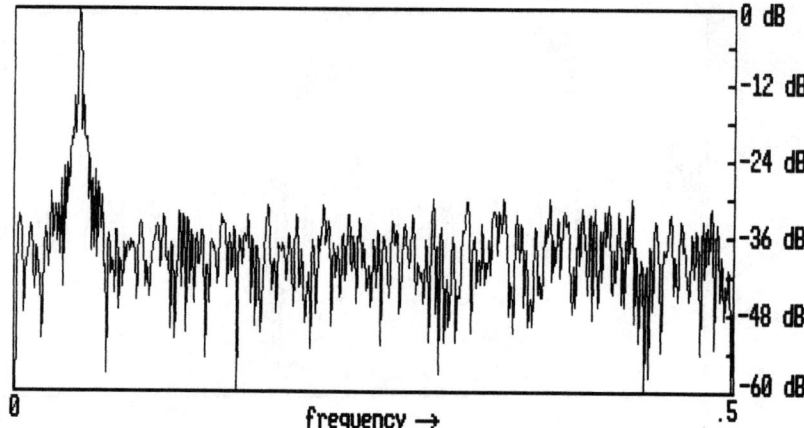

Figure 3.102   Line spectrum after windowing and padding.

The background white noise in the original time-domain data had a standard deviation of 0.145. The presence of the sine wave increased the standard deviation to 0.7138. After the filtering, the middle 412 points of the transformed time-domain data had a standard deviation of 0.144. This observation, coupled with the figures, suggests that the frequency-domain filtering with padding and windowing was successful in removing the sine wave from the original data.

The combination of windowing and padding is useful whenever the analyst wants to use the fast Fourier transform to compute the line spectrum of a data set that is not a power of 2. In this case the original data might best be tapered to zero at both ends and then padded until the number of points is a power of 2. As a rough rule of thumb, approximately 10% of the data at each end should be tapered. Note however, that many software packages, such as MATLAB™, are able to default to a slower fast Fourier transform algorithm when the number of points is not a power of 2.

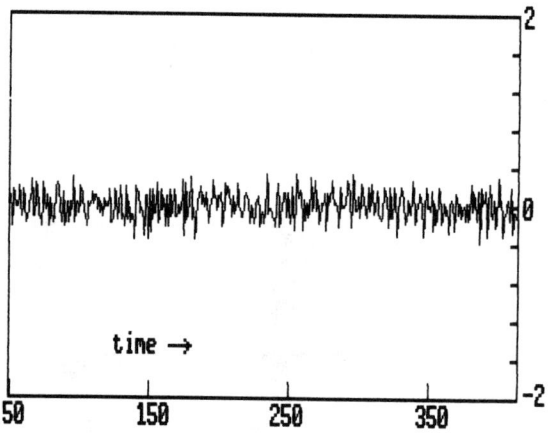

Figure 3.103   Untapered portion of filtered data in time domain.

3.12.5 Consistency of the Line Spectrum Estimate

In Section 1.3 it was mentioned that the average of N samples, x_i, $i = 1, \ldots, N$, of a stationary stochastic sequence is an estimate of the population mean, which in turn is the expected value of x. One would expect that as N increases the average would become a better estimate of the mean. Alternatively, one would expect the variance of the estimate about the true mean to decrease to zero as N increases to infinity. In this case the estimate would be considered *consistent*.

For each of the harmonics of the fundamental frequency, the line spectrum can be considered as an estimate of a true spectrum, and there is a variance associated with such an estimate based on N samples. If the line spectrum were a consistent estimator, this variance would go to zero as N goes to infinity. For the case of the x_i being a white noise sequence, it can be shown (Appendix 4.A, Marple, 1987) that this is not the case. In fact, as N increases, a plot of the line spectrum of white noise appears to become more "hashy." As N increases, the fundamental frequency

$$f_1 = \frac{1}{Nh}$$

decreases, as does the spacing between the harmonics of the fundamental frequency. The total harmonic power of the white noise signal, which was shown earlier to be proportional to the variance of the white noise signal, is independent of N. As N increases, the total variance, which is constant, is divided among more harmonics. Since the distribution of the variance among the harmonics is known to be erratic for small N, it follows that it will become even more erratic as the number of harmonics increases. This is somewhat analogous to a histogram becoming more erratic as the size of the cells are decreased.

When the analyst is studying empirical data resulting from sampling an industrial process, there are two things that somewhat ameliorate this problem. First, although the variance of the line spectrum estimate does not decrease to zero as N increases without limit, it also does not diverge, and since the ability to decipher periodic components buried in noisy signals increases as N increases, it often is beneficial to work with as many samples as possible. Second, in the situation where it is felt that the line spectrum is too erratic for good analysis, the cumulative line spectrum will often provide sufficient insight because of the smoothing afforded by the summing.

When the analyst is using the Fourier line spectrum to study the results of simulations, as was done in Section 3.9, the noise associated with the line spectrum can be reduced somewhat by making several simulation runs and averaging the line spectra over the runs. Similar approaches can also be used with empirical data. For example, N samples could be divided into subsets of, say, M groups, each having L samples, such that $ML = N$. The line spectrum would be calculated for each subset and an average could be taken of the M line spectra. Although the variance associated with the average line spectrum would be smaller, the ability to locate peaks in the spectrum would not be as great since the record length for each subset is decreased by a factor of M. This is called the Bartlett method. The Welch method extends this idea by using overlapping segments and applying windows. One of the simplest approaches is to apply a moving average (or some other type of low-pass filter) to the line spectrum. These methods are summarized in Chapter Five of Marple (1987).

3.12.6 Effect of the Sampling Interval

The sampling interval must also be chosen with care. Essentially, one wants the peaks associated with the frequencies of interest to lie in the center portion of the line spectrum. Some rough guidelines can be derived for choosing the sampling frequency by requiring that for the maximum sampling frequency the peak associated with the frequency of interest, f_a, should lie no farther to the left of the Nyquist interval than one-eighth of the Nyquist frequency:

$$\frac{1}{8}\left(\frac{f_s^{\max}}{2}\right) = f_a$$

For the minimum sampling frequency, the peak associated with the frequency of interest should lie no farther to the right than seven-eighths of the Nyquist frequency:

$$\frac{7}{8}\left(\frac{f_s^{\min}}{2}\right) = f_a$$

These two conditions result in

$$\frac{16}{7} f_a < f_s < 16 f_a$$

which can be used as a rough guideline for choosing f_s.

In reality, the frequency of the hidden periodic component is not known ahead of time. Thus, sometimes the sampling interval has to be based on the analyst's knowledge of the process's dynamics so that the Nyquist interval will only include frequencies that will have an impact on the process. For example, if the process under analysis has an effective time constant of approximately 1 hour, it would not make sense to sample process variables at 1.0 Hz, because even though there may be harmonic power near the Nyquist frequency of 0.5 Hz it is not likely that this power would have any effect on a process variable. Here it might make sense to try a sampling interval on the order of 1 to 5 minutes.

In summary, the reader should note that the resolving power of the Fourier method is a function of the record length. Assuming that the data are not padded, then for a record having length L, where $L = Nh$, values of the line spectrum can be calculated at zero frequency, at the fundamental frequency, $1/(Nh) = 1/L$, and at multiples of the fundamental from $2/(Nh)$ up to the Nyquist frequency, which is $1/(2h)$. Decreasing the sampling interval h only serves to increase the Nyquist frequency, allowing one to search for harmonic strength at higher frequencies and perhaps discover the sources of aliases. It does not increase the resolving power.

3.13 SUMMARY

The goal in this chapter was to convince the reader of the wide utility of the Fourier line spectrum in analyzing noisy processes. The discrete-time Fourier series was introduced at an elementary level, but with sufficient detail so that the reader can understand how and why it works. The Fourier line spectrum and cumulative line spectrum were derived from the Fourier series and compared with the autocorrelation by means of several numerical examples. It was shown how the Fourier line spectrum could be interpreted as the distribution of the variance of a noisy signal. The color numbers were suggested as a way of compressing the information contained in the cumulative line spectrum. The Nyquist frequency and the concept of aliasing were shown to play an important part in the analysis of discrete time-domain data. Remembering that the line spectrum is derived from a discrete-time Fourier series aided in interpreting data containing sharp shifts.

Differencing data can be useful in making stochastic sequences stationary as long as one is aware that such an operation on the data causes harmonic power in the higher frequencies to be amplified. By means of simulation, the line spectrum was applied to processes under control, where it was seen to be helpful in evaluating the effect of changing tuning constants, especially when the line spectra from several simulations were averaged. The exponential form of the discrete-time Fourier series pair was derived, and the results were applied to the development of surgically precise filters, which were compared to other filters that have already been discussed.

We never cease to be amazed at the wide applicability of the Fourier line spectrum as a tool for process analysis. In our work experience we have used it to gain insight into processes as diverse as ceramic kilns, large glass-melting tanks, and manufacturing equipment for optical fibers. In studying this diversity of processes we have dealt with time constants that range from fractions of a second to days.

The material in this chapter has only scratched the surface of digital signal processing and classical spectral estimation. Modern spectral estimation methods have not been mentioned. Interested readers are encouraged to take a look at the books by Marple (1987), Kay (1988), Oppenheim and Schafer (1989), Hamming (1989), and Karl (1989).

3.14 DISCUSSION OF THE REFERENCES

The Fast Fourier transform was first discussed in
COOLEY, J. W., and TUKEY, J. W., "An Algorithm for the Machine Calculation of Complex Fourier Series," *Math Computation,* vol. 19, 1965, pp. 297–301.

The concept of orthogonality and many other issues related to applied mathematics in the discrete time domain appear in the following text. This text is designed as an introduction to digital signal processing for the nonspecialist:

HAMMING, R, W., *Digital Filters,* 3rd ed., Prentice-Hall, Englewood Cliffs, N.J., 1989.

Even though the following text, written by statisticians, does not make extensive use of the fast Fourier transform, it is still a widely referenced text, and notation in this book derives heavily from it:

JENKINS, G. M., and WATTS, D. G., *Spectral Analysis and Its Applications*, Holden-Day, Oakland, Calif., 1968.

An extensive review of spectral analysis, where nonFourier series approaches are also discussed, appears in

KAY, S. M., and MARPLE, S. M., "Spectrum Analysis—A Modern Perspective," *Proceedings of the IEEE,* vol. 69, no. 11, 1981, pp. 1380–1419.

The authors of the above review have each written their own textbooks on the subject of spectral estimation, and both books have a chapter on the classical spectral estimation, which can be used as a reference to Chapter Three of this book:

KAY, STEVEN, *Modern Spectral Estimation: Theory and Applications*, Prentice-Hall, Englewood Cliffs, N.J., 1988.

MARPLE, S. L., *Digital Spectral Analysis with Applications,* Prentice-Hall, Englewood Cliffs, N.J., 1987.

Engineering counterparts to the Jenkins and Watts text are

OPPENHEIM, A. V., and SCHAFER, R. W., *Digital Signal Processing,* Prentice-Hall, Englewood Cliffs, N.J., 1975.

and more recently

OPPENHEIM, A. V., and SCHAFER, R. W., *Discrete Time Signal Processing,* Prentice-Hall, Englewood Cliffs, N.J., 1989.

Although the above two texts comprehensively and exhaustively cover the field of digital signal processing, they may be a bit overwhelming to the practicing control engineer or statistician. An interesting introductory text to digital signal processing that emphasizes discussion rather than derivation and that complements the two Oppenheim and Schafer texts nicely is

KARL, JOHN H., *An Introduction to Digital Signal Processing,* Academic Press, New York, 1989.

CHAPTER 4

INTRODUCTION TO THE CONTINUOUS FREQUENCY DOMAIN USING THE Z-TRANSFORM

The first chapter reviewed the fundamentals of process dynamics using the first-order-with-deadtime (FOWDT) model as the benchmark. Control algorithms were the subject of the second chapter. In those two chapters all analysis was done in the discrete time domain. In the third chapter most of the discussion took place in the discrete frequency domain, and the reader was introduced to the Fourier line spectrum, which was used as an empirical tool for analyzing the dynamics of processes. In this chapter the discussion will move to the continuous frequency domain by means of the Z-transform. This is done for two reasons. First, the Z-transform by itself provides the analyst with a potent tool for the study of discrete-time dynamic systems. Second, the Z-transform provides a link to Chapter Five, where much of the foregoing material in this book will be tied together by means of the power spectral density and noise transmission curves.

There is a price of admission, however: use must be made of some relatively advanced mathematical tools that, although familiar to many engineers, are sophisticated enough to warrant a review. The equations for the FOWDT model were developed in Section 1.6. The PI control algorithm equations were developed in Section 2.3. To facilitate the journey toward the power spectral density and noise transmission curves in Chapter Five, these and other equations will be reformulated in the Z-transform domain.

Having rewritten these equations in this new domain, the dynamics of a control loop will be dealt with in terms of transfer functions. Unlike most texts on control theory, the Z-transform will be considered as a tool independent of the Laplace transform. However, the two transforms will be compared so that the reader can appreciate each in its own right.

Although the development of the noise transmission functions in Chapter Five is the main reason for introducing the Z-transform, it will also be used in this chapter to continue the discussion of filters that was started in Chapter One. Second, the Z-transform will be used to derive the PI tuning rules that were presented in Chapter Two. In the process of deriving these tuning rules, a Dahlin-like control algorithm will also be derived and shown to be quite similar to the Smith predictor. Third, the transition from the discrete time domain to the continuous frequency domain will be seen to be facilitated by the Z-transform. Finally, the Z-transform will be used to analyze the stability of controlled systems.

It is hoped that the reader will not be put off by the use of slightly more sophisticated mathematical tools so that by the end of this chapter the reader will be convinced of the utility of the Z-transform as a powerful analytical tool for discrete time control systems.

4.1 BACKSHIFT OPERATOR AND TRANSFER FUNCTIONS

In Section 1.6 the FOWDT model equations were written as follows:

$$C_i = C_{i-1} \exp(-h/T) + G [1 - \exp(-h/T)] M_{i-1-d}$$

or

$$C_i = AC_{i-1} + G(1 - A)M_{i-1-d} \tag{4.1}$$

where $A = \exp(-h/T)$
         $h =$ sampling/control interval
         $T =$ process time constant
         $G =$ process gain

Note that C_i was introduced as a sample of C at the time t_i, where $t_i = t_{i-1} + h$, but M_i represented a staircase function, meaning that M was defined to be held constant at the value M_i over the time interval between t_i and t_{i+1}. In other words, C could be changing during the time interval of length h, and C_i is simply a snapshot of C at time t_i. However, M is assumed to be constant over the time interval h, and for $t_i \leq t < t_{i+1}$, $M(t) = M_i$. It is important to note that this is equivalent to putting the snapshots of the manipulated variable $M_1, M_2, M_3, \ldots$, through a *zero-order hold*. As a result, the process does not "see" isolated spikes having the values $M_1, M_2, M_3, \ldots$. Instead, it sees a series of steps in the manipulated variable that changes every h seconds but that remains constant during the time between changes.

The backshift operator, represented by z^{-1}, is defined to have the ability to shift a quantity backward in time by h seconds:

$$C(t - h) = z^{-1}C(t) \tag{4.2}$$

where h is the sampling interval. Using indexes, Equation (4.1) can be written as

$$C_{i-1} = z^{-1}C_i \qquad (4.3)$$

Equation (4.3) can be generalized to

$$C_{i-n} = z^{-n}C_i \qquad (4.4)$$

Usually, the operator z^{-1} is applied to a discrete-time domain equation so that all time arguments are the same, and then the time arguments are dropped. For example, Equation (4.1) becomes

$$C = z^{-1}CA + G(1 - A)z^{-1-d}M \qquad (4.5)$$

Now, C in Equation (4.5) is treated as an algebraic variable and solved for

$$C = \frac{G(1 - A)Mz^{-1-d}}{1 - Az^{-1}} \qquad (4.6)$$

Shortly, the reason behind these manipulations will be more clear.

In Section 2.3, the PI control algorithm was written as follows:

$$\Delta M_i = IhE_i + P\,\Delta E_i \qquad (4.7)$$

$$M_i = M_{i-1} + \Delta M_i \qquad (4.8)$$

Before applying the z^{-1} operator to Equations (4.7) and (4.8), it will be beneficial to develop the notation for the delta operator Δ. Applying z^{-1} to

$$\Delta M_i = M_i - M_{i-1} \qquad (4.9)$$

and dropping the time argument gives

$$\Delta M = (1 - z^{-1})M \qquad (4.10)$$

Since this relationship applies to any quantity, Equation (4.10) can be written as

$$\Delta = 1 - z^{-1} \qquad (4.11)$$

With this new relationship in hand, Equation (4.7) can be rewritten as follows:

$$(1 - z^{-1})M = IhE + P(1 - z^{-1})E$$

$$M = \frac{IhE}{(1 - z^{-1})} + PE \qquad (4.12)$$

Equation (4.12) is a compact way of saying that the manipulated variable consists of a term proportional to the error, *PE,* and a term proportional to the sum of the errors, $IhE/(1 - z^{-1})$. Therefore, while the operator $(1 - z^{-1})$ takes the *difference* of the quantity that it operates on, the operator $1/(1 - z^{-1})$ computes the *sum* of the quantity that it operates on; that is, these two operators are inverses and they are the discrete analogs of differentiation and integration.

Before proceeding, a quick review of just what has been accomplished by introducing this backshift operator z^{-1} is in order. First, the equations are no longer in the time domain and, second, simple rules of algebra can be applied to the new equations. By replacing C_i with C, a new variable is being introduced. The quantity without any subscript or time argument represents the original variable in a new domain, which is called the Z-domain. In other words, the controlled variable C is no longer represented by $C(t)$ or C_i, which are quantities whose independent variable is time or a time index. Instead, C is now a function of z^{-1} or z. The notation so far has indeed been unfortunate. Strictly speaking, an argument should be appended to C showing this new variable's dependence on z rather than t or i and it should also be distinguished by using boldface italics. Thus, for the time domain one would write $C(t)$ or just C; for the Z-domain one would write $C(z)$ or just C. With this new notation, Equations (4.6) and (4.12) become

$$C(z) = \frac{G(1 - A)z^{-1-d}M(z)}{1 - Az^{-1}} \tag{4.13}$$

$$M(z) = \frac{Ih}{(1 - z^{-1}) + P} E(z) \tag{4.14}$$

Later this concept will be expanded and the reader will see that these equations result from applying the Z-transform. So far, no reason has been given for introducing the backshift operator other than allowing one to apply the simple rules of algebra to the transformed equations and solve directly for the quantities of interest.

Equations (4.13) and (4.14) can also be written as ratios:

$$\frac{C(z)}{M(z)} = \frac{G(1 - A)z^{-1-d}}{1 - Az^{-1}} = G_p(z) \tag{4.15}$$

$$\frac{M(z)}{E(z)} = \frac{Ih}{(1 - z^{-1}) + P} = G_c(z) \tag{4.16}$$

where $G_p(z)$, which is $C(z)/M(z)$, is called the process transfer function and $G_c(z)$, which is $M(z)/E(z)$, is called the control algorithm transfer function. (Note that G without a subscript has been used as the process gain. In Chapter Two, g_m was used as the Smith predictor reference model gain and, in this chapter, G's with subscripts will refer to transfer functions.) Equation (4.15) says that, in the Z-domain, the controlled variable $C(z)$ can be obtained by multiplying the manipulated variable in the Z-domain $M(z)$ by the process transfer function $G_p(z)$; that is, $G_p(z)$ transfers the dynamics of M into C. Equation (4.16) says that the manipulated variable $M(z)$ can be obtained by multiplying the error between the setpoint and the controlled variable $E(z)$ by the control algorithm transfer function $G_c(z)$; that is, G_c transfers the dynamics of E into M.

If Equations (4.15) and (4.16) are combined with the definition of the controller error,

$$E(z) = R(z) - C(z)$$

then, after some straightforward algebra, the transfer function relating the controlled variable to its set point can be obtained:

$$C(z) = G_p M(z) = G_p G_c E(z) = G_p G_c [R(z) - C(z)]$$

$$C(z)(1 + G_p G_c) = G_p G_c R(z)$$

$$\frac{C(z)}{R(z)} = \frac{G_p G_c}{1 + G_p G_c} \tag{4.17}$$

Although it is not apparent now, Equation (4.17) could be used to investigate the dynamic relationship between the controlled variable and its set point. Note that when two transforms are multiplied together it is assumed that the input to the second element from the first element is a discrete signal sampled or computed every h seconds. This will be the case for every instance in this book. When this is not the case, one cannot determine the transform of the combined elements by multiplication.

The effect of disturbances or noise can be brought into the picture by referring back to Figure 2.3 in Section 2.2.1, which will be reproduced here as Figure 4.1. The disturbance is represented by $N(t)$ in the time domain and $N(z)$ in the Z-domain. By looking at the flow diagram in Figure 4.1, the following relations can be written:

$$E(z) = R(z) - C(z)$$

$$M(z) = G_c E(z)$$

$$C_p(z) = G_p M(z)$$

$$C(z) = C_p(z) + N(z)$$

Some more straightforward algebra gives one of the noise transmission transfer functions, $E_T(z)$:

$$E(z) = R(z) - C(z), \qquad R(z) \to 0 \text{ from now on}$$

$$E(z) = -[C_p(z) + N(z)]$$

$$= -G_p M(z) - N(z) \tag{4.18}$$

$$= -G_p G_c E(z) - N(z)$$

$$\frac{E(z)}{N(z)} = -(1 + G_p G_c)^{-1} = E_T(z)$$

Note that $R(z)$ is set to zero since it is not changing. In other words, it is assumed that an arbitrary constant can be subtracted from both the set point and controlled variable such that the new set point is zero. Since the immediate interest is in the dynamic relationship between $E(z)$ and $N(z)$, the value of $R(z)$ will be assumed equal to zero and the term will drop out. The transfer function in Equation (4.18), $E(z)/N(z)$ or $E_T(z)$, contains the desired dynamic relationship between disturbances or noise and the control error.

As a default situation, the reader should note that, in the case of no feedback control, both P and I would be zero, as would G_c. Consequently, Equation (4.18) would become

$$\frac{E(z)}{N(z)} = -1$$

or

$$E_i = -N_i$$

which is consistent with a visual analysis of Figure 4.1.

A second transmission transfer function of interest, $M_T(z)$, can be derived as follows:

$$\frac{M(z)}{N(z)} = \frac{M(z)}{E(z)}\frac{E(z)}{N(z)}$$

$$= G_c E_T(z)$$

$$= M_T(z)$$

$$= -\frac{G_c}{1 + G_c G_p} \tag{4.19}$$

This transmission transfer function, $M_T(z)$, can be used to show how dynamic information on a disturbance represented by N affects the dynamic behavior of the manipulated variable M.

Much of this and the following chapter will be devoted to studying these transmission functions so as to yield insight into how the control system (represented by G_c) is able to turn the disturbances $N(z)$ experienced by the process (represented by G_p) into an acceptable error function $E(z)$.

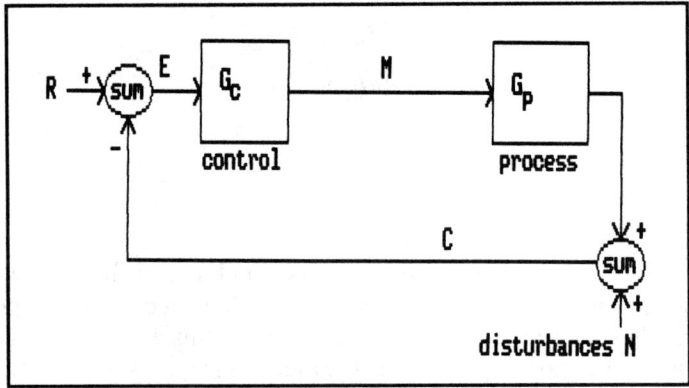

Figure 4.1   Relation between R, C, E, and N.

4.2 THE Z-TRANSFORM

In the previous section, two basic working equations were transformed from the time domain, where they were originally derived, into the Z-domain, where they became algebraic equations subject to simple algebraic operations, and it was hinted that they might have other attractive properties. Later it will show that these Z-domain equations can be converted into the frequency domain where the power spectrum approach can be used for analysis. But, before this is done, this transformation from the time domain to the Z-domain should be put on a sounder footing. This section will not attempt to be even remotely comprehensive. There are many good texts that the reader can refer to (for example, Franklin and Powell, 1980, and Saucedo and Shiring, 1968) for more detail and background.

By definition, a variable $C(t)$, or C_i, in the time domain can be transformed into a new variable $C(z)$ in the Z-domain by the following equation:

$$C(z) = C(0) + C(h)z^{-1} + C(2h)z^{-2} + C(3h)z^{-3} + \cdots$$

or

$$C(z) = C_0 + C_1 z^{-1} + C_2 z^{-2} + C_3 z^{-3} + \cdots$$

or

$$C(z) = \sum_{i=0}^{\infty} C_i z^{-i}$$

or

$$C(z) = Z\{C_i\}$$

where $Z\{\ \}$ denotes taking the Z-transform of whatever is inside the braces. This definition assumes that all values of C for time less than zero are zero. The fact that the summation index ranges from 0 to ∞ causes some authors to refer to the above operation as the unilateral Z-transform. Had the indexes ranged from $-\infty$ to ∞, the operation would have been called the bilateral Z-transform.

This definition can be used to derive the shifting feature that was introduced in the previous section:

$$Z\{C_{i-1}\} = \sum_{i=0}^{\infty} C_{i-1} z^{-1} = z^{-1} \sum_{i=0}^{\infty} C_{i-1} z^{-(i-1)} = z^{-1} \sum_{k=-1}^{\infty} C_k z^{-k}$$

where the last term uses a change in variable $k = i - 1$. Since C_{-1} is by definition zero, the last sum becomes

$$Z\{C_{i-1}\} = z^{-1} \sum_{k=0}^{\infty} C_k z^{-k} = z^{-1} Z\{C_i\} = z^{-1} C(z)$$

In almost all the work in this book, it can be assumed, without loss of generality, that all values of C are zero for time less than zero.

The inverse operation, that of transforming $C(z)$ back to the time domain to get $C(t)$ or C_i, is given by

$$C_i = C(ih) = \frac{1}{2\pi j} \oint dz\, C(z)z^{i-1} = Z^{-1}\{C(z)\}$$

This expression contains a complex contour integral, which, if discussed in detail, would require mathematics well beyond the scope of this book. Fortunately, this integral will not be needed in this book.

The two quantities C_i and $C(z)$ can be referred to as a Z-transform pair, just as c_m and F_m were referred to as a discrete Fourier transform pair in Section 3.10. Note that, while the time-domain index i takes on real integer values,

$$\ldots 0, 1, 2, 3, \ldots$$

the Z-domain variable z is a continuous complex variable having a real and an imaginary part. Therefore, the time-domain index can be thought of as taking on discrete values along a line, while the Z-domain variable z takes on continuous values in a plane where the x-axis of the plane is the real component of z and the y-axis is the imaginary component of z. Note that the parameter z is unitless, so the units of C_i and $C(z)$ are the same.

On the whole, the use of the Z-transform is not complicated; it is in the details where the mathematics can make the subject difficult for the beginner. One usually starts with an equation such as the FOWDT model in the time domain. Solving that equation for C_i as a function of the parameter i is possible but rather difficult, because C_{i-1} as well as C_i is present; that is, simple algebra will not suffice. So, instead of solving the equation in the time domain, each term in the equation is transformed into the Z-domain. Now only $C(z)$ occurs and it can be solved for algebraically.

If one is interested in finding C_i, the expression for $C(z)$ must be inverted. The inversion usually does not come from applying the complex inversion integral mentioned above. Instead, the expression for $C(z)$ is algebraically grouped into components that have known inverses. Then, by inspection, the expression for C_i can be determined. To do this effectively, one must be familiar with several simple time-domain functions and their Z-transforms. Therefore, the Z-transforms of a few commonly used time-domain functions will be developed. No attempt will be made to actually equip the reader to solve time-domain equations using the Z-transform since the Z-transform is of interest for other reasons.

First, consider the Z-transform of a constant, say 1.0. Applying the definition of the Z-transform, where, for all $i \geq 0$, C_i equals 1.0, gives

$$Z\{1.0\} = 1 + z^{-1} + z^{-2} + z^{-3} + \cdots$$

$$= \frac{1}{1 - z^{-1}}$$

$$= \frac{z}{z - 1}$$

The conversion from the infinite series to a closed form results from applying long division to

$$\frac{1}{1 - z^{-1}}$$

Since it is assumed that $C_i = 0$ for $i = -1, -2, -3, \ldots$, then, in effect, $z/(z - 1)$ is the Z-transform of a unit step change; that is,

$$Z\{U(t)\} = \frac{z}{z - 1} \tag{4.20}$$

where

$$U(t) = 0, \quad \text{for } t < 0$$

$$U(t) = 1, \quad \text{for } t \geq 0$$

Next, consider the exponential function $\exp(-at)$. Here $C_i = \exp(-aih)$, where h is the sampling interval. Applying the definition of the Z-transform gives

$$Z\{\exp(-aih)\} = \sum_{i=0}^{\infty} \exp(-aih)z^{-i} = \sum_{i=0}^{\infty} [\exp(-ah)z^{-1}]^i$$
$$= \frac{1}{1 - \exp(-ah)z^{-1}} = \frac{z}{z - \exp(-ah)} \tag{4.21}$$

where long division methods have been applied to

$$\frac{1}{1 - \exp(-ah)z^{-1}}$$

to obtain another infinite series.

The following example will demonstrate the utility of these two Z-transforms. Consider the FOWDT model with no deadtime:

$$C_i = AC_{i-1} + G[1 - A]M_{i-1}$$

which is a finite difference equation in the discrete time domain. The Z-transform of this time-domain equation is

$$C(z) = Az^{-1}C(z) + Bz^{-1}M(z) \tag{4.22}$$

where $B = G[1 - A]$. Require that $C_0 = 0$ and that M take on a constant value of unity at time zero. Therefore, in the time domain, M has been given a unit step change at time zero. Solving Equation (4.22) for C will yield the step change response of C. Since M is a unit step, Equation (4.20) can be combined with Equation (4.22) to yield

$$C(z) = Az^{-1}C(z) + \frac{Bz^{-1}}{1 - z^{-1}} \tag{4.23}$$

Solving Equation (4.23) for $C(z)$ gives

$$C(z) = \frac{Bz^{-1}}{(1 - Az^{-1})(1 - z^{-1})}$$

$$= \frac{Bz}{(z - A)(z - 1)}$$

(4.24)

Note that, of the two roots in the denominator of Equation (4.24), the first, at $z = A$, is dependent on the process while the second, at $z = 1$, is dependent on the nature of M. (To simplify the algebra, assume that the process gain G is unity so that $B = 1 - A$.)

To invert the above expression for $C(z)$, the right side of Equation (4.24) is expanded into groups that can be recognized as Z-transforms of known time-domain functions. This can be done by applying partial fractions and getting

$$C(z) = \frac{1}{1 - z^{-1}} - \frac{1}{1 - Az^{-1}}$$

(4.25)

Equation (4.20) indicates that the first term on the right side of Equation (4.25) is the Z-transform of unity. Equation (4.21) indicates that the second term is the Z-transform of

$$\exp(-hi/T)$$

Therefore, each term in Equation (4.25) can be inverted, giving

$$C_i = 1 - \exp(-ih/T)$$

or

$$C_i = 1 - (A)^i$$

which is the step response of C to a unit step change in M.

For future reference, note that in Equation (4.24) the denominator had two roots, one at $z = 1.0$, and one at $z = A$, where $A = \exp(-h/T)$. The time-domain solution for C_i turned out to be a special case of

$$C_i = a_1(r_1)^i + a_2(r_2)^i$$

(4.26)

where $r_1 = 1.0$ and $r_2 = A$, and where the coefficients a_1 and a_2 could have been determined from an initial coondition, $C_0 = 0$, and a final condition, $C_\infty = 1.0$. This latter condition results from the fact that, after a unit step change in M when the gain is unity, the ultimate value of C will be unity. The initial condition is

$$0 = a_1 + a_2 A^0 = a_1 + a_2$$

and the final condition is

$$C_\infty = 1 = a_1 + 0$$

These two equations can be solved to give $a_1 = -a_2 = 1$. Later this example will be referred to in the discussion of stability.

The Z-transform can be useful in determining the *final value* of a process variable, that is, the value of the process variable after all transients have settled out and time has increased without limit. The final value property can be stated as follows: If $Y(z)$ is the Z-transform of Y_i, $i = 0, 1, 2, \ldots$, then the final value of Y_i, symbolized by Y_∞, can be determined from

$$Y_\infty = \lim_{i \to \infty} \{Y_i\} = \lim_{z \to 1} \{(1 - z^{-1})Y(z)\} \qquad (4.27)$$

For example, the transfer function for a FOWDT process is

$$\frac{C(z)}{M(z)} = \frac{G(1 - A)z^{-1-d}}{1 - Az^{-1}} = G_p(z)$$

If M is given a unit step change, its Z-transform will be

$$M(z) = \frac{1}{1 - z^{-1}}$$

and the Z-transform for the response will be

$$C(z) = G_p(z)M(z) = \frac{G(1 - A)z^{-1-d}}{(1 - Az^{-1})(1 - z^{-1})}$$

The final value that C tends to as time goes to infinity can be obtained by multiplying $C(z)$ by $(1 - z^{-1})$ and letting z go to unity:

$$C_\infty = \lim_{z \to 1} \left\{ \frac{G(1 - A)z^{-1-d}}{1 - Az^{-1}} \right\} = G$$

This result makes sense, because for a unit step in M the ultimate change in C would be equal to the process gain.

4.3 SOME APPLICATIONS OF THE Z-TRANSFORM

This section will consist of three side trips where the Z-transform will be used to develop some equations that have been referred to in previous chapters. First, the Z-transform will be used to develop the simple but useful third-order process model that was referred to in Chapters One and Two. Models will also be derived for the integrator process and the inverse responding process. Next, the relationship between the Z-transform, the Laplace transform, and the time domain will be discussed, and, finally, the Z-transform will be used to derive the tuning rules that were presented in Chapter Two. As a by-product of deriving the tuning rules, a Dahlin-like control algorithm quite similar to the Smith predictor will be obtained.

4.3.1 Developing Process Models

In this subsection the Z-transform will be used to derive three process models.

Simple third-order process model.   The Z-transform can now be used to discuss process models other than the FOWDT model, which has the time-domain form of

$$C(t) = AC(t - h) + G(1 - A)M(t - h - D) \tag{4.28}$$

where G = process gain
A = $\exp(-h/T)$
h = sampling interval
T = process time constant

The Z-domain form of Equation (4.28) is

$$\begin{aligned} C(z) &= Az^{-1}C(z) + G[1 - A]z^{-1-d}M(z) \\ &= Az^{-1}C(z) + Bz^{-1-d}M(z) \end{aligned} \tag{4.29}$$

where $B = G(1 - A)$. Equation (4.29) shows C to be dependent on itself shifted by *one* time unit and is called a *first*-order model.

A *third*-order model could be written as

$$C(z) = A_1z^{-1}C(z) + A_2z^{-2}C(z) + A_3z^{-3}C(z) + Bz^{-1-d}M(z)$$

or in the time domain as

$$C(t) = A_1C(t - h) + A_2C(t - 2h) + A_3C(t - 3h) + BM(t - h - D) \tag{4.30}$$

where C is dependent on itself shifted one, two, and three time units. Using simple algebra, Equation (4.30) can be rearranged to

$$C(z)(1 - A_1z^{-1} - A_2z^{-2} - A_3z^{-3}) = Bz^{-1-d}M(z)$$

or

$$\frac{C(z)}{M(z)} = \frac{Bz^{2-d}}{z^3 - A_1z^2 - A_2z - A_3} \tag{4.31}$$

just as Equation (4.29) can be rearranged to

$$C(z)(1 - Az^{-1}) = Bz^{-1-d}M(z)$$

or

$$\frac{C(z)}{M(z)} = \frac{Bz^{-d}}{z - A} \tag{4.32}$$

Rearranging Equation (4.31) as

$$C(z)(z^3 - A_1z^2 - A_2z - A_3) = Bz^2z^{-d}M(z) \tag{4.33}$$

allows it to be factored as

$$C(z)[(z - r_1)(z - r_2)(z - r_3)] = Bz^2z^{-d}M(z) \qquad (4.34)$$

where r_1, r_2, and r_3 are the roots, which in general can be complex numbers. The relationship between the roots r_1, r_2, r_3 and the parameters A_1, A_2, A_3 can be determined by collecting powers of z in Equations (4.33) and (4.34) and equating coefficients of z^0, z^1, z^2, and z^3:

$$A_1 = r_1 + r_2 + r_3$$

$$A_2 = -(r_1r_2 + r_2r_3 + r_1r_3)$$

$$A_3 = r_1r_2r_3$$

For those readers not comfortable with complex variables, these products of the roots show that, in order for the coefficients A_1, A_2, and A_3 to be real, the roots either have to all be real or there must be one real root and two complex roots that are complex conjugates. In the latter case, one root, r_3, would be real while the other two roots must satisfy

$$r_1 = a + jb$$

$$r_2 = a - jb$$

The FOWDT model, Equation (4.32), when written in the form of Equations (4.33) and (4.34), looks like

$$C(z)(z - A) = Bz^{-d}M(z)$$

$$C(z)(z - r) = Bz^{-d}M(z)$$

$$A = r$$

showing that the first-order root

1. is A or $\exp(-h/T)$,
2. depends on the interval h and the process time constant T, and
3. lies on the positive real axis inside the unit circle in the Z-plane.

This exponential relationship between the root and the time constant shows that as the root moves toward the unit circle the time constant becomes larger.

For the third-order model, the roots can also be specified in a similar manner. If desired, the three roots could be required to depend on three time constants, according to

$$r_i = \exp(-h/T_i), \qquad i = 1, 2, 3$$

where all rooots would lie on the positive real axis inside the unit circle in the Z-plane. It makes no sense for these real roots to lie beyond unity since no time constant would correspond to this case.

Consider the case where there is one real root and a conjugate pair; that is,

$$r_1 = a + jb$$

$$r_2 = a - jb$$

$$r_3 = c$$

Stability, which will be discussed later in more detail, requires that these roots lie on or within the unit circle in the Z-plane, that is, that the magnitude of the roots be less than or equal to unity. Since the magnitude of a quantity, which may be complex, is the square root of the sum of the squares of its real and imaginary parts, the following must be true:

$$|c| \le 1$$

and

$$\sqrt{(a^2 + b^2)} \le 1$$

In general, as the roots move nearer to the unit circle, the effective time constant becomes larger. If the roots are outside the unit circle, the values of the time constants are meaningless because the model is unstable. As will be shown later, an unstable process has a positive step change response that will not settle out to a new steady-state value but will increase at an increasing rate indefinitely.

When the roots appear as complex conjugates inside the unit circle, the model usually has an underdamped oscillatory step change response. An example of the step change response of a model with the roots

$$r_1 = 0.8 + j.5$$

$$r_2 = 0.8 - j.5$$

$$r_3 = 0.9$$

is shown in Figure 4.2. Note that although feedback control is not active the response variable, C, exhibits over- and undershoot. This is the same model (model 12) that was discussed in Sections 1.7.4 and 2.6.6.

For the most part, this book will deal with process models that have roots on the positive real axis inside the unit circle since, in the absence of feedback control action, most industrial processes are stable and do not exhibit underdamped behavior in response to steps in the manipulated variable.

Equations have been developed that allow the determination of the coefficients A_1, A_2, A_3 from the roots. Now, the final value property, Equation (4.27), will be used to determine a relationship for the remaining coefficient B. Let M be a unit step change

$$M(z) = \frac{z}{z - 1}$$

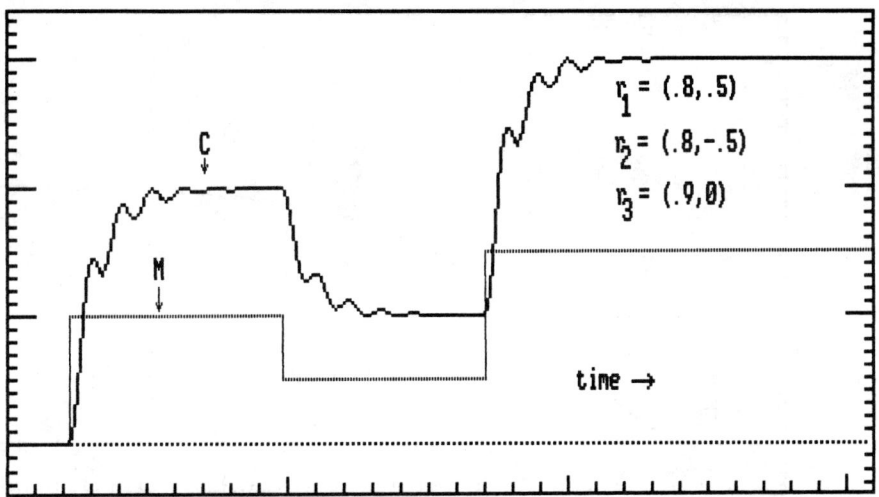

Figure 4.2 Step change response for a third-order process with complex conjugate roots.

The response to this step change is given by

$$C(z) = \frac{Bz^{-1-d}z}{(1 - A_1 z^{-1} - A_2 z^{-2} - A_3 z^{-3})(z - 1)}$$

Application of the final value property gives

$$C_\infty = \frac{B}{1 - A_1 - A_2 - A_3}$$

In the previous section, it was shown that, when the manipulated variable is given a unit step change, C_∞ is equal to the process gain G, so the value of B can be determined from

$$B = G(1 - A_1 - A_2 - A_3)$$

Note the similarity of this expression to the one for the FOWDT model:

$$B = G(1 - A)$$

Thus the Z-transform has provided a simple way of determining expressions for the four coefficients in a third-order model. Figure 4.3 shows the step change response for a third-order model (model 4) with the following parameters:

$$T_1 = 25, \quad T_2 = 26, \quad T_3 = 27, \quad G = 2, \quad d = 0, \quad h = 1$$

$$r_1 = \exp(-\tfrac{1}{25}) = 0.9607894, \qquad r_2 = \exp(-\tfrac{1}{26}) = 0.9622687$$

$$r_3 = \exp(-\tfrac{1}{27}) = 0.9636405, \qquad A_1 = 2.886699$$

$$A_2 = -2.777674, \qquad A_3 = 0.8909218, \qquad B = 0.0001072884$$

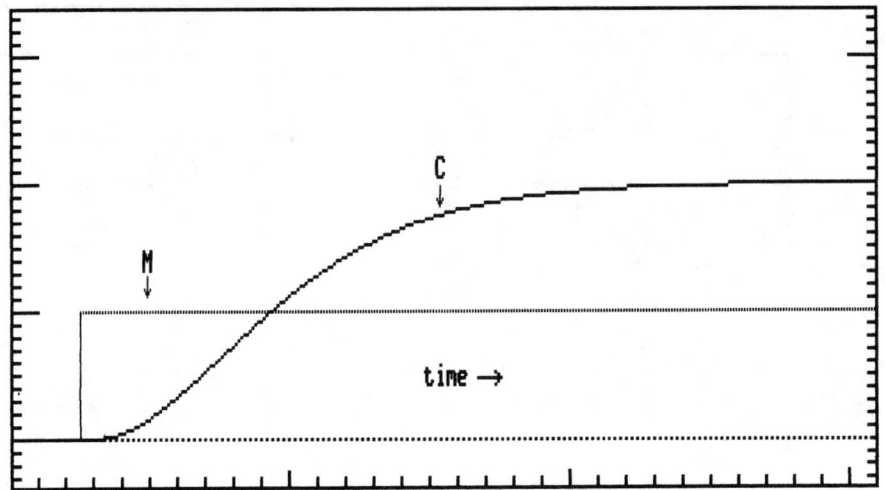

Figure 4.3 Step change response of a third-order model with all real roots.

Note that even with no explicit deadtime in the model it exhibits an effective deadtime and an inflection point in a manner similar to the three-tank model discussed in Section 1.6.2 and whose step change response is shown in Figure 1.30.

Unlike the discrete-time FOWDT process model, which was derived by starting with a first-order differential equation, this discrete-time third-order model does not relate nicely to a third-order differential equation. However, its parameters can be derived from process concepts such as time constants and process gains, and it can be extremely useful as a test process model for control algorithms.

This third-order model [Equation (4.30)] can be extended to include additional displaced values of the manipulated variable to give

$$C(t) = A_1 C(t - h) + A_2 C(t - 2h) + A_3 C(t - 3h)$$
$$+ B_1 M(t - h - D) + B_2 M(t - 2h - D) + B_3 M(t - 3h - D)$$

and

$$C(z) = A_1 z^{-1} C(z) + A_2 z^{-2} C(z) + A_3 z^{-3} C(z)$$
$$+ B_1 z^{-1-d} M(z) + B_2 z^{-2-d} M(z) + B_3 z^{-3-d} M(z)$$

or

$$\frac{C(z)}{M(z)} = \frac{B_1 z^{-1-d} + B_2 z^{-2-d} + B_3 z^{-3-d}}{1 - A_1 z^{-1} - A_2 z^{-2} - A_3 z^{-3}}$$

Although this model has no significantly different dynamic properties, it does relate to a third-order differential equation, as will be shown in Section 4.3.2. Note that, unlike the simple third-order model, this model's numerator as well as its denominator can have roots.

Integrator process.   Reconsider the limiting case of the integrator process that was discussed in Section 1.7.3, where it was shown that an integrator model can be obtained from a FOWDT model by letting the gain G and time constant T increase without limit while keeping their ratio, G/T, constant. Starting with the FOWDT model and replacing A and B by their definitions gives

$$C(z)(z - A) = Bz^{-d}M(z)$$

$$C(z)\,[z - \exp(-h/T)] = G\,[1 - \exp(-h/T)]\,z^{-d}M(z)$$

Letting T and G increase without limit gives

$$C(z)(z - 1) = \frac{Gh}{T}\,z^{-d}M(z)$$

$$C(z) = \frac{(Gh/T)z^{-d}M(z)}{z - 1} \tag{4.35}$$

$$C(z) = \frac{(Gh/T)z^{-d}M(z)z^{-1}}{1 - z^{-1}}$$

where the approximation

$$\exp(-h/T) \approx \frac{1 - h}{T}$$

has been used. This approximation requires

$$\frac{h}{T} \ll 1$$

which is acceptable because T is increasing without limit. Since the quantity Gh/T is a constant, Equation (4.35) shows that C is effectively the integral of M (after the deadtime has been applied) because of the presence of the summation operator

$$\frac{1}{(1 - z^{-1})}$$

Notice that the denominator of the transform of the integrator has a root at unity, that is, on, not within, the boundary of the unit circle.

Inverse responding process.   The inverse responding process was discussed in Sections 1.7.4 and 2.6.6, where it was suggested that the difference of two first-order models could explain the dynamic behavior. Such a process model can be written as

$$\frac{C(z)}{M(z)} = \frac{Bz^{-1}}{1 - Az^{-1}} - \frac{B'z^{-1}}{1 - A'z^{-1}}$$

where

$$
\begin{aligned}
B &= G(1 - A) \\
A &= \exp(-h/T) \\
B' &= G'(1 - A') \\
A' &= \exp(-h/T') \\
G, G' &= \text{process gains} \\
T, T' &= \text{process time constants}
\end{aligned}
$$

The gain and time constant of the second or primed component, G' and T', would be less than that of the first or unprimed component. After a little algebra, one can rewrite the model as follows:

$$\frac{C(z)}{M(z)} = \frac{B_1 z + B_2}{z^2 - A_1 z - A_2}$$

or

$$\frac{C(z)}{M(z)} = \frac{B_1 z^{-1} + B_2 z^{-2}}{1 - A_1 z^{-1} - A_2 z^{-2}}$$

or

$$C_i = A_1 C_{i-1} + A_2 C_{i-2} + B_1 M_{i-1} + B_2 M_{i-2}$$

where

$$
\begin{aligned}
A_1 &= A + A' \\
A_2 &= -AA' \\
B_1 &= B - B' \\
B_2 &= B'A - BA'
\end{aligned}
$$

If the numerical values

$$
\begin{aligned}
A &= 0.95, & T &= 19.5 \\
A' &= 0.7, & T' &= 2.8 \\
G &= 3 \\
G' &= 1
\end{aligned}
$$

are chosen, then the model coefficients for this second-order discrete-time model are

$$A_1 = 1.65, \quad A_2 = -0.665, \quad B_1 = -0.15, \quad B_2 = 0.18$$

which describe the process model (model 13) that was discussed in Sections 1.7.4 and 2.6.6.

4.3.2 Relationships between the Time Domain, the Laplace Transform, and the Z-transform

In constructing process models, one usually starts with first principles by applying some or all of the three conservation laws augmented by various constitutive equations and ends up with differential equations. As a simple example, the FOWDT model was derived in Section 1.6, where the result was an ordinary first-order differential equation. That differential equation was solved for the discrete time domain, and the continuous time domain was effectively ignored for the rest of the book. This has been acceptable since the control algorithms were derived in the discrete time domain and since all the analysis techniques dealt with equally spaced time-domain data.

This approach is in contrast to most control theory texts, which spend a good deal of time in the continuous time domain and make extensive use of the Laplace transform before introducing the Z-transform. To help bridge the gap between this book and conventional control theory textbooks, this section will briefly review how the Laplace transform and the Z-transform are related. Since the Laplace transform is not actually used in this book and since there is a fair amount of mathematical detail in the following paragraphs, the reader may wish to give this section only a cursory reading. On the other hand, the Laplace transform has played a critical part in the development of the control literature, and any serious student of the subject will want to be aware of how this tool can be applied to continuous time-domain control problems.

In the continuous time domain the FOWDT process model is described as

$$T\frac{dC}{dt} + C = GM(t - D) \tag{4.36}$$

In the discrete time domain the FOWDT model is described by

$$C_i = AC_{i-1} + BM_{i-1-d}$$

where

$$A = \exp(-h/T)$$
$$B = G(1 - A)$$
$$D = hd$$
$$t_i = t_{i-1} + h$$

The Z-transform was used to convert the above finite difference equation to an algebraic equation. A similar thing can be done in the continuous time domain by using the Laplace transform, which is defined for the generic variable $C(t)$ as

$$\mathbf{C}(s) = \int_0^\infty \exp(-st)C(t)\,dt = L\{C(t)\} \tag{4.37}$$

Note that the integration with respect to time is carried out over the interval (0, ∞) and as a result this operation is sometimes called the unilateral Laplace transform. In some texts the integration is carried out over the interval $(-\infty, \infty)$ and the operation is called the bilateral Laplace transform. The integration uses a weighting factor of $\exp(-st)$. The Z-transform uses summation and a weighting factor of z^{-i}. Thus, by comparison, $\exp(-st)$ corresponds to z^{-i} and $\exp(-sh)$ corresponds to z^{-1}.

Note that in order to keep the argument of the exponential unitless the units of s must be second^{-1}. Note also that if the units of $C(t)$ are volts, the units of $C(s)$ are volts-second.

Only the Laplace transforms of a few simple functions are needed to allow the comparisons with the Z-transform. First, the Laplace transform of a derivative is

$$L\left\{\frac{dC}{dt}\right\} = sC(S) - C(0) \tag{4.38}$$

If the initial value of C is zero, then multiplying the Laplace transform of C by s in the s-domain is the same thing as taking the derivative of C in the time domain. Since the derivative is only defined on the continuous time domain, there is no comparable Z-transform. However, taking the derivative in the continuous time domain is somewhat analogous to taking the difference in the discrete time domain. Therefore, multiplying by s in the s-domain is comparable to multiplying by

$$1 - z^{-1}$$

in the Z-domain.

Second, the Laplace transform of a time shifted variable is

$$L\{M(t - \alpha)\} = \exp(-s\alpha)M(s) \tag{4.39}$$

where α is not restricted to be a multiple of a sampling interval. Earlier, the Z-transform of a time-shifted variable was shown to be

$$Z\{M(t - nh)\} = z^{-n}M(z)$$

where the time shifting was restricted to be a multiple of the sampling interval h.

Third, the Laplace transform of a unit step change at time $t = \beta$ is

$$L\{U(t - \beta)\} = \frac{\exp(-\beta s)}{s} \tag{4.40}$$

where $U(t) = 0$ for $t < 0$, $U(t) = 1$ for $t \geq 0$. As a special case, the unit step change at time zero has a Laplace transform of $1/s$. The Z-transform of a unit step change at time zero was shown earlier to be

$$Z\{U(t)\} = \frac{1}{1 - z^{-1}}$$

Since multiplying the Z-transform of C by $1/(1 - z^{-1})$ in the Z-domain is the same thing as summing C in the discrete time domain, Equation (4.39) suggests that multiplying the Laplace transform of C by $1/s$ in the s-domain is the same thing as integrating C in the continuous time domain. In fact, the following can be written:

$$L\left\{\int_0^t C(u)\,du\right\} = \frac{\mathbf{C}}{s}$$

Fourth, the Laplace transform of the exponential time function is

$$L\{\exp(-\alpha t)\} = \frac{1}{s + \alpha} \tag{4.41}$$

and the Z-transform of the exponential time function is

$$Z\{\exp(-\alpha t)\} = \frac{1}{1 - \exp(-\alpha h)z^{-1}}$$

With these four rules, the Laplace transform can be applied to Equation (4.36), yielding

$$T[s\mathbf{C} - C(0)] + \mathbf{C} = G\exp(-sD)\mathbf{M}$$

which can be solved for $\mathbf{C}$ to give

$$\mathbf{C} = \frac{G\exp(-sD)\mathbf{M}}{Ts + 1} + \frac{TC(0)}{Ts + 1} \tag{4.42}$$

Thus a differential equation has been replaced with an algebraic equation, just as a difference equation was replaced with an algebraic equation earlier in this chapter.

Frequently, to simplify the algebra, the initial value of $C(t)$ is assumed to be zero, and in this chapter that assumption was made when dealing with the Z-transform. However, for the time being that assumption will not be made. If M is a unit step change at $t = 0$, then Equation (4.41), combined with Equation (4.40) becomes

$$\mathbf{C} = \frac{G\exp(-sD)}{s(Ts + 1)} + \frac{TC(0)}{Ts + 1} \tag{4.43}$$

(The reader should note the parallels with Section 4.2.)

To solve for $C(t)$, Equation (4.43) has to be inverted. As with the Z-transform, partial fractions are applied to give

$$\mathbf{C} = G\exp(-sD)\left[\frac{1}{s} - \frac{T}{Ts + 1}\right] + \frac{TC(0)}{Ts + 1} \tag{4.44}$$

An expression for $C(t)$ can be found by matching up the components in Equation (4.44) with Laplace transforms of known time-domain functions. Using the above four rules, Equation (4.44) becomes

$$C(t) = GU(t - D)\{1 - \exp[-(t - D)/T]\} + C(0) \exp(-t/T) \quad (4.45)$$

Note that Equation (4.45) is the same as Equation (1.29).

Setting $C(0)$ to zero in Equation (4.42) gives

$$\frac{\mathbf{C}}{\mathbf{M}} = G_p(s) = \frac{G \exp(-sD)}{Ts + 1} \quad (4.46)$$

where $G_p(s)$ is the Laplace transform of the transfer function for the FOWDT process model. The notation is unfortunate and needs some clarification. Uppercase G's with subscripts refer to transfer functions, not process gains, and it should be clear that $G_p(s)$ is not the $G_p(z)$ of this chapter with z replaced with s.

There is one significant but subtle difference between the continuous and discrete time-domain approaches. The above process model places few restrictions on the nature of M in the time domain. It can be a series of steps (a staircaselike function), or a continuously varying function. As long as one can write the Laplace transform of M, an expression for $\mathbf{C}(s)$ can be obtained. Then $\mathbf{C}(s)$ can be inverted to yield $C(t)$. For digitally based control systems, M is almost always a series of discrete values (or spikes) put out by a control algorithm. However, the process sees a series of steps, not a series of isolated spikes. Thus the receiver sees a different signal than the sender. In between the control algorithm and the process is a zero-order hold, which receives the discrete values of M every h seconds and holds them until the next new value is sent. By definition, the zero-order hold consists of a unit step at time zero, followed by a negative unit step h seconds later. In the time domain, the zero-order hold is written as

$$h_0(t) = U(t) - U(t - h)$$

Thus $h_0(t)$ is a function that is zero for negative time, unity for h seconds, and then zero for all remaining time; that is, it "holds" its input value at time zero for h seconds. The Laplace transform can be written using Equation (4.39):

$$L\{h_0(t)\} = H_0(s)$$

$$= \frac{1}{s} - \frac{\exp(-sh)}{s} \quad (4.47)$$

$$= \frac{1 - \exp(-sh)}{s}$$

Therefore, when the process input is a staircaselike function, the conversion of transfer functions of differential equations in the continuous time domain to Z-transform transfer functions of difference equations in the discrete time domain requires special consideration. Attention must be given the product of $H_0(s)$ and the appropriate Laplace transform. For example, the Laplace transform of the FOWDT model is

$$\frac{\mathbf{C}}{\mathbf{M}} = G_p(s) = \frac{G \exp(-sD)}{Ts + 1}$$

which corresponds to

$$T\frac{dC}{dt} + C = GM(t - D)$$

To convert from the continuous time domain to the discrete time domain where the process sees M as a staircase function, the zero-order hold must be included:

$$H_0 G_p(s) = \frac{[1 - \exp(-hs)]}{s} \frac{G \exp(-shd)}{Ts + 1} \tag{4.48}$$

where $D = hd$. A partial fraction expansion of Equation (4.48) is

$$H_0 G_p(s) = [1 - \exp(-hs)] \, G \exp(-shd) \left\{ \frac{1}{s} - \frac{T}{Ts + 1} \right\} \tag{4.49}$$

To convert Equation (4.49) from the s-domain to the Z-domain, the following steps are followed. First, replace $\exp(-hs)$ by z^{-1} and $\exp(-shd)$ by z^{-d} to get

$$(1 - z^{-1})Gz^{-d} \left\{ \frac{1}{s} - \frac{T}{Ts + 1} \right\}$$

Next, match up $1/s$ and $T/(Ts + 1)$ with their counterparts in the z-domain and get

$$\frac{C}{M} = \frac{G(1 - A)z^{-1}z^{-d}}{1 - \exp(-h/T)z^{-1}} \tag{4.50}$$

Using the shifting property of z^{-1}, Equation (4.50) can be written in the discrete time domain as

$$C_i = \exp(-h/T)C_{i-1} + G[1 - \exp(-h/T)]M_{i-1-d} \tag{4.51}$$

which is the discrete time-domain version of the FOWDT model. Thus, initially, the process model was a differential equation:

$$T\frac{dC}{dt} + C = GM(t - D)$$

After using the Laplace transform, the zero-order hold, and the Z-transform, the result is

$$C_i = AC_{i-1} + BM_{i-1-d}$$

where $A = \exp(-h/T)$ and $B = G[1 - \exp(-h/T)]$.

If the mass balance method, used in Section 1.6 for the FOWDT model, were to be applied to the three-tank process mentioned in Section 1.7.2, the result would be a third-order differential equation having the form

$$T_1 T_2 T_3 \frac{d^3 C}{dt^3} + (T_1 T_2 + T_2 T_3 + T_1 T_3) \frac{d^2 C}{dt^2} + (T_1 + T_2 + T_3) \frac{dC}{dt} + C = GM$$

$$(4.52)$$

where the three time constants T_1, T_2, and T_3 can be related to properties of the three tanks. Applying the Laplace transform to Equation (4.52) gives

$$\frac{C}{M} = \frac{G}{T_1 T_2 T_3 s^3 + (T_1 T_2 + T_2 T_3 + T_1 T_3)s^2 + (T_1 + T_2 + T_3)s + 1} \qquad (4.53)$$

To derive the discrete time-domain finite difference equation for this third-order model, one would multiply the transfer function in Equation (4.53) by $H_0(s)$ and follow the same procedure that was followed for the FOWDT model. The result, after some involved algebraic bookkeeping, is

$$C_i = A_1 C_{i-1} + A_2 C_{i-2} + A_3 C_{i-3} + B_1 M_{i-1} + B_2 M_{i-2} + B_3 M_{i-3}$$

where the coefficients are

$$A_1 = W_1 + W_2 + W_3$$

$$A_2 = W_1 W_2 + W_1 W_3 + W_2 W_3$$

$$A_3 = W_1 W_2 W_3$$

$$B_1 = A_h + B_h + C_h$$

$$B_2 = A_h(W_2 + W_3) + B_h(W_1 + W_3) + C_h(W_1 + W_2)$$

$$B_3 = A_h W_2 W_3 + B_h W_1 W_3 + C_h W_1 W_2$$

$$W_i = \exp(-h/T_i), \qquad i = 1, 2, 3$$

$$A_h = \frac{G(1 - W_1)}{(1 - T_3/T_1)(1 - T_2/T_1)}$$

$$B_h = \frac{G(1 - W_2)}{(1 - T_1/T_2)(1 - T_3/T_2)}$$

$$C_h = \frac{G(1 - W_3)}{(1 - T_1/T_3)(1 - T_2/T_3)}$$

Upon comparing this approach with the one in Section 4.3.1, one sees that this latter approach allows a direct relationship with an ordinary differential equation, which in turn is developed from a physical model. However, for the purposes of testing control algorithms, there is no reason to go to all this trouble when the dynamic performances of the models are effectively the same.

Note that the algebra of this latter approach would have to be modified considerably for the case where the dynamics are underdamped. However, to use the method of Section 4.3.1, one simply changes the location of the roots and applies the straightforward algebra. It should be noted that software packages such as MATLAB™ provide a relatively painless path between the s-domain and the Z-domain and make the development of models in either domain almost transparent to the user.

This concludes this side trip into the s-domain. Since the rest of the book will deal almost exclusively with the discrete time domain, knowledge of the Z-transform will be sufficient for the reader to assimilate the material in the remainder of this text.

4.3.3 Derivation of the PI Tuning Rules

As a third example of a useful application of the Z-transform, the tuning rules of Section 2.5 will be derived by following a procedure that originated with Dahlin (1968). First, the equation for the controlled system that relates the controlled variable C to the target R will be written using the Z-transform. This equation will contain, first, the transfer function for the process, which is known because FOWDT models will be dealt with exclusively, and, second, the transfer function of the control algorithm, which is to be determined. Next, C will be required to follow R in a manner described by a unity-gain FOWDT model that has a specifiable time constant. Writing this specification in the Z-domain will give a second equation in terms of C and R. Eliminating C and R from these equations will provide a means of solving for the unknown control algorithm transfer function, which will serendipitously look like a PI control algorithm with the gains defined in terms of process parameters and the specifiable time constant. The reader is forewarned that some of the algebraic manipulations in what follows are relatively convoluted.

Development of the initial equations.   The Z-domain description of the process is

$$C(z) = Az^{-1}C(z) + G[1 - A]z^{-1-d}M(z)$$
$$= Az^{-1}C(z) + Bz^{-1-d}M(z)$$

where $A = \exp(-h/T)$, $B = G(1 - A)$, and $d = D/h$. The process transfer function $G_p(z)$ is therefore

$$\frac{C(z)}{M(z)} = G_p(z) = \frac{Bz^{-1-d}}{1 - Az^{-1}}$$

The manipulated variable is related to the error and controlled variable as follows:

$$M(z) = G_c(z)E(z) \qquad (4.54)$$

$$E(z) = R(z) - C(z) \qquad (4.55)$$

where $G_c(z)$ represents the transfer function of the control algorithm to be determined. Eliminating $M(z)$ and $E(z)$ between Equations (4.54) and (4.55) gives

$$(1 + G_p G_c)C = G_p G_c R \tag{4.56}$$

where the arguments of each element do not appear in order to simplify the book-keeping.

Specify that the response of C to changes in R be a first-order, unity-gain response defined in the time domain by

$$T_d \frac{dC}{dt} + C = R(t - D) \tag{4.57}$$

where T_d represents a specifiable time constant describing the response of C to R. The value of T_d is a measure of how aggressive the control action should be. The above equation can also be written in the discrete time domain as

$$C_i = A_d C_{i-1} + (1 - A_d)R_{i-1-d} \tag{4.58}$$

where $A_d = \exp(-h/T_d)$. In the Z-domain, Equation (4.58) can be written as

$$\frac{C(z)}{R(z)} = Q(z) = \frac{(1 - A_d)z^{-1-d}}{1 - A_d z^{-1}} \tag{4.59}$$

where $Q(z)$ has been introduced as the ratio of C to R. For example, if a step change was given to the target R, one would expect to see C respond in a manner described by unity gain, a specifiable time constant T_d, and a deadtime $D = dh$. It would be unrealistic to expect C to respond to R with a deadtime less than that of the process, but the flexibility of choosing the time constant will be retained.

At this point there are two equations relating C to R, one from the definition of how the process and control algorithm are related and one from the specification of how C is to respond to R:

$$(1 + G_p G_c)C = G_p G_c R \quad \text{and} \quad C = QR$$

If these two equations are combined to eliminate C, the quantity R also drops out, giving

$$G_c = \frac{Q}{G_p(1 - Q)} \tag{4.60}$$

where G_p and Q are known and the goal is to find G_c.

Tuning equations for a process without deadtime.   Setting $d = 0$ in Equation (4.60) gives

$$G_c = \frac{(1 - Az^{-1})(1 - A_d)}{B(1 - z^{-1})} = \frac{M}{E} \tag{4.61}$$

which is written in terms of process parameters A and B and a specifiable parameter A_d. Since this expression looks rather useless, some further algebra must be used to get the control algorithm in a more familiar form:

$$(1 - z^{-1})M = \frac{1 - A_d}{B}(1 - z^{-1})E + \frac{1 - A_d}{B}(1 - A)z^{-1}E \qquad (4.62)$$

Remembering that the operator $(1 - z^{-1})$ is the difference operator, Equation (4.62) says that the change in M is equal to a coefficient, $[(1 - A_d)/B]$, times the change in the error plus a coefficient, $[(1 - A_d)/B](1 - A)$, times the error delayed by one sample. The first term on the right side is the proportional component and, except for the one-sample delay, the second term is the integral component. Since the basic form for the PI control algorithm is

$$(1 - z^{-1})M = P(1 - z^{-1})E + IhE \qquad (4.63)$$

by comparison one can write expressions for the control gains:

$$P = \frac{1 - A_d}{B} \qquad (4.64)$$

$$I = \frac{1 - A_d}{B}\frac{1 - A}{h} \qquad (4.65)$$

Furthermore, if A, B, and A_d are replaced by their definitions, the result is

$$P = \frac{1 - \exp(-h/T_d)}{G[1 - \exp(-h/T)]} \qquad (4.66)$$

$$I = \frac{1 - \exp(-h/T_d)}{hG} \qquad (4.67)$$

The one-sample delay in the integral term of Equation (4.62) has been removed for the sake of convenience and common sense; it would be no benefit to have a control algorithm with an unnecessary delay in it.

   Similar equations can be derived for the integrator process without deadtime, where

$$G_p = \frac{(Gh/T)z^{-1}}{1 - z^{-1}} \qquad (4.68)$$

Combining Equations (4.68) and (4.60), while using Equation (4.59) for the definition of $Q(z)$, one arrives at

$$G_c = \frac{M}{E} = \frac{1 - A_d}{Gh/T} \qquad (4.69)$$

which is a proportional-only control algorithm. The process supplies the integral component. For the case where $h \ll T_d$, Equation (4.69) yields the following expression for the proportional control gain:

$$P = \frac{T}{GT_d} \qquad (4.70)$$

showing that the gain increases as the desired time constant decreases. In practice, a small amount of integral control is sometimes added to augment the integrating feature of the process.

Dahlin control algorithm for processes with deadtime, the Box–Jenkins algorithms, and the Smith predictor.   In the preceding the following expression [Equation (4.60)] was derived for the as yet undetermined control algorithm:

$$G_c = \frac{Q}{G_p(1 - Q)} \tag{4.71}$$

Upon replacing Q and G_p by their definitions in terms of z^{-1}, the process parameters A and B, and the specifiable parameter A_d, Equation (4.71) becomes

$$
\begin{aligned}
G_c &= \frac{M}{E} \\
&= \frac{(1 - Az^{-1})(1 - A_d)}{B[1 - A_d z^{-1} - (1 - A_d)z^{-1-d}]}
\end{aligned} \tag{4.72}
$$

where, unlike the development of the previous subsection, the deadtime has been included. After some algebra, Equation (4.72) becomes

$$(1 - z^{-1})M = (1 - A_d)(z^{-1-d} - z^{-1})M \tag{4.73}$$

$$+ \frac{(1 - A_d)}{B}(1 - z^{-1})E + \frac{(1 - A_d)(1 - A)}{B}z^{-1}E$$

which has an interesting interpretation. The reader will recognize the left side as the change in the manipulated variable. The second term on the right side is the proportional component and the third term, if the one-sample delay is removed, is integral component. With this in mind, Equation (4.73) can be written as

$$(1 - z^{-1})M = (1 - A_d)(z^{-1-d} - z^{-1})M + P(1 - z^{-1})E + IhE \tag{4.74}$$

where

$$P = \frac{1 - \exp(-h/T_d)}{G[1 - \exp(-h/T)]} \tag{4.75}$$

$$I = \frac{1 - \exp(-h/T_d)}{Gh} \tag{4.76}$$

Note that the z^{-1} operator has been removed from the last term of Equation (4.73). Therefore, with the removal of this term, Equation (4.74) becomes a PI control algorithm augmented by an extra term,

$$(1 - A_d)(z^{-1-d} - z^{-1})M$$

which is a scaled difference between the last value of the manipulated variable and the value one deadtime earlier. This extra term deals with the deadtime. The expressions for the control gains P and I in Equations (4.75) and (4.76) are the same as those for the case without deadtime considered in the last subsection [Equations (4.62) and (4.63)].

The form of Equation (4.74) is similar to the Box–Jenkins algorithm that was presented without derivation in Section 2.8:

$$(1 - z^{-1})M = (1 + \beta)(z^{-1-d} - z^{-1})M + [P(1 - z^{-1}) + Ih]E$$

where β is a parameter describing the stochastic disturbance and P and I are control gains that have a different dependence than that given in Equations (4.75) and (4.76). Thus the Dahlin control algorithm has a form similar to that of Box–Jenkins, but the coefficients are different. This difference in coefficients is a consequence of the different conditions that are the basis for the derivation of the Dahlin algorithm and the Box–Jenkins algorithm. In the latter case, the control algorithm is designed to drive the controlled variable to target in $1 + d$ steps, a requirement that sometimes causes excessive activity in the manipulated variable if the time constant is large relative to the control interval. On the other hand, the Dahlin approach includes a specifiable time constant that defines the approach of the controlled variable to the target. This time constant can be chosen to prevent the excessive activity of the Box–Jenkins approach.

With proper choice of the P and I gains, the Dahlin control algorithm given in Equation (4.74) is almost identical to the Smith predictor, which was derived in Section 2.7. Start with the PI control algorithm in the Smith predictor, which is

$$(1 - z^{-1})M = [(1 - z^{-1})P + Ih](R - C - C_p' + C_p) \qquad (4.77)$$

where

$$R = \text{set point}$$
$$C = \text{controlled variable}$$
$$C_p' = \text{predicted controlled variable without deadtime}$$
$$C_p = \text{predicted controlled variable with deadtime}$$

Since C_p' and C_p are model based, they can be written as

$$C_p' = \frac{Bz^{-1}}{1 - Az^{-1}} \qquad (4.78)$$

$$C_p = \frac{Bz^{-1-d}}{1 - Az^{-1}} \qquad (4.79)$$

Combining Equations (4.77), (4.78), and (4.79) gives

$$(1 - z^{-1})M = [(1 - z^{-1})P + Ih](R - C) \qquad (4.80)$$

$$+ \frac{P + Ih - Pz^{-1}}{1 - Az^{-1}} B(z^{-1} - z^{-1-d})M$$

where the $(z^{-1} - z^{-1-d})M$ term pops up again.

Comparing Equations (4.80) and (4.74) suggests that, with the proper choice of P and I, the Smith predictor and the Dahlin control algorithm can be made identical. If the term in the Smith predictor from Equation (4.80) (where the subscript s has been appended to P and I to denote Smith predictor)

$$\frac{P_s + I_s h - P_s z^{-1}}{1 - Az^{-1}} B(z^{-1} - z^{-1-d})M$$

is set equal to the corresponding term in the Dahlin control algorithm [Equation (4.74)],

$$(1 - A_d)(z^{-1-d} - z^{-1})M$$

then the Smith predictor control gains must satisfy

$$P_s = \frac{A(1 - A_d)}{B}$$

$$I_s = \frac{(1 - A_d)}{hG}$$

From the Dahlin analysis [Equations (4.75) and (4.76)], the corresponding control gains (with a subscript d for Dahlin) are

$$P_d = \frac{1 - A_d}{B}$$

$$I_d = \frac{1 - A_d}{hG}$$

Therefore, with the exception of the factor A in the expression for P_s, the two algorithms are identical. When the process time constant is large, the factor A will be nearly unity and the difference will be negligible.

PI tuning rules for processes with deadtime.   Start by specifying that the control algorithm have the form

$$G_c = \frac{(1 - Az^{-1})(1 - A_d)}{B(1 - z^{-1})} - \frac{M}{E} \tag{4.81}$$

which was derived previously for the process without deadtime [Equation (4.61)]. In addition, there is an expression for the FOWDT process transfer function (which includes the deadtime),

$$G_p(z) = \frac{Bz^{-1-d}}{1 - Az^{-1}} \tag{4.82}$$

and the relation between C and R derived from the loop diagram,

$$(1 + G_p G_c)C = G_p G_c R \tag{4.83}$$

Combining these last three equations to eliminate G_p and G_c gives a relationship between C and R:

$$\frac{C}{R} = \frac{z^{-1-d}(1 - A_d)}{1 - z^{-1} + z^{-1-d}(1 - A_d)} \tag{4.84}$$

which has a form similar to the relation containing a specifiable parameter that was used previously [Equation (4.59)]. Note that the order of things has been reversed. Here the beginning point is the desired control algorithm form along with a relation between C and R, which will presently yield a value for the parameter A_d. Earlier, the starting point was a relation between C and R containing a specifiable parameter from which a control algorithm was derived.

The behavior of the relationship between C and R that has just been derived in Equation (4.84) depends on the roots of the denominator, which are the values of z, say z_r, that satisfy

$$1 - z_r^{-1} + z_r^{-1-d}(1 - A_d) = 0 \tag{4.85}$$

Equation (4.85) can be arranged to

$$A_d = z_r^d(z_r - 1) + 1 \tag{4.86}$$

Therefore, a value for z_r can be specified and (4.86) will yield the parameter A_d, which in turn defines the relationship between C and R. Note that if $d = 0$ then $A_d = z_r$. However, when $d = 0$, there is no deadtime. For this case, it was shown previously that

$$A_d = \exp(-h/T_d)$$

Therefore, if

$$z_r = \exp(-h/T_d)$$

then Equation (4.86) yields

$$A_d = [\exp(-h/T_d)]^d [\exp(-h/T_d) - 1] + 1$$

By definition, $dh = D$, so

$$A_d = [\exp(-D/T_d)] [\exp(-h/T_d) - 1] + 1 \tag{4.87}$$

The tuning rules derived previously [Equations (4.64) and (4.65)] were

$$P = \frac{1 - A_d}{B}$$

$$I = \frac{1 - A_d}{B} \frac{1 - A}{h}$$

so, replacing A_d by the new expression [Equation (4.81)] yields

$$P = \exp(-D/T_d) \frac{1 - \exp(-h/T_d)}{G[1 - \exp(-h/T)]}$$

$$I = \exp(-D/T_d) \frac{1 - \exp(-h/T_d)}{Gh}$$

which are the tuning rules presented in Section 2.5.

4.4 FREQUENCY RESPONSE FROM THE Z-TRANSFORM

The Z-transform provides a straightforward path to the continuous frequency domain where the frequency response of transfer functions can be studied. The general method will be applied to the FOWDT model and then used to compare some of the filtering techniques that have been mentioned in earlier chapters. Finally, the sampling interval will be decreased to an infinitesimal value to show how the FOWDT model and the first-order filter evolve into their continuous or analog counterparts.

4.4.1 Application to the FOWDT Model

Earlier, the FOWDT model with no deadtime was presented in the discrete time domain and the Z-domain as follows:

$$C_i = AC_{i-1} + G(1 - A)M_{i-1}$$

$$C(z) = Az^{-1}C(z) + Bz^{-1}M(z)$$

$$= \frac{Bz^{-1}M(z)}{1 - Az^{-1}}$$

$$= G_p(z)M(z)$$

In Sections 4.2 and 4.3 the manipulated variable M was given a step change, and the Z-transform was used to find the response of the controlled variable C. In this section the manipulated variable will be a sine wave.

For linear systems, whenever the system input is a sinusoidal function, the steady-state system output (after all the transients have died away) will also be a sinusoidal function. The steady-state output will in general have a different amplitude and a different phase relative to the input sinusoid, but it will always be a sinusoid. This feature of linear systems is one of the main reasons why they are so attractive as models for real processes: a process can be characterized by the steady-state amplitude ratio of the output sinusoid to the input sinusoid and by the steady-state phase of the output sinusoid relative to the input sinusoid.

These two characteristics make up the frequency response. The process is said to have phase *lag* when the phase of the process output is negative relative to that of the input. When the output phase is positive relative to the input, the process is said to have phase *lead*. In Chapter Five the magnitude will be of primary interest, but as will be seen later in this chapter, the phase can play an important part in the stability analysis of controlled systems.

The steady-state frequency response of a dynamic system can be determined from the Z-transform transfer function by making a simple change of variable, a second feature of linear systems as amazing and beneficial as the orthogonality of sinusoids mentioned in Chapter Three. For example, start with the transfer function of the deadtime-less FOWDT model

$$G_p(z) = \frac{C(z)}{M(z)} = (1 - Az^{-1})^{-1}G(1 - A)z^{-1}$$

$$= \frac{G(1 - A)}{z - A}$$

(4.88)

To obtain the frequency response, the following change of variable is made:

$$z = \exp(j\Omega).$$

This transformation effectively contrains the variable z to lie on the unit circle in the complex Z-plane where the new variable Ω ranges from 0 to 2π. As a result, the transfer function is now a function of the angular rotation Ω, which has units of radians:

$$G_p[\exp(j\Omega)] = G_p'(j\Omega) = \frac{B}{\exp(j\Omega) - A}$$

where $B = G(1 - A)$. Note that a prime has been added to $G_p'(j\Omega)$ to indicate that it is not just $G_p(z)$ with z replaced by $j\Omega$. Using Euler's formula,

$$\exp(j\Omega) = \cos(\Omega) + j\sin(\Omega)$$

$G_p'(j\Omega)$ can be written as

$$G_p'(j\Omega) = \frac{B}{\cos(\Omega) + j\sin(\Omega) - A}$$

$$= \frac{B}{[\cos(\Omega) - A] + j\sin(\Omega)}$$

The denominator is a complex number having a magnitude (which is the square root of the sum of the squares of the real and imaginary parts) of

$$\sqrt{[\cos(\Omega) - A]^2 + \sin(\Omega)^2}$$

and a phase (which is the angle whose tangent is the ratio of the imaginary part to the real part) of

$$\arctan\left\{\frac{\sin(\Omega)}{\cos(\Omega) - A}\right\}$$

Therefore, the magnitude of $G'_p(j\Omega)$ is

$$| G'_p(j\Omega) | = \sqrt{\frac{B^2}{(\cos(\Omega) - A)^2 + \sin(\Omega)^2}}$$

and the phase (or argument) of $G'_p(j\Omega)$ is

$$\arg\{G'_p(j\Omega)\} = -\arctan \frac{\sin(\Omega)}{\cos(\Omega) - A}$$

This magnitude and phase describe the steady-state response of the controlled variable to the manipulated variable when the latter is a unity-amplitude, zero-phase sine wave.

First, note that this result has been obtained by dealing only with the transfer function; that is, there was no need to invert a Z-transform. Second, note that the phase of $G'_p(j\Omega)$ is equal to the phase of its numerator, B, which is zero, minus the phase of the denominator. The next two paragraphs should clear up this last manipulation. Readers already familiar with complex algebra may want to skip ahead.

Quantities can be analyzed in terms of their real and imaginary parts. For example, the variable z can be written as

$$z = x + jy$$

$$= \text{Re}\{z\} + j \text{ Im}\{z\}$$

Alternatively, a quantity can be analyzed in terms of its magnitude, $| z |$, and its argument, $\arg(z)$ (or angle or phase), as

$$z = | z | \exp[j \arg(z)]$$

where $| z | = \sqrt{x^2 + y^2}$, and $\arg(z) = \arctan(y/x)$. Graphically, this means that the quantity z can be represented in the z-plane as a vector from the origin to the point (x, y) having the length of $| z |$ and the angle relative to the positive x-axis of $\arg(z)$. For example, the transformation $z = \exp(j\Omega)$ means that z can be represented by a vector having unit length and an angle (or phase) of Ω. From this point of view, Euler's formula shows that the vector $\exp(j\Omega)$ has a real part of $\cos(\Omega)$ and an imaginary part of $\sin(\Omega)$.

For example, the real constant 1.0 can be considered a quantity having a magnitude of unity and an angle of zero or multiples of 2π; that is,

$$1.0 = | 1 | \exp(j0)$$

$$= | 1 | \exp(jk2\pi), \quad k = 1, 2, \ldots$$

Likewise, the real constant -1.0 can be considered a quantity having unity magnitude and an angle of $\pm\pi$ or $\pm 180°$:

$$-1.0 = | 1 | \exp(j\pi) = | 1 | \exp(-j\pi)$$

or

$$-1.0 = \exp(jk\pi), \quad k = \pm 1, \pm 2, \pm 3, \ldots$$

Therefore, the act of negating a quantity is equivalent to changing its phase by π and leaving its magnitude unaffected. Note that adding 2π to the $\arg(z)$ of the quantity -1.0 will leave it unaffected.

Finally, the result of dividing one complex variable, z_1, by another, z_2, is

$$\frac{z_1}{z_2} = \frac{|z_1|\,\exp[j\,\arg(z_1)]}{|z_2|\,\exp[j\,\arg(z_2)]}$$

$$= \frac{|z_1|}{|z_2|}\,\exp\{j[\arg(z_1) - \arg(z_2)]\}$$

which explains the manipulation referred to above.

The main interest is in how the magnitude and phase of transfer functions depend on frequency. To show this dependence, the angular rotation Ω in the example transfer function, is replaced with the frequency f, which has units of cycles per second or hertz (Hz):

$$\Omega = 2\pi f h$$

$$|G_p''(f)| = \sqrt{\frac{B^2}{1 - 2A\,\cos(2\pi f h) + A^2}} \qquad (4.89)$$

$$\arg\{G_p''(f)\} = -\arctan\frac{\sin(2\pi f h)}{\cos(2\pi f h) - A} \qquad (4.90)$$

where f ranges from 0.0 to $1/(2h)$ Hz, which defines the Nyquist interval. The double prime will be used to indicate the two steps of transformation: $z \to \Omega \to f$, so $G_p''(f) = G_p'(j2\pi f h)$.

Figure 4.4 shows $|G_p''(f)|$ plotted versus frequency when the process gain G is 1.0, the process time constant is 20 seconds, and the sampling interval h is 1 second. The shape of the curve should make sense to the reader. The FOWDT model without deadtime is effectively a first-order low-pass filter that passes low frequencies and attenuates high frequencies, with the attentuation being greater as the frequency increases.

For example, if M (the process or filter input) were a constant, that is, a sinusoidal signal with zero frequency, then, since the process gain is unity, the effect of M on C (the process or filter output) would be one to one, and the magnitude of C would be equal to that of M. When the frequency of M becomes greater than zero, that is, when M actually becomes a sinusoid, the effect of M on C is attenuated. If M were a sinusoid with a frequency equal to the Nyquist frequency, its impact would be negligible. This would make sense physically, since a process with a time constant of 20 seconds would have too much inertia to be able to closely follow an input sinusoid with a frequency of 0.5 Hz and a period of 2 seconds.

The abscissa and ordinate were plotted on linear scales in Figure 4.4 since that was often done in Chapter 3 when line spectra were being analyzed. However, it is frequently enlightening to plot the two axes on logarithmic scales. Here the abscissa is the log of the frequency. Since the log of zero frequency is not defined,

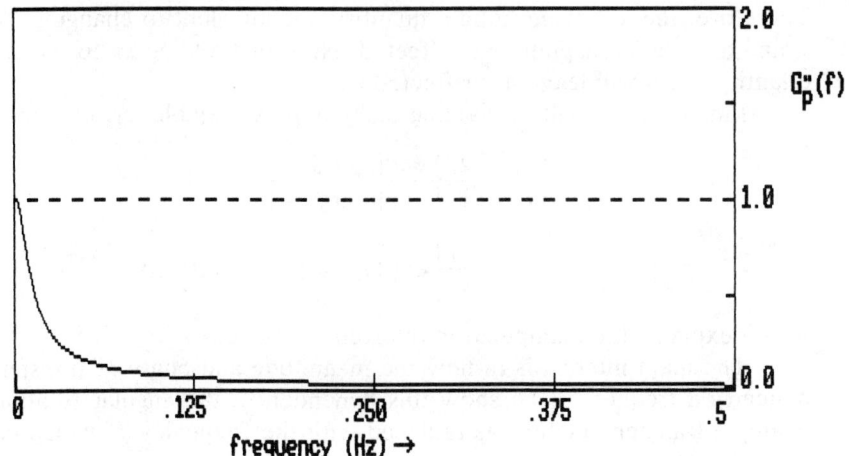

Figure 4.4   Frequency response for a first-order process (linear plot).

the maximum frequency is chosen to be the Nyquist frequency, and the minimum frequency is chosen to be four decades lower: $f_{Ny}/10^4$. The ordinate is the log of the amplitude ratio multiplied by 20.0:

$$20 \log | G_p''(f)|$$

so now the ordinate has units of decibels. Note that in Chapter Three $10 \log(R_k/R_{max})$ was plotted and the units of decibels were also used. The presence of the factor of 10 rather than 20 is due to the fact that the harmonic strength R_k is proportional to the square of the amplitude. The frequency response is replotted using log scales in Figure 4.5. The two types of plotting emphasize different portions of the relationship between amplitude ratio and frequency.

Now it is time to move on to the FOWDT model that does have deadtime. Start with

$$C_i = AC_{i-1} + BM_{i-1-d}$$

$$\frac{C(z)}{M(z)} = G_p(z)$$

$$= \frac{Bz^{-1-d}}{1 - Az^{-1}}$$

$$= \frac{Gz^{-d}(1 - A)}{z - A}$$

which is the same as the case without deadtime except for the presence of z^{-d} in the numerator of the transfer function.

After the transformation of

$$z = \exp(j2\pi fh)$$

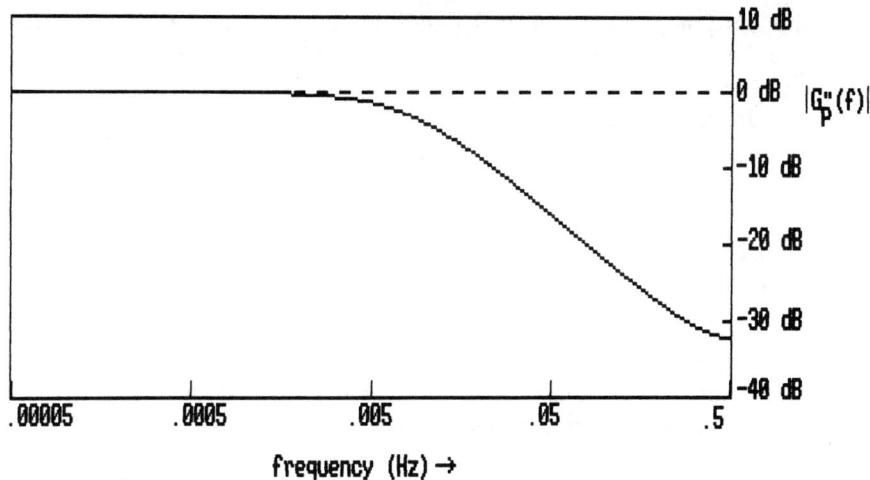

Figure 4.5   Frequency response for a first-order process (log–log plot).

the quantity z^{-d} becomes $\exp(-j2\pi f\,dh)$ or $\exp(-j2\pi fD)$, which has a magnitude of unity and a phase of $-2\pi fD$. This means that the magnitude of the transfer function is unchanged from that given in Equation (4.89); that is, the addition of a deadtime has no impact on the magnitude of the transfer function. However, because of the z^{-d} term, the phase lag is increased over that of the model without deadtime by $2\pi fD$, an amount that increases linearly with frequency:

$$\arg\{G''_p(f)\} = -\arctan \frac{\sin(2\pi fh)}{\cos(2\pi fh) - A} - 2\pi fD \qquad (4.91)$$

This equation makes sense because, in passing a sine wave through a pure deadtime, one would expect the output sine wave to have the same magnitude as the input but to lag it in phase more and more as the frequency increased.

4.4.2 Application to Filtering, Smoothing, and Differencing

In Section 3.11, three filtering methods that reduce noise were discussed. The first and most widely used is the discrete first-order filter, which can be written in the time domain as

$$Y_k = \alpha X_k + (1 - \alpha)Y_{k-1} \qquad (4.92)$$

where the filtering coefficient α is related to the time constant of the filter, T_f, by

$$\alpha = 1 - \exp(-h/T_f)$$

This last relationship was developed in Section 1.9. Note that, unlike the FOWDT process model, the index of X in Equation (4.92) is k and not $k - 1$, since the filter output is effectively calculated at the same instant that the filter input is read from the process. Note also that the filter coefficient α corresponds to $1 - A$ in the FOWDT model in the previous section.

In the Z-domain, the discrete first-order filter becomes

$$\frac{Y(z)}{X(z)} = H(z) = \frac{\alpha}{1 - (1 - \alpha)z^{-1}} \tag{4.93}$$

$$= \frac{\alpha z}{z - (1 - \alpha)}$$

where $H(z)$ is the transfer function of the filter. Comparing Equation (4.93) with Equation (4.92), indicates that the only difference is the appearance of z in the numerator of the former equation, which is a consequence of the different index of X in the discrete time-domain equations.

Making the substitution of

$$z = \exp(j2\pi fh) = \cos(2\pi fh) + j\sin(2\pi fh)$$

allows one to move to the frequency domain where, after collecting the real and imaginary parts of $H''(f)$, the magnitude and the phase of the transfer function can be written as

$$|H''(f)| = \frac{\alpha}{\sqrt{1 - 2(1 - \alpha)\cos(2\pi fh) + (1 - \alpha)^2}} \tag{4.94}$$

$$\arg\{H''(f)\} = 2\pi fh - \arctan\frac{\sin(2\pi fh)}{\cos(2\pi fh) - (1 - \alpha)} \tag{4.95}$$

The two primes indicate that there has been a two-step transition from z to Ω to f; that is, $z = \exp(j\Omega)$ and $\Omega = 2\pi fh$. The algebraic details of deriving Equations (4.94) and (4.95) have been skipped since these equations are almost identical to Equations (4.89) and (4.90) in the previous section.

For the case of $T_f = 20$, the frequency domain plot of $|H''(f)|$ has the same shape as the curves in Figures 4.4 and 4.5, since the FOWDT process model, without a deadtime, is a first-order filter. Note how the distribution of the attenuation over the Nyquist interval starts at unity (no attenuation) and gradually moves toward, but never quite reaches, zero (complete attenuation). Compare this gradual change in attenuation to the sharp discontinuity in attenuation for the low-pass frequency-domain filter that was designed in Section 3.11.

Figure 4.6 shows how the phase of $H''(f)$ starts at zero at low frequencies and decreases to a minimum of slightly more than $-90°$ at approximately 0.05 Hz and then returns to zero phase at the Nyquist frequency. The dependence of phase for this filter on the sampling interval will be shown in Section 4.4.3.

The second filter is a three-point central smoother, which can be written in the time domain as

$$Y_k = \frac{X_{k-1} + X_k + X_{k+1}}{3}$$

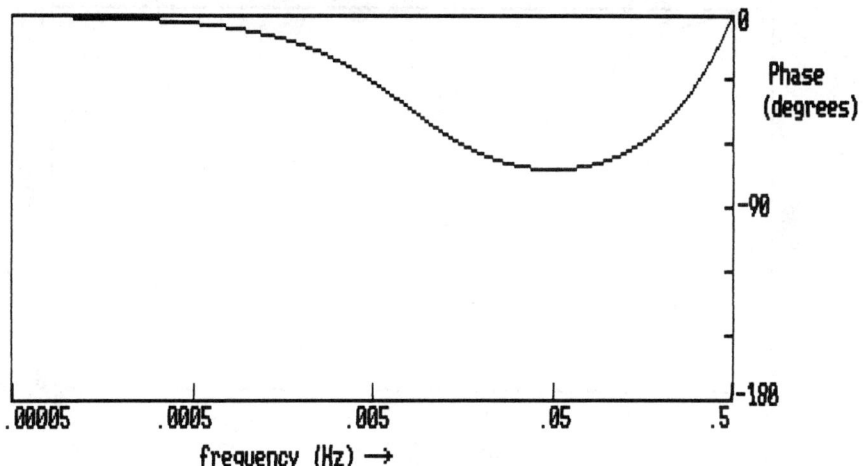

Figure 4.6   Phase of a first-order filter.

and can be written in the Z-domain as

$$\frac{Y(z)}{X(z)} = H(z) = \frac{z^{-1} + 1 + z}{3}$$

Using $z = \exp(j2\pi fh)$, one can move to the frequency domain and get

$$H''(f) = \frac{2\cos(2\pi fh) + 1}{3}$$

Note that, due to the symmetry about the kth point, the transfer functon has no imaginary component and therefore zero phase. (Whenever the variable z only occurs as $z + z^{-1}$, it means there will be no imaginary parts.) Without plotting $|H''(f)|$, one can see that there is a frequency f_0 such that $H''(f_0) = 0$, which means that

$$2\cos(2\pi f_0 h) = -1$$

The solutions of this equation are $f_0 = 0.333333/h, 0.666666/h, \ldots$. Since only $0.33333/h$ hertz lies in the Nyquist interval, which is bounded by the Nyquist frequency of $0.5/h$, one can be certain that the smoothing filter will completely remove power at that frequency. This is to be expected, since a sinusoid having a period equal to $3h$ would be obliterated by such a three-point smoother. Figure 4.7 shows how other frequencies in the Nyquist interval would be attenuated.

This analysis suggests that smoothing formulas of this type should be used with care. Note that central smoothing formulas using five, seven, or more points would have similar characteristics except that the number of obliterated frequencies in the Nyquist interval would increase. For example, a seven-point smoother would obliterate power at a frequency having a period of $7h$ seconds and all its harmonics; that is, power at the following frequencies in the Nyquist interval would be obliterated: $1/(7h)$, $2/(7h)$, $3/(7h)$.

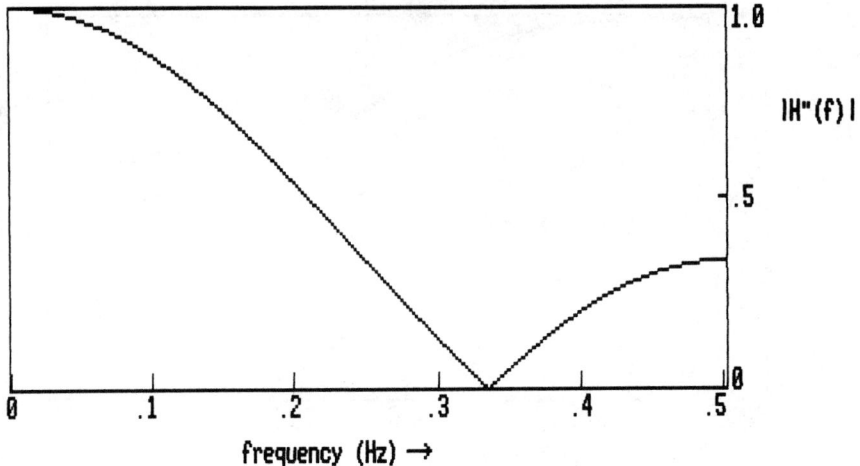

Figure 4.7   Frequency response for a three-point central smoother.

The third filter is the double-sweep filter, where the left-to-right filter is

$$Y_1 = X_1, \qquad Y_k = \alpha X_k + (1 - \alpha) Y_{k-1}, \qquad k > 1$$

$$\frac{Y(z)}{X(z)} = \frac{\alpha}{1 - (1 - \alpha)z^{-1}}$$

and the right-to-left filter is

$$W_N = X_N, \qquad W_k = \alpha X_k + (1 - \alpha) W_{k+1}, \qquad k < N$$

$$\frac{W(z)}{X(z)} = \frac{\alpha}{1 - (1 - \alpha)z}$$

The two intermediate filter outputs are averaged to give

$$V_k = \frac{W_k + Y_k}{2}$$

$$\frac{V(z)}{X(z)} = \frac{1}{2} \left\{ \frac{\alpha}{1 - (1 - \alpha)z} + \frac{\alpha}{1 - (1 - \alpha)z^{-1}} \right\}$$

$$= \frac{1}{2} \frac{2\alpha - \alpha\beta(z + z^{-1})}{1 + \beta^2 - \beta(z + z^{-1})}$$

where $\beta = 1 - \alpha$. Once again the variable z only occurs as $z + z^{-1}$, indicating that there is no imaginary part and therefore no phase shift. Transforming to the frequency domain using the substitution $z = \exp(2\pi f h)$ gives

$$H''(f) = \frac{1}{2} \frac{2\alpha - 2\alpha\beta \cos(2\pi f h)}{1 + \beta^2 - 2\beta \cos(2\pi f h)}$$

The magnitude $| H''(f)|$ is plotted in Figure 4.8 along with that of the discrete first-

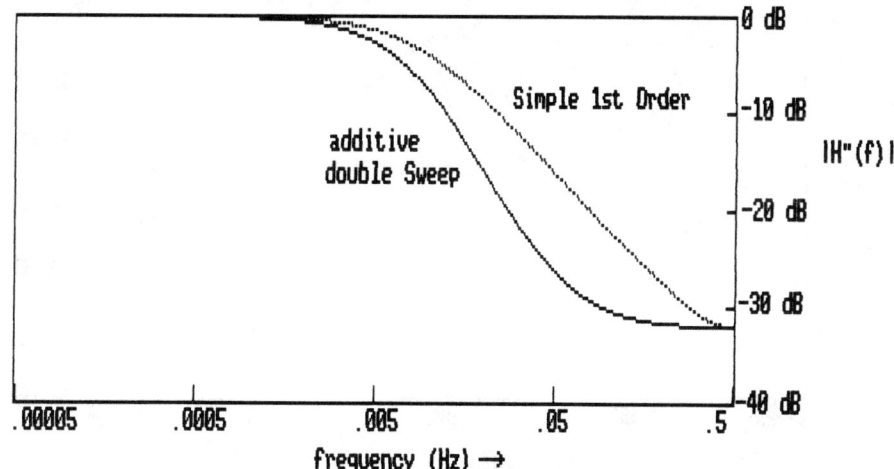

Figure 4.8 Comparison of a first-order filter with a double-sweep filter having the same filtering time constant.

order filter having the same filtering time constant. The double-sweep filter is called *additive* in order to differentiate it from another double-sweep filter, which will be discussed below.

Note that for low frequencies both filters behave similarly, but at the middle frequencies the double-sweep filter attenuates more. As the frequency approaches the Nyquist frequency, the performances of the two filters approach each other.

An alternative form of the double-sweep filter can be obtained by applying the two first-order filters in cascade fashion. On the first sweep the process data represented by X are transformed into Y as follows:

$$Y_1 = X_1, \qquad Y_k = \alpha X_k + (1 - \alpha) Y_{k-1}, \qquad k > 1$$

$$\frac{Y(z)}{X(z)} = \frac{\alpha}{1 - (1 - \alpha) z^{-1}}$$

On the second sweep the one-time filtered data represented by Y are transformed into W as follows:

$$W_N = Y_N, \qquad W_k = \alpha Y_k + (1 - \alpha) W_{k+1}, \qquad k < N$$

$$\frac{W(z)}{Y(z)} = \frac{\alpha}{1 - (1 - \alpha) z}$$

By simple algebra, the transformation between X and W (the filter output) is

$$\frac{W(z)}{X(z)} = \frac{\alpha^2}{[1 - (1 - \alpha) z^{-1}][1 - (1 - \alpha) z]}$$

$$= \frac{\alpha^2}{1 - (1 - \alpha)(z^{-1} + z) + (1 - \alpha)^2}$$

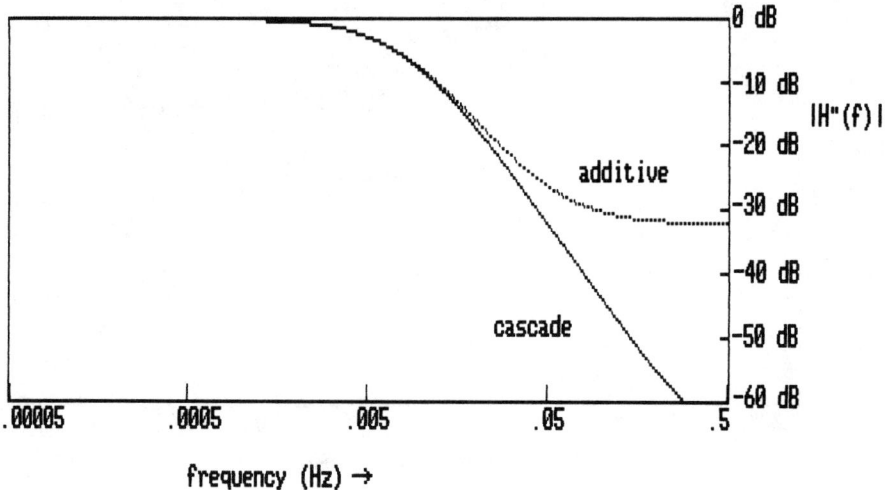

Figure 4.9 Comparison of additive and cascade double-sweep filters.

which shows that this approach also has no phase shift. Transforming to the frequency domain using the substitution $z = \exp(2\pi f h)$ gives

$$H''(f) = \frac{\alpha^2}{1 - 2\beta \cos(2\pi f h) + \beta^2}$$

where $\beta = 1 - \alpha$. The cascade double-sweep filter will exhibit more attenuation than the previous double-sweep filter because it is really a second-order filter. Figure 4.9 compares the attenuation of the two double-sweep filters when the filtering coefficient is the same.

Note that all three of these filters are low-pass filters, and all three have a gradual drop off in magnitude as frequency increases. This is to be compared to the frequency-domain filtering technique discussed in Section 3.11, where the change in the filter magnitude with frequency was a step.

Differencing data has been mentioned as a method of removing low-frequency power and thereby turning nonstationary stochastic sequences into stationary sequences, which then may yield to analysis with the line spectrum and autocorrelation. Since the differencing operation is also an example of a discrete filter (a high-pass filter), it will also be studied in the frequency domain. First, consider the backward difference

$$Y_k = \frac{X_k - X_{k-1}}{h}$$

$$\frac{Y(z)}{X(z)} = H(z) = \frac{1}{h}(1 - z^{-1})$$

Having the transformation $z = \exp(2\pi fh)$ gives

$$|H''(f)| = \frac{1}{h}\sqrt{2 - 2\cos(2\pi fh)}$$

[2]Note that $|H''(0)| = 0$, which is the desired feature of removing all power at zero frequency.

Figure 4.10 shows how $|H''(f)|$ increases with frequency, taking on a maximum value of $2/h$ at the Nyquist frequency. Therefore, if the sampling interval is small, this differencing formula can amplify high frequencies significantly, which in some cases can muddy the analysis.

As an alternative, analysts sometimes use central differencing, which operates on the data as follows:

$$Y_k = \frac{X_{k+1} - X_{k-1}}{2h}$$

$$\frac{Y(z)}{X(z)} = H(z) = \frac{1}{2h}(z - z^{-1})$$

Transforming to the continuous frequency domain gives

$$|H''(f)| = \frac{1}{h}\sin(2\pi fh)$$

Like the backward difference, this operation removes all power at zero frequency, but unlike the backward difference, it also removes all power at the Nyquist frequency and has a maximum value of $1/h$ at the middle of the Nyquist interval,

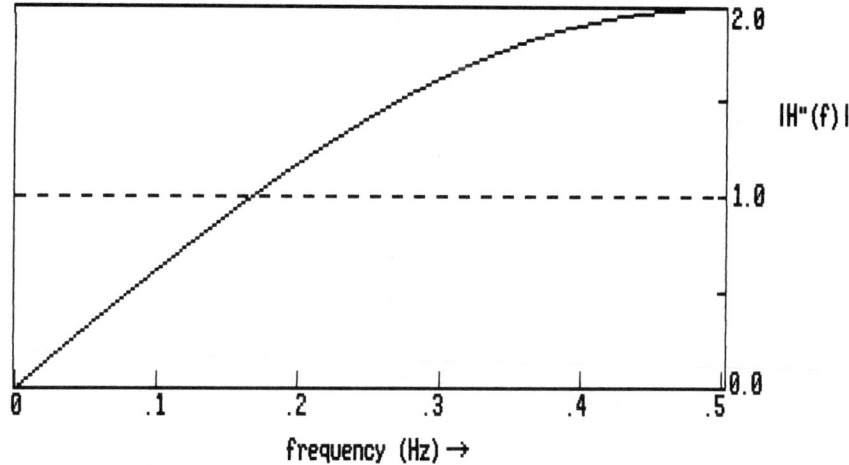

Figure 4.10 Frequency response of the backward difference.

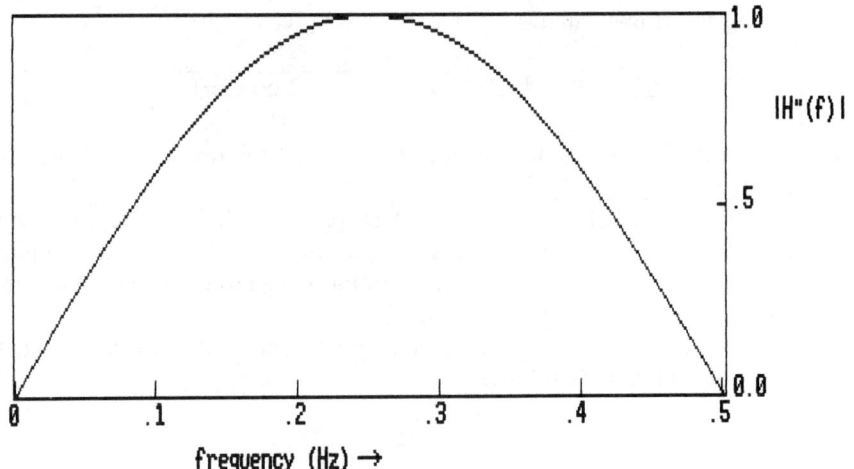

Figure 4.11 Frequency response of the central difference.

that is, where $f = 1/(4h)$. Figure 4.11 shows how this operation amplifies the middle frequencies. (Note that in most differenting operations the computing interval is set equal to unity even though the sampling interval may not be unity.)

These last two paragraphs suggest that one should difference data only with care. These comments complement those in Section 1.10, where passing reference was made to the consequences of differencing data when trying to fit dynamic data to FOWDT models.

The time-domain transient characteristics of filters, such as the time constant and gain, were studied in Chapter One. In that chapter and in Chapter Three, the expected value approach was used to determine the effect of the filters on white noise. In Chapter Three the filters were fed white noise and the empirical line spectrum was calculated. Finally, in this chapter, the Z-transform was used to reveal the theoretical frequency response. In the next section, the discrete time-domain versions of the first-order process and the first-order filter will be related to their counterparts in the continuous time domain.

4.4.3 Effect of Shrinking the Sampling Interval

The filter shown in Figures 4.4, 4.5, and 4.6 has a time constant of 20 seconds and a sampling interval of 1 second, and the transfer function was plotted from 0.00005 Hz to the Nyquist frequency of 0.5 Hz. The sampling interval will now be changed to 0.1 second, while keeping the time constant at 20 seconds. The transfer function will be plotted over the same frequency interval of 0.00005 to 0.5 Hz, even though the Nyquist frequency is now 5.0 Hz.

Figure 4.12 shows the effect of changing the sampling interval on the magnitude by comparing the two 20-second time constant filters using logarithmic scales for the frequency and the transfer function magnitude. Note that over the

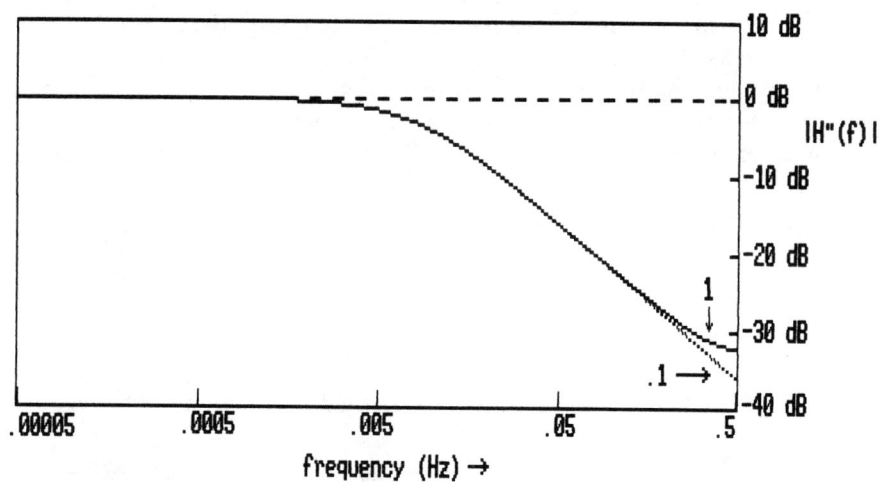

Figure 4.12   Effect of sampling interval on the magnitude of a FO filter.

frequency range shown in Figure 4.12 the transfer function of the filter with the 0.1-second sampling interval can be approximated by a horizontal straight line of unit magnitude and a sloping straight line that decreases with increasing frequency.

Figure 4.13 shows the effect of changing the sampling interval on the phase. Note that decreasing the sampling interval appears to bring the phase nearer to $-90°$ as the frequency increases. In effect, the 20-second time constant filter with the 0.1 second sampling interval behaves as though the sampling frequency were almost infinite, at least over the frequency range from 0.00005 to 0.5 Hz.

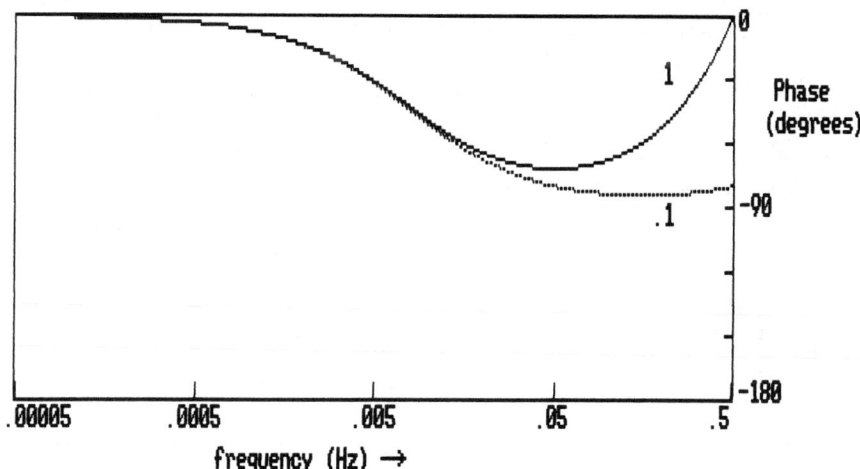

Figure 4.13   Effect of sampling interval on the phase of a FO filter.

To support this last contention, reexamine Equations (4.94) and (4.95) as h is decreased to an infinitesimal value. This situation would correspond to an infinite sampling frequency and would cause the filter to behave as though it were continuous. From Taylor's series expansions for small h/T_f, the following approximations can be written:

$$\alpha = 1 - \exp(-h/T_f) \approx \frac{h}{T_f}$$

$$\cos(2\pi fh) \approx \frac{1 - (2\pi fh)^2}{2}$$

$$\sin(2\pi fh) \approx 2\pi fh$$

Therefore, under the condition of small h/T_f, Equation (4.95) becomes

$$|H''(f)| \approx \frac{h/T_f}{\sqrt{1 - 2(1 - h/T_f)[1 - (2\pi fh)^2/2] + (1 - h/T_f)^2}}$$

and Equation (4.93) becomes

$$\arg\{H''(f) \approx 2\pi fh - \arctan\left\{\frac{2\pi fh}{1 - (2\pi fh)^2/2 - (1 - h/T_f)}\right\}$$

If, when all the factors are multiplied out, terms containing h^4 are neglected while constants and terms containing h^2 are retained, the result is

$$|H''(f)| = \frac{1}{\sqrt{(2\pi fT_f)^2 + 1}} \tag{4.96}$$

$$\arg\{H''(f)\} = -\arctan\{2\pi fT_f\} \tag{4.97}$$

where the right sides are identical to the expressions one would obtain for a continuous or analog filter with a time constant of T_f. (See the next subsection where this contention will be supported.) Note that in deriving Equation (4.97) the $2\pi fh$ term was discarded when compared to the arctan term since h is infinitesimal while the arctan term is finite.

Taking the logarithm of both sides of Equation (4.96) and letting f shrink to an infinitestimal value gives

$$\lim_{f \to 0} \log|H''(f)| = \log\{1\} = 0$$

which is a straight horizontal line at 0 dB on the log–log magnitude versus frequently plot. For large f, Equation (4.96) gives

$$\lim_{f \to \infty} \log|H''(f)| = -\log\{\sqrt{(2\pi fT_f)^2}\} = -\log(2\pi fT_f)$$

which, on the log–log magnitude versus frequency plot, is a straight line with a negative slope. These two straight lines meet at a "corner," where $2\pi fT_f = 1$

and $|H''(f)| = 1/\sqrt{2}$. When the magnitude is plotted in decibels, the value of the transfer function at the corner value is

$$20 \log \left(\frac{1}{\sqrt{2}}\right) = -3.01 \text{ dB}$$

so this corner point is often referred to as the 3-dB point. The frequency at this corner is sometimes called the *corner frequency* or *crossover frequency* and has the value $1/(2\pi T_f)$. Note that, as frequency increases without limit, Equation (4.97) indicates that the maximum phase lag is 90°, which corresponds to the behavior of the phase in Figure 4.13.

Briefly refer to the discussion in Section 3.4 on aliasing, where it was suggested that a low-pass analog filter should be applied to remove harmonic power at frequencies greater than the Nyquist frequency. This is the same as saying that the corner frequency of such an analog filter should be approximately equal to the Nyquist frequency. Keep in mind that for discrete-time low-pass filters, the maximum frequency that can be attenuated is the Nyquist frequency. For low-pass analog filters the maximum frequency that can be attenuated is in principle infinite, and the attenuation increases with frequency.

Now, the sampling interval will be decreased for the equations representing the first-order process without deadtime [Equations (4.89) and (4.90)]. Compared with the first-order filter, the magnitude of the first-order process is unchanged. However, the phase changes to the different numerator, which is a consequence of the index on M being $k - 1$ instead of k, as in the case of the first-order filter. Figure 4.14 shows the effect of changing the sampling interval on the phase. This suggests that, as the sampling interval decreases, the phase at high frequencies

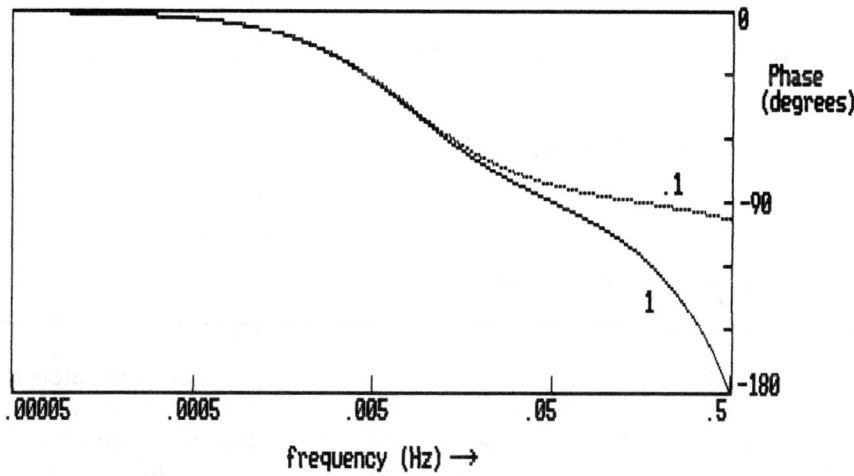

Figure 4.14   Effect of sampling interval on the phase of a FO filter.

approaches $-90°$ but from below, as compared to from above, which was the case in Figure 4.13 for the first-order filter. In other words, when the sampling interval is large there is a deadtime effect resulting from the index on M being $k - 1$, which causes the phase lag to increase with frequency, and therefore the phase lag is greater than $90°$ for large frequencies. When the sampling interval decreases, the deadtime effect becomes less, and, in the limiting case, when the process becomes continuous, the effect is negligible and the maximum phase lag is $90°$.

4.4.4 Frequency Response for the Continuous Time Domain

In Section 4.3.2, the s-domain transfer function for the continuous time-domain, first-order differential equation was found to be

$$\frac{C(s)}{M(s)} = G_p(s) = \frac{G}{Ts + 1}$$

Since the frequency response for a Z-domain transfer function can be obtained by a simple change of variable, one might suspect that something similar can be done for s-domain transfer functions. In fact, this can be done and the change of variable is even simpler than with the Z-transform. The magnitude and angle (or phase) of $G_p(s)$ can be obtained by letting $s = j\phi$, which will give

$$G_p(j\phi) = \frac{G}{Tj\phi + 1}$$

$$= \frac{G(-Tj\phi + 1)}{(T\phi)^2 + 1}$$

where the numerator and denominator have been multiplied by

$$-Tj\phi + 1$$

The magnitude squared of $G_p(j\phi)$ is

$$|G_p(j\phi)|^2 - \frac{1}{(T\phi)^2 + 1}$$

while the phase of $G_p(j\phi)$ is

$$\arg\{G_p(j\phi)\} = -\arctan(T\phi)$$

The units of the parameter s are second^{-1}, and it can be considered as a complex-valued frequency. The units of ϕ are second^{-1}, so ϕ can be considered a real-valued frequency. If $\phi = 2\pi f$, then the above expressions are seen to be identical with those derived in the previous subsection [Equations (4.96) and (4.97)].

For the Z-transform, the substitution $z = \exp(j2\pi fh)$ constrains z to the unit circle in the complex Z-plane. For the Laplace transform, the substitution $s = j2\pi f$ constrains s to the imaginary axis in the complex s-plane.

Since the primary interest in this book is in the discrete time domain, the frequency response of continuous time systems will not be pursued.

4.5 STABILITY OF CONTROLLED SYSTEMS

Passing references have been made to the idea of stability several times in this chapter, and in this section a more quantitative definition will be given. First, one must determine whether a system is stable or not (the yes-no question); then, if the system is found to be stable, one can determine the degree of stability. Regarding the first question of yes-no stability, several different definitions of stability will be presented and criticized in order to make the reader aware of some of the subtleties. A reasonable starting point is the following. Essentially, but crudely, a process for which there is no feedback control activity (an open-loop configuration) could be considered stable when a bounded change in the manipulated variable (process input) causes a bounded change in the controlled variable (process output). Conversely, an unbounded response of the controlled variable to a bounded change in the manipulated variable would mean instability.

This definition needs further discussion because if a positive step change (a bounded input) is applied to the example process, which was referred to in Section 1.7.3 as an integrator and which will be considered a stable process, the output will increase indefinitely (an unbounded output) but at a constant rate. Thus, for the case of positive process gains and positive step changes in the manipulated variable, an uncontrolled process might be defined to be unstable if the controlled variable were to respond in an ever-increasing manner with an increasing rate of change. This modification would allow the integrator to be defined as being stable, since for this kind of process the response of the controlled variable to a positive step change in the manipulated variable is a ramp with a constant rate of change. However, by this definition, a double integrator would be considered unstable.

If stability was defined in terms of responses to sinusoidal inputs, then a process would be considered stable if the response to a constant amplitude sinusoid was another sinusoid with a constant amplitude. In this case, a process would be considered unstable if the response to a constant-amplitude sinusoid was another sinusoid with an ever-increasing amplitude. This definition would allow an integrator to be considered stable because the sinusoidal response of an integrator is another sinusoid with the same amplitude but with a different phase. However, all of the foregoing definitions of stability depend on the type of input variation.

Similar problems arise in dealing with a process under feedback control, that is, a *controlled* process (a closed loop configuration). Using the approach of the previous paragraph, a closed loop system would be considered stable if a constant-amplitude sinusoidal variation in the set point or disturbance would cause a sinusoidal variation in the controlled variable having constant amplitude.

To make the stability of a process or system independent of the type of disturbing function, a definition will be presented in terms of either the roots of the denominator or the open-loop magnitude and phase of the appropriate transfer function.

4.5.1 Root Location

The definition of stability can be made more precise by stating it in terms of where the roots of the denominator of the appropriate Z-transform transfer function lie. These roots of the denominator of the transfer function are also called the *poles* of the transfer function. The following will demonstrate that an uncontrolled process in an open-loop configuration can be considered stable if the poles of the process transfer function lie on or within the boundary of the unit circle.

Referring back to Section 4.2, consider an uncontrolled first-order process without deadtime having a transfer function of

$$\frac{C(z)}{M(z)} = G_p(z) = \frac{G(1 - A)z}{z - A}$$

The time-domain analytical solution of $C(t)$ for the case when M was a unit step change showed that C depended on the pole of the process transfer function and on the pole of $M(z)$. In general, the stability of the response of C to M, no matter what kind of time domain behavior M has, will depend on the pole(s) of the process transfer function. If the poles have an absolute value less than or equal to unity, that is, if the poles do not lie outside the unit circle in the Z-plane, C will be stable.

The example in Section 4.2 showed that the time domain behavior of C had the form [Equation (4.26)]

$$C_i = a_1(r_1)^i + a_2(r_2)^i \tag{4.98}$$

where one root, r_1, came from the denominator of the process transfer function and the other, r_2, came from $M(z)$. As time evolves, the exponents of the roots in Equation (4.98) increase without limit, so a bounded response requires that the roots have magnitudes less than or equal to unity. Since the root coming from the transfer function is A and since $A = \exp(-h/T)$, the response of C to M, no matter what M is, will be stable (unless M itself is unstable), because the absolute value of A is less than unity and anytime such a quantity, whether real or complex, is taken to higher and higher powers, the result ultimately approaches zero. Therefore the process described by the above transfer function, having the pole at $z = A$, is stable.

If the third-order process model mentioned in Section 4.3.1 were studied, there would be three poles associated with the process, two of them complex conjugates. In this case the time-domain solution for C when M is a unit step change would have the form

$$C_k = v^k[a_1 \cos(ku) + a_2 \sin(ku)] + a_3(r_3)^k + a_4(r_4)^k \qquad (4.99)$$

where the first two poles, r_1 and r_2, are complex conjugates defined as

$$r_1 = p + jq, \qquad r_2 = p - jq$$

with

$$v^2 = p^2 + q^2 \quad \text{and} \quad u = \arctan \frac{q}{p}$$

Poles r_1, r_2, and r_3 (real) come from the process transfer function, while r_4 comes from the Z-transform of M. The presence of the sine and cosine terms means that the time-domain behavior of C will be oscillatory; but if $|r_1| = |r_2| < 1$, those terms will die away with increasing time and, if $|r_3| \leq 1$, the process represented by the third order model will be stable. If $|r_1| = |r_2| = |r_3| = 1$, then C will exhibit sustained but stable bounded oscillations plus an offset. [The derivation of Equation (4.99) is not given because it would add little to the understanding of the stability concept. It will be left to the interested reader to derive this equation using the simple collection of tools presented so far in this chapter.]

This example supports the contention that stability requires that the poles of the transfer function lie inside or on the unit circle in the Z-domain. With the exception of the integrator process, which, as mentioned in Section 4.3.1, has a pole on the unit circle, most of this book will be directed toward processes having poles inside the unit circle.

In Section 4.1, for the case of a process embedded in a conventional feedback control loop, the transfer function relating C to its set point R was shown to be

$$\frac{C(z)}{R(z)} = \frac{G_p G_c}{1 + G_p G_c} \qquad (4.100)$$

The stability of this system, consisting of a process described by G_p and a control algorithm described by G_c, also depends on the location of the poles of the transfer function. For the case where the process is FOWDT and the control algorithm is PI, the components of Equation (4.100) are

$$G_p = \frac{Bz^{-d}}{z - A}$$

and

$$G_c = [P(z - 1) + Ihz]/(z - 1)$$

Therefore,

$$C(z)/R(z) = \frac{B[P(z - 1) + Ihz]}{(z - A)(z - 1) + B[P(z - 1) + Ihz]z^{-d}}$$

If the denominator of $C(z)/R(z)$ is denoted by

$$W(z) = z^d(z - A)(z - 1) + B[P(z - 1) + Ihz]$$

then, for this system to be stable, the roots r of the polynomial $W(z)$ that satisfy the equation

$$W(r) = r^d(r - A)(r - 1) + B[P(r - 1) + Ihr] = 0 \qquad (4.101)$$

must lie inside or on the unit circle in the Z-domain. If, for example, the parameters are $h = 1$, $T = 40$, and $d = 40$, i.e., a process having a time constant equal to the deadtime of 40, then stability would require that all the roots of the above 42nd-order polynomial, $W(z)$, lie inside or on the unit circle in the Z-plane.

Although the degree of stability depends on where inside the unit circle the roots lie, the yes-no question of stability only depends on whether the roots lie inside or outside the unit circle. Therefore it may not make sense to address this latter question by employing a root-finding algorithm, especially when the number of roots can be high. Instead, the problem can be straightforwardly solved by applying the following contour integral which is derived in most advanced engineering mathematics texts (for example, see Wylie, 1966):

$$\oint W'(z)/W(z)\, dz = 2\pi j(n - p) \qquad (4.102)$$

where $W'(z)$ is the derivative with respect to z of the polynomial $W(z)$, n is the number of roots of $W(z)$ which lie *inside* the closed contour in the Z-plane around which the integration is done and p is the number of poles of $W(z)$ in the same area. Since $W(z)$ is the denominator of transfer function, there will be no poles, so $p = 0$.

If the closed contour is chosen to be the unit circle, then the system represented by the polynomial $W(z)$ will be stable if n equals the order of the polynomial, that is, if all the roots of the polynomial lie inside the unit circle. Since the system could have roots on the unit circle and still be stable, this method is not completely general, but it can provide a useful tool to verify stability. In other words, if this method says the system is stable, one can be sure it is. If it says it is unstable, one will not know if the instability indication is a consequence of roots on the unit circle or outside of the unit circle. As will be shown below, when it is known that there is a root at unity, measures can be taken to get around that problem.

For those readers not familiar with contour integration in the complex plane, this procedure may not seem attractive but the following simplification may modify that. Change the variable of integration from z to v by the following transformation:

$$z = \exp(jv) = z(v)$$

where v varies from $-\pi$ to π as z moves around the unit circle. Substituting this variable into the integrand of Equation 4.102 gives

$$2\pi n = \int_{-\pi}^{\pi} \frac{\exp(jv)W'[z(v)]}{W[z(v)]} \, dv$$

so that the contour integral now becomes a simple one-dimensional definite integral. The integration limits are chosen to be $-\pi$ to π instead of 0 to 2π because many of the roots lie near the point $1.0 + j0$ in the complex Z-plane, and better accuracy will be achieved by not starting or stopping the numerical integration in that neighborhood. Since the result of the integration is an integer multiple of 2π, round-off or truncation error will not have a significant impact, and this integral can be evaluated numerically by any simple method, such as the trapezoidal rule. The only calculational subtlety comes in evaluating $\exp(jv)$, $W'[z(v)]$, and $W[z(v)]$, which are complex quantities with real and imaginary parts. Programming a digital computer to carry out this numerical integration can be simple if a version of FORTRAN is used that allows complex variables. Similarly, higher-level languages such as MATLAB™ provide a simple means of evaluating the integral.

For the case of a FOWDT process where the integral control gain is zero, Equation (4.101) simplifies to

$$W(r) = r^d(r - A)(r - 1) + BP(r - 1) = 0 \qquad (4.103)$$

where one of the roots is unity. Applying the above stability-determining algorithm to this polynomial will cause problems since one of the roots lies exactly on the unit circle. This difficulty is bypassed by applying the stability-determining algorithm to Equation (4.103) after the factor $(r - 1)$ has been removed, that is, to

$$W(r) = r^d(r - A) + BP$$

The method presented in this section has dealt only with the yes-no stability question. If the system is found to be stable, the degree of stability can be assessed by the closeness of the transfer function poles to the unit circle. However, this method gives no quantitative measure for the degree of stability. The method of the next section will address this problem.

4.5.2 Magnitude and Phase in the Frequency Domain

An alternative approach to stability makes use of the magnitude and phase concepts of Section 4.4 along with the transfer function concept of Section 4.1. The effect of the set point R on the controlled variable C when the control loop is *open* can be studied with

$$C(z) = G_p(z)M(z)$$

$$= G_p G_c E(z)$$

$$= G_p G_c R(z).$$

In the last equality, $E(z)$ was replaced with $R(z)$ and not $R(z) - C(z)$ because the loop is open and C is not being fed back to the summing point, where it would be subtracted from the set point R. If R is a sinusoid with unit amplitude and if there are no disturbances, then, as a consequence of G_pG_c, C will also be a sinusoid with, in general, differing magnitude and phase. When the loop is *closed* at the summing point, C is subtracted from R (or $-C$ is added to R). Therefore R and $-C$, both of which are sinusoids, are being combined.

If the magnitude of C just before the summing point is unity, then two signals of equal magnitude will be combined at the summing point. If the phase of C is $-\pi$ radians relative to R, then $-C$ will have a phase of -2π radians, since negating a quantity is equivalent to subtracting π radians from its phase. Thus, since a phase of -2π is equivalent to a phase of zero, C will be in phase with $-R$, and if the magnitude of C is unity, the result of combining $-C$ and R will be a new signal with twice the magnitude of R. When this new signal propagates around the now closed loop, there will be another doubling at the summing point, and the amplitude of all the signals circulating around the loop will soon become unbounded.

This heuristic discussion suggests that instability will be imminent when G_cG_p, which is called the *open-loop* transfer function, transforms a unity magnitude sinusoid into another unity magnitude sinusoid with a phase of $-\pi$ radians. Under these conditions the system is said to be *marginally stable*. In other words, marginal stability occurs when G_cG_p operates on 1.0 and gives -1.0, that is,

$$G_cG_p = -1$$

In the Z-domain, this condition requires finding a value of z, say z_c, that satisfies

$$G_c(z_c)G_p(z_c) + 1 = 0$$

That is, z_c is a pole of C/R in Equation (4.100).

The substitution $z = \exp(j2\pi fh)$ moves the problem to the frequency domain, where the search is now for a frequency f_ϕ, called the *phase crossover frequency* or *critical frequency*, at which the phase of the open-loop transfer function, $G_c''(f_\phi)G_p''(f_\phi)$, first becomes $-\pi$ radians. If, for this frequency, the open-loop magnitude is unity then the system is marginally stable. One can assess the degree of stability by studying how the open-loop phase and magnitude approach $-\pi$ and unity respectively as the control gains are increased.

These concepts can be quantified using the *gain margin* and the *phase margin*. The former quantity is defined in terms of the ratio of the value of the open-loop magnitude when the system is marginally stable (that is, unity) to the open-loop magnitude at the phase crossover frequency f_ϕ:

$$\frac{1}{[G_c(f_\phi)G_p(f_\phi)]}$$

The gain margin is this ratio written in terms of decibels:

$$\text{Gain Margin} = -20 \log_{10} | G_c(f_\phi)G_p(f_\phi)|$$

For stable systems, the gain margin is the factor by which the open-loop gain must be increased in order to make to system marginally stable. A rule of thumb is to choose control gains such that the gain margin is not less than 6 dB.

The phase margin is defined as the difference between 180° and the open-loop phase lag at the frequency (called the gain crossover frequency) where the open-loop magnitude equals unity. Thus, if the control gains were chosen to make the open-loop phase equal to $-130°$ at the frequency where the open-loop magnitude equaled one, then the phase margin would be 50°. A rule of thumb is to choose control gains that keep the phase margin greater than 45°.

Note that for stable systems, the phase crossover frequency where the open-loop phase equals $-180°$ and gain crossover frequency where the magnitude equals unity are not the same. Only for marginally stable systems are they the same.

As an elementary example of how one might use these concepts, consider the case where there is proportional-only control:

$$G_c = P$$

Most processes act like low-pass filters so that for low frequencies the magnitude of G_p is just less than its steady-state gain, which would be G if the process were FOWDT. Also, as was seen in the previous sections, the phase lag of G_p is small at small frequencies. Since $G_c = P$, both G_p and G_cG_p will have the same phase. However, for large values of P, the magnitude of G_cG_p may be greater than unity at low frequencies. As the frequency increases, the phase lag of G_cG_p will usually increase and the magnitude of G_cG_p will decrease.

Instability will occur if, as the frequency increases, the phase lag reaches π radians before the open-loop magnitude dips below unity. If this happens, then the choice of the control gain P is too aggressive and must be decreased. The value of the control gain that just causes the magnitude of G_cG_p to equal unity at the frequency that the phase of G_cG_p equals $-\pi$ is called the critical control gain and is denoted by P_c. If P is lower than P_c then the controlled process will be stable but if P is near P_c the performance will be stable but oscillatory. Consistent with the discussion on gain margin, a rule of thumb is to choose P equal to one-half of P_c. Some of these ideas are illustrated in Figure 4.15.

Although proportional-only control was chosen to make the arguments simple, the goal of control design by this method is to choose control strategies or control gains such that the critical frequency and the critical control gains are as high as possible. With this simple argument in hand, the reader can see that process characteristics such as deadtime, which add phase lag, will cause this critical condition to occur at lower frequencies and at lower control gains. Similarly, the addition of filters in the loop will also cause the phase lag to increase and therefore lower the critical frequency.

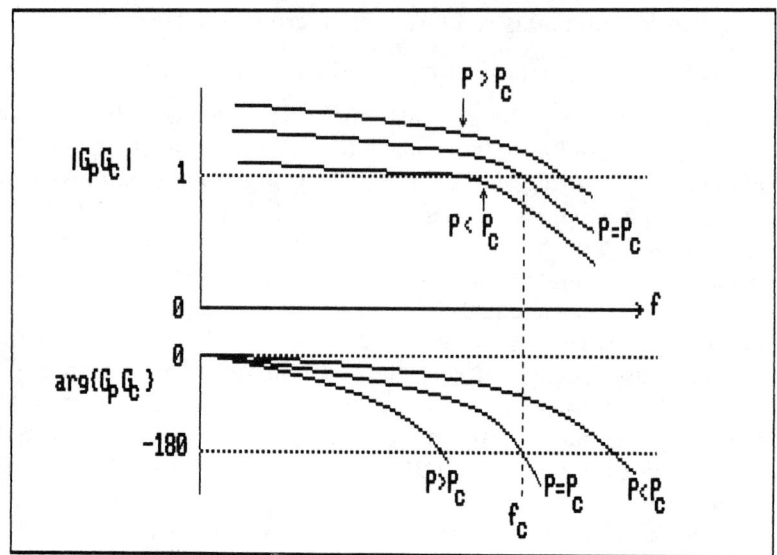

Figure 4.15   Critical frequency and critical gain.

Even though the magnitude/phase approach to control design will not be pursued in the remainder of the book, it is important that the process analyst be aware that conditions that cause the phase of G_pG_c to approach $-\pi$ may cause instabilities. To help drive this idea home, consider the following example of position control. The process consists of a device whose position is to be kept near a moving target by adjusting its velocity. The control algorithm consists of making a change in the device's velocity every h seconds where the velocity change is to be proportional to the position error.

From the discussion in Chapter Two, the control algorithm is recognized to be integral-only, that is, the output of the control algorithm is effectively the integral of the input error. Likewise from first principles, the process itself is recognized to be an integrator because the device's output, its position, is the integral of its input, the velocity. Consequently, the open-loop transfer function G_pG_c consists of the transforms of two integrators.

By definition, an integrator has a phase of $-\pi/2$ radians. For example, if the integrator's input were $\sin(\alpha t)$, the output would be the integral of input, or $-(1/\alpha)\cos(\alpha t)$, which, using trigonometry, can be written as $(1/\alpha)\sin(\alpha t - \pi/2)$; that is, the output, except for the gain factor $(1/\alpha)$, is the same as the input but with a phase of $-\pi/2$ radians. The phase of G_pG_c is $-\pi$ since both G_p and G_c have a phase of $-\pi/2$. Therefore, no matter what the controller gain is, this controlled system will have stability problems in the sense that the phase margin will be unacceptably low.

There are extensive graphical methods for analyzing controlled processes from the magnitude/phase point of view and they are especially well suited for designing servomechanism systems where the main challenge is a varying set point rather than a controlled variable contaminated with noise. For process control, the classic text for this approach in the continuous time domain is the book by Harriott (1964). For the discrete time domain, the book by Saucedo and Schiring (1968) is recommended. This approach will not be pursued here since methods using the power spectral density will be presented in Chapter Five for studying processes contaminated by noise. However, no matter what method is being used to design or analyze controlled systems, it is always a good idea to augment, if possible, the approach with a phase margin calculation.

4.6 SUMMARY

The initial motive for introducing the Z-transform in this chapter was to provide a steppingstone to the power spectral density that will be used in the next chapter. However, once the Z-transform was in hand, it was found to be extremely useful in developing simple process models, deriving the tuning rules of Chapter Two, analyzing filters, and studying the stability of controlled systems. It should have struck the reader that the real utility of the Z-transform is not just its ability to aid in searching for closed solutions to time-domain difference equations that describe controlled systems. Instead, it is the straightforward path to the frequency domain provided by the Z-transform that makes it so useful. Also, the backshifting feature of z^{-1} allows one to easily convert a Z-transform version of a control strategy into a working time-domain algorithm.

The reader should realize that the Z-transform has not been "covered" as much as it has been "uncovered." Any serious student of applied control theory will want to read more about the Z-transform in texts like the ones referenced in the next section.

4.7 DISCUSSION OF THE REFERENCES

A straightforward introduction to the Z-transform as well as an interesting approach to process control is provided in
FRANKLIN, G. F., and POWELL, J. DAVID, *Digital Control of Dynamic Systems*, Addison Wesley, Reading, Mass., 1980,

and
SAUCEDO, R., and SCHIRING, E. E., *Introduction to Continuous and Digital Control Systems*, Macmillian, New York, 1968.

Probably the best text for applying frequency-domain servomechanism design methods in the continuous time domain to process control is
HARRIOTT, P., *Process Control*, McGraw-Hill, New York, 1964.

Two of the many basic texts on advanced engineering mathematics that deal with the complex contour integral used in Section 4.4 are

WYLIE, C. R., *Advanced Engineering Mathematics*, 3rd ed., McGraw-Hill, New York, 1966,

and

HILDEBRAND, F. B., *Advanced Calculus with Applications*, 2nd ed., Prentice Hall, Englewood Cliffs, N.J., 1976.

Readers of the text by Box and Jenkins, which has been referenced in previous chapters, will notice that their shift operator B is equivalent to z^{-1}.

The approach used in Section 4.3.1 to derive the tuning rules is quite similar to that given in

DAHLIN, E. B., *Designing and Tuning Digital Controllers, Instruments and Control Systems*, 41, 77–83 (1968).

CHAPTER 5

THEORETICAL SPECTRAL ANALYSIS OF NOISY PROCESSES

In Chapter Four the mathematical foundation for the development of this chapter's main tool, the power spectral density, was laid. This tool will provide a theoretical method for evaluating the variance of the error between the controlled variable and the set point as a function of the variance of the disturbance. The Fourier line spectrum, which was introduced in Chapter Three and used to analyze sampled data coming from noisy processes, is an empirical function of the harmonics of the data set's fundamental frequency. On the other hand, the power spectral density is a theoretical continuous function of frequency, even though the time-domain function on which it is based is only defined at discrete points in time.

In Chapter Three, line spectra were calculated using data from simulations based on FOWDT models (as well as a few others) subject to white and colored stochastic disturbances. In this chapter it will be shown how to calculate power spectral densities when the process model, the control algorithm, and the stochastic disturbance models are known. For zero-mean stochastic sequences, the area under the power spectral density curve is the variance, and the noise transmission curves will be seen to be related to power spectral densities. Therefore, these methods will enable the process analyst to determine, without resorting to simulation, whether or not the deviation of a disturbance is decreased by the action of a control algorithm.

When simulation methods are used, standard deviations and line spectra can be estimated, but they are subject to sampling error because each simulation is a look at only one realization of a stochastic process. In dealing with the power spectral density, mathematical models of both the process and the stochastic disturbances are used, and the uncertainty associated with sampling a realization of a stochastic process is avoided.

The tools of this chapter will be applied to the PID control algorithm and to the Smith predictor, which was mentioned in Chapter Two. The internal model control approach will be introduced and analyzed. This method has not been mentioned earlier since its construction requires knowledge of the Z-transform. The Box–Jenkins control algorithms will be discussed again, this time using a hybrid notation combining time-domain quantities with the Z-transform. The chapter will conclude with comments on how knowledge of the process and noise models can be used to determine off-line estimates of the PID control gains.

5.1 POWER SPECTRAL DENSITY

The power spectral density of a time-domain function will first be defined in terms of a Z-transform and in terms of a chain rule that relates the power spectral density of the process output to the power spectral density of the process input. To develop this latter relationship, more mathematics will be involved, although the derivation will be far from rigorous.

A second definition in terms of the discrete-time Fourier transform of the autocorrelation function will be presented in summary fashion. The introduction of yet another transform should augment the first definition of power spectral density rather than overwhelm the reader. A short summary of all the transforms used in this book is presented in the subsection following this second definition. If the readers are at all interested in this subject, they are urged to look at Chapter Four of the book by Astrom (1970), at Appendix D in the book by Franklin and Powell (1980), at Chapters Eight and Ten of the first text by Oppenheim and Schafer (1975), Chapter One in the second text by Oppenheim and Schafer (1989), and Chapters Two and Four in Marple (1987).

5.1.1 Power Spectral Density from the Z-transform

For the sake of generality, let $Y(t)$ or Y_i be the time-domain description of a process output, and let $X(t)$ or X_i be the process input. Previously in this book, C has been used as the controlled variable or process output and M as the manipulated variable or process input. In the Z-domain the relationship between X and Y would be written

$$Y(z) = H(z)X(z)$$

where $H(z)$ would represent the transfer function that transfers the dynamics of X into Y. In the running example in Chapter Four, $H(z)$ corresponds to $G_p(z)$.

As was pointed out in Section 4.4, z, $Y(z)$, $X(z)$, and $H(z)$ are complex variables having real and imaginary parts; that is,

$$z = x + jy = \text{Re}\{x\} + j\,\text{Im}\{z\}$$

$$H(z) = \text{Re}\{H(z)\} + j\,\text{Im}\{H(z)\}$$

$$Y(z) = \text{Re}\{Y(z)\} + j\,\text{Im}\{Y(z)\}$$

A complex variable, such as z, can also be represented in terms of its magnitude, $|z|$, and its angle, $\arg\{z\}$:

$$z = |z|\exp(j\,\arg\{z\})$$

where $|z| = \sqrt{x^2 + y^2}$ and $\arg\{z\} = \arctan(y/x)$. Likewise, one can write

$$H(z) = |H(z)|\exp[j\,\arg\{H(z)\}]$$

where

$$|H(z)| = \sqrt{\text{Re}\{H(z)\}^2 + \text{Im}\{H(z)\}^2}$$

$$\arg\{H(z)\} = \arctan\frac{\text{Im}\{H(z)\}}{\text{Re}\{H(z)\}}$$

This latter way of representing a complex variable is called the exponential or polar form.

The complex conjugate of a function of a complex variable, such as $H(z)$, is symbolized by $H^*(z)$ and defined by

$$H^*(z) = \text{Re}\{H(z)\} - j\,\text{Im}\{H(z)\}$$

or in the exponential form as

$$H^*(z) = |H(z)|\exp[-j\,\arg\{H(z)\}]$$

That is, the complex conjugate of a quantity has the same real part, but its imaginary part has the opposite sign. If a complex variable is multiplied by its complex conjugate, the exponential form shows that the angular arguments cancel each other, giving the magnitude squared:

$$H(z)H^*(z) = |H(z)|^2$$

which no longer has an imaginary part.

The power spectral density of a quantity is defined as the product of its Z-transform and the complex conjugate of its Z-transform scaled by the sampling interval h. Strictly speaking, this definition only makes sense when the Z-transform of the quantity of interest exists. From the definition of the Z-transform, one should perceive that stochastic quantities such as white noise sequences have no Z-transform. However, if the Z-transform exists, the power spectral density of Y is therefore defined as

$$P_Y(z) = hY(z)Y^*(z) = h\,|Y(z)|^2$$

Note that, as a consequence of this definition, P_Y has no imaginary part; that is, it is a real variable. Note also that, if the engineering units of the signal Y were volts, the units of the Z-transform $Y(z)$ would also be volts, and the units of the power spectral density would be volts^2-second or volts^2 per hertz.

Since Y can be described in terms of its relationship with X as

$$Y(z) = H(z)X(z)$$

it follows that the power spectral density of Y can be written in terms of its dependence on the power spectral density of X as

$$P_Y(z) = hY(z)Y^*(z)$$

$$= h[H(z)X(z)][H^*(z)X^*(z)] \tag{5.1}$$

$$= [H(z)H^*(z)]h[X(z)X^*(z)]$$

$$= |H(z)|^2 P_X(z)$$

Therefore, the power spectral density of a process output can be determined if the process transfer function and the power spectral density of the process input are known. It is important to note that if the input to the process is a stochastic quantity such as a white noise sequence, which will be shown to have a power spectral density but does not have a Z-transform, Equation (5.1) will still allow one to find the power spectral density of the process output.

For the case where the signals are cascaded, the above expression can be applied in chain-rule fashion. For example, if $Y_3(t)$ is the output of a process whose input is $Y_2(t)$ which in turn is the output of a process whose input is $Y_1(t)$ then, if $H_{32}(z)$ and $H_{21}(z)$ are the appropriate transfer functions, the following can be written:

$$Y_3(z) = H_{32}(z)Y_2(z) = H_{32}(z)H_{21}(z)Y_1(z)$$

$$P_3(z) = |H_{32}(z)|^2 P_2(z) = |H_{32}(z)|^2 |H_{21}(z)|^2 P_1(z)$$

Using the substitution

$$z = \exp(j2\pi fh)$$

these power spectral densities can be related to some of the things that were done in Chapter Three. For example, Equation (5.1) becomes

$$P_Y''(f) = |H''(f)|^2 P_X''(f)$$

where the double primes should remind the reader of the following substitution progression:

$$z \rightarrow \exp(j\Omega) \rightarrow \exp(j2\pi fh)$$

In Section 3.5 it was shown that, for a zero-mean sequence, a sum of the components of the line spectrum components over the Nyquist interval was equal to the variance. In the continuous frequency domain, for a zero-mean stochastic sequence, the integral of the power spectral density over the Nyquist interval is equal to the variance:

$$V_Y = \int_0^{f_{Ny}} 2P_Y''(f) \, df \tag{5.2}$$

(More of a justification for this expression will be given in the next subsection.) Hence the power spectral density can be considered as a continuous distribution of the variance over the Nyquist interval. Taking the derivative of both sides of Equation (5.2) gives

$$\left. \frac{dV_Y(a)}{da} \right|_{a=f} = 2P_Y''(f)$$

which says that $2P_Y''(f) \, df$ is an estimate of the contribution to the variance coming from the frequency range f to $f + df$.

The goal of this subsection has been to present a compact but formal definition of the power spectral density that will suffice for the purposes of this book. The approach was chosen because the definition needed only knowledge of the Z-transform and some complex algebra. In the next subsection, this definition will be complemented with a significantly more involved approach using the discrete-time Fourier transform and the autocorrelation function. Following this alternative definition, a subsection is devoted to the Blackman–Tukey approach to spectral estimation. Neither of these subsections is critical to the overall flow of the book, so it is suggested that the reader give them reduced attention on the first reading so as to not get bogged down in mathematical details.

5.1.2 Power Spectral Density from the Autocorrelation Function

This section contains an alternative definition of the power spectral density that requires a little more mathematics but is generally accepted as a more rigorous approach. Since there is more interest in what the power spectral density can do and in what can be done with the power spectral density, rather than where it came from, the discussion will be relatively superficial.

In Section 1.3.8, the autocorrelation function for a lag k for an infinite sequence $X_1, X_2, X_3, \ldots$, was defined as the expected value of the quantity $X_i X_{i+k}$:

$$A_X(k) = E\{X_i X_{i+k}\}$$

and it was noted that $A_X(0) = V_X$ if X had a zero mean. (Remember that the autocorrelation function is not the same as the autocorrelation that was introduced in Section 1.3.1, where a finite set of data was dealt with.)

The actual values of the indexes used in denoting the sequence X_i are unimportant, so for the sake of being consistent with the digital signal processing literature, let the index take on any integer value, positive, negative, and zero; that is, deal with

$$X_i, \quad -\infty < i < \infty$$

In the alternative definition, the power spectral density is the discrete-time Fourier *transform* of the autocorrelation function. Unfortunately, the discrete-time Fourier transform has not been defined, but a fair amount of time has been spent on the discrete-time Fourier *series pair* in Section 3.10. The discrete-time Fourier transform can be considered as an extension of the discrete-time Fourier series pair for a discrete-time sequence of infinite extent. From Section 3.10, it is known that if the elements of the finite time-domain sequence are

$$X_1, X_2, \quad \ldots, X_N$$

(a subset of the above described infinite sequence) and if the elements of the discrete-time Fourier series coefficients are

$$F_{-n}, F_{-n+1}, \quad \ldots, F_0, F_1, \quad \ldots, F_{n-1}, \quad \text{where } 2n = N,$$

then these two sets of quantities are related according to

$$F_i = F_{\text{DTFS}}\{X_k\} = \frac{1}{N} \sum_{k=1}^{N} X_k \exp\left(\frac{-j2\pi ki}{N}\right)$$

$$X_k = F_{\text{DTFS}}^{-1}\{F_i\} = \sum_{i=-n}^{n-1} F_i \exp\left(\frac{j2\pi ki}{N}\right)$$

Here the finite discrete time-domain sequence $X_k, k = 1, \ldots, N$, is transformed into the finite discrete frequency-domain sequence $F_i, i = -n, -n + 1, \ldots,$ 0, $\ldots, n - 1$, where $N = 2n$.

If a time-domain sequence (of infinite extent) is transformed into the frequency domain, a new quantity $X^F(f)$, the discrete-time Fourier transform,

$$X^F(f) = F\{X_k\} = h \sum_{k=-\infty}^{\infty} X_k \exp(-j2\pi fhk)$$

is obtained, where the index k ranges from $-\infty$ to ∞ (hence some authors refer to the above operation as the *bilateral* discrete-time Fourier transform). If the units of X_k are volts, then the units of $X^F(f)$ are volts-second or volts per hertz. Unlike the discrete-time Fourier series pair, the discrete-time Fourier transform is a continuous function of the frequency f. The notation of Marple (1987), which is a slight deviation from most of the digital signal processing literature (see Oppenheim and Schafer, 1989, for example), is being followed because it will yield a power spectral density that has units of *power per hertz*. The variance, which has units of power, can then be obtained by integrating, with respect to frequency, the power spectral density over the Nyquist interval.

If X_k is zero for $k < 0$, then the discrete-time Fourier transform $X^F(f)$ is the same as the unilateral Z-transform with the change of variable

$$z = \exp(j2\pi fh)$$

and a scaling factor of h; that is,

$$X^F(f) = hX(z)|_{z=\exp(j2\pi fh)}, \qquad X_k = 0, \, k < 0$$

The other member of the discrete-time Fourier transform pair is given by

$$X_k = F^{-1}\{X^F(f)\} = \int_{-f_{Ny}}^{f_{Ny}} \exp(j2\pi fhk)X^F(f) \, df$$

Thus, to transform the infinite sequence X_k into the continuous frequency domain, an infinite sum is used. To transform the quantity $X^F(f)$ back to the time domain, an integral is used rather than a sum, as is the case with the discrete-time Fourier series. Note the similarities with the Z-transform, which used an infinite sum to transform X_k from the discrete time domain to the z-plane and an integral to make the inverse transformation.

With this notation, the alternative definition of the power spectral density as the discrete-time Fourier transform of the autocorrelation function can be written as

$$P_X''(f) = F\{A_X(i)\} = h \sum_{i=-\infty}^{\infty} A_X(i) \exp(-j2\pi fih) \qquad (5.3)$$

The other member of the discrete-time Fourier transform pair is

$$A_X(k) = \int_{-f_{Ny}}^{f_{Ny}} \exp(j2\pi fhk)P_X''(f) \, df \qquad (5.4)$$

When the lag k is set to zero, the autocorrelation function equals the variance (if the stochastic sequence has a zero mean)

$$A_X(0) = V_X = \int_{-f_{Ny}}^{f_{Ny}} P_X''(f) \, df$$

Since the power spectral density is an even function, that is,

$$P_X''(-f) = P_X''(f)$$

it follows that

$$V_X = 2 \int_0^{f_{Ny}} P_X''(f) \, df \qquad (5.5)$$

The definition of the power spectral density given in the previous section has the advantage of making algebraic manipulations apparently more straightforward. However, it leaves a lot to be desired when one is interested in power spectral densities of stochastic sequences that have no Z-transform. Since those types of sequences have autocorrelation functions, which in turn often have discrete-time Fourier transforms, the alternative definition allows the analysis of noisy processes to proceed with a little more comfort. Although Equations (5.3) and (5.4) are being used as a definition of the power spectral density, they can be derived from first principles and are often referred to as the Wiener–Khintchine theorem.

In the case where the power spectral density of a composite variable such as

$$v = X + Y$$

is sought, the first definition of the power spectral density would suggest

$$P_v(z) = hv(z)v^*(z) = h[X(z) + Y(z)][X^*(z) + Y^*(z)]$$

$$= h \mid X(z) \mid^2 + hX(z)Y^*(z) + hX^*(z)Y(z) + h \mid Y(z) \mid^2$$

If X and Y are both white noise sequences and therefore independent of each other, intuition would suggest that $X(z)Y^*(z)$ and $X^*(z)Y(z)$ would drop out. In terms of Z-transforms, there would be no basis for such a supposition. However, using the second definition, one could write

$$P_v''(f) = F\{A_v(i)\}$$

The autocorrelation function of v is the expected value of $v_k v_{k+i}$ or

$$A_v(i) = E\{v_k v_{k+i}\}$$

$$= E\{(X_k + Y_k)(X_{k+i} + Y_{k+i})\} \tag{5.6}$$

$$= E\{X_k X_{k+i}\} + E\{X_k Y_{k+i}\} + E\{Y_k X_{k+i}\} + E\{Y_k Y_{k+i}\}$$

If X and Y are white noise sequences, then the cross-correlation functions $E\{X_k Y_{k+i}\}$ and $E\{Y_k X_{k+i}\}$ would be zero and Equation (5.6) becomes

$$A_v(i) = E\{X_k X_{k+i}\} + E\{Y_k Y_{k+i}\}$$

$$= A_X(i) + A_Y(i)$$

Therefore, the power spectral density for the sum of two white noise sequences is

$$P_v''(f) = P_X''(f) + P_Y''(f)$$

The Z-transform-based definition of the power spectral density will be used in the remainder of this chapter only because it makes algebraic manipulations so much easier. The reader should keep in mind the more rigorous definition when following the manipulations.

5.1.3 Blackman–Tukey Spectral Estimate

Equation (5.3) suggests an alternative to the Fourier line spectrum for estimating the power spectral density when a finite set of discrete time-domain data is being studied. If the autocorrelation function $A_X(k)$ is replaced with the *un*normalized autocorrelation

$$R_X(k) = r_X(k)s_X^2 = \frac{1}{N} \sum_{i=1}^{N-k} [X(i) - \overline{X}][X(i + k) - \overline{X}]$$

then the Blackman–Tukey (BT) estimate of the power spectral density results:

$$P_X^{\mathrm{BT}}(f) = \sum_{k=-L}^{L} R_X(k)q_k \exp(-j2\pi fhk)$$

where q_k are weights that define a *lag window* in a manner similar to the data windows discussed in Section 3.12.3. The maximum lag index L is sometimes chosen to be approximately $N/10$, and there are a variety of windows to chose from.

This approach will not be pursued since the method presented in Chapter Three, which combines the fast Fourier transform with the Fourier line spectrum as derived from the discrete-time Fourier series, gives the most efficient and most widely used estimate of the power spectral density. Also, the BT approach is not as effective in revealing hidden periodic components in noise as is the line spectrum. Kay (1988) compares the line spectrum (which, except for a scaling factor, is the same as the periodogram when the frequencies chosen for computation are the harmonics of the fundamental) and the BT approach for several different kinds of time series in Section 4.6 of his text.

5.1.4 Short Review of All the Transforms Used So Far

Counting the discrete-time Fourier transform, which has just been used in Section 5.1.3, four different kinds of transforms have been dealt with in this book. For those readers who are practicing engineers and statisticians and who are not daily users of applied mathematics, trying to keep track of all these transforms may be a bit discouraging. To help put things in perspective, the four transforms are compared from the following points of view:

1. The nature of the time domain where the data originates (discrete or continuous)
2. The extent of the time domain (finite or infinite)
3. The functional form of the time domain data (empirical or analytical)
4. The nature of the frequency domain to which the time-domain data are transformed (discrete or continuous)
5. The main use (in this book) of the transform

The discrete-time Fourier series pair, introduced in Chapter Three, provides a way of changing a *finite* sequence of data (of length N) in the *discrete* time domain into a *finite* sequence of harmonic strengths in the *discrete* frequency domain. It was shown that this transform could be used to study experimental data as well as the results of simulations. The magnitudes squared of the harmonic strengths make up the components of the Fourier line spectrum. As a consequence of the discrete-time Fourier series approach, the time-domain data can also be looked at as being infinite in extent but with period N.

The unilateral Z-transform, introduced in Chapter Four, provides a way of changing a function represented by an *infinite* sequence in the *discrete* time domain into a new function defined on the *continuous* complex z-plane. It was shown that this transform could be applied to time-domain functions whose analytical form is known, as compared to the discrete-time Fourier series, which could be applied to empirically determined quantities. To get the frequency response associated with this time-domain function whose analytical form is known, the domain of definition of the Z-transform is constrained to be a unit circle in the complex z-plane by a change of variable. In Chapter Four the Z-transform was applied to a wide variety of problems. In this chapter it was used to develop the power spectral density, and it will be used in the rest of the chapter to aid in developing working equations.

The companion to the Z-transform is the unilateral Laplace transform, which provides a way of changing a function defined on the *continuous* time domain into a new variable defined on the *continuous* complex s-plane. The Laplace transform was shown to be useful in changing time-domain differential equations into s-domain algebraic equations and in developing transfer functions in a manner analogous to the Z-transform. To get the frequency response associated with the continuous time-domain function, the domain of definition of the Laplace transform is constrained to be the imaginary axis in the s-plane. In this book the primary interest is in the discrete time domain, so the unilateral Laplace transform has received little attention.

The bilateral discrete-time Fourier transform, introduced in this chapter, provides a way of changing a function represented by an *infinite* sequence defined on the *discrete* time domain into a new function defined on the *continuous* frequency domain. It was shown that this transform could be useful in defining the power spectral density.

5.2 POWER SPECTRAL DENSITY OF STOCHASTIC PROCESSES DRIVEN BY WHITE NOISE

In Section 1.3 and in Chapter Three, autoregressive and moving average stochastic sequences were discussed. The zero-mean autoregressive stochastic sequence Y_k was described in the time domain as

$$Y_k = \alpha Y_{k-1} + w_k$$

where w_k represents a zero-mean white noise sequence with a variance V_w.

The power spectral density for white noise is constant; that is, all frequencies in the Nyquist interval are represented by equal power. Therefore, if $P''_w(f)$ is the power spectral density of white noise, then

$$P''_w(f) = A$$

where A is a constant that can be determined from the requirement that the integral of this power spectral density must equal the variance:

$$V_w = 2 \int_0^{f_{Ny}} A \, df = 2Af_{Ny} \rightarrow A = \frac{V_w}{2f_{Ny}}$$

$$P''_w(f) = \frac{V_w}{2f_{Ny}} = hV_w$$

The Z-transform of the defining equation for the autoregressive sequence is

$$Y(z) = \alpha z^{-1} Y(z) + W(z)$$

where $W(z)$ represents the Z-transform of the white noise sequence even though it does not exist. Solving for $Y(z)$ gives

$$Y(z) = \frac{W(z)}{1 - \alpha z^{-1}} = H(z)W(z)$$

$$H(z) = \frac{1}{1 - \alpha z^{-1}}$$

where $H(z)$ represents the transfer function between w_k and Y_k.

Using the chain rule, the power spectral density of Y_k can be determined as a function of $H(z)$ and the variance of the white noise sequence that generates Y_k:

$$P_Y(z) = |H(z)|^2 P_w(z)$$

$$= \left| \frac{z}{z - \alpha} \right|^2 \frac{V_w}{2f_{Ny}}$$

$$= \left| \frac{z}{z - \alpha} \right|^2 hV_w$$

If z is replaced by $\exp(j2\pi fh)$ and some moderately tedious algebra is carried out, one comes up with

$$P''_Y(f) = \frac{V_w/(2f_{Ny})}{1 - 2\alpha \cos(2\pi fh) + \alpha^2} \tag{5.7}$$

$$= \frac{hV_w}{1 - 2\alpha \cos(2\pi fh) + \alpha^2}$$

The reader should note that a similar function [Equation (4.89)] was derived in Section 4.4 for the frequency response of a first-order process.

Consider a moving average stochastic sequence

$$Y_k = w_k + \beta w_{k-1}$$

After transforming to the Z-domain, the above expression becomes

$$Y(z) = (1 + \beta z^{-1})W(z) = H(z)W(z)$$

$$H(z) = 1 + \beta z^{-1}$$

Now that the transfer function for the moving average stochastic sequence has been determined, the power spectral density can be written as

$$P_Y(z) = |H(z)|^2 P_w(z)$$

$$= \frac{|1 + \beta z^{-1}|^2 V_w}{2f_{Ny}}$$

$$= \frac{(1 + \beta z^{-1})(1 + \beta z)V_w}{2f_{Ny}}$$

$$= \frac{[1 + \beta(z + z^{-1}) + \beta^2]V_w}{2f_{Ny}}$$

Replacing z by $\exp(j2fh)$ takes the analysis to the frequency domain:

$$P_Y''(f) = \frac{[1 + 2\beta \cos(2\pi fh) + \beta^2]V_w}{2f_{Ny}}$$

Some liberties have been taken in using the Z-transform of white noise even though it does not exist. The reader interested in more rigor is directed toward the text by Jenkins and Watts (1968). In any case, this example illustrates how the power spectral density of a process variable can be determined formally from the transfer function and the power spectral density of the process input.

In general, twice the integral of the power spectral density over the Nyquist interval equals the variance; however, in the case of the autoregressive stochastic sequence, there is a simpler way to obtain an expression for its variance and it was used in Section 1.3.2 to obtain

$$V_Y = \frac{V_w}{1 - \alpha^2}$$

Note that α must be less than unity, and as α approaches unity, or as the autoregressive sequence approaches becoming a random walk, the variance increases rapidly. When α equals unity and the sequence becomes a random walk, there is no expression for the variance since the sequence is nonstationary.

For the sake of variety, the variance of the moving average stochastic sequence will be derived as follows:

$$V_Y = 2 \int_0^{f_{Ny}} P_Y''(f) \, df$$

$$= \frac{2V_w}{2f_{Ny}} \int_0^{f_{Ny}} [1 + 2\beta \cos(2\pi fh) + \beta^2] \, df$$

The second term in the integrand vanishes, leaving

$$V_Y = (1 + \beta^2)V_w$$

5.3 WHITE NOISE VARIANCE RATIO

The goal in this section is to derive an expression relating the variance of a zero-mean white noise process input, w, to the variance of the zero-mean process output, Y, where the type of process has yet to be specified. This can be done by combining three equations that have been developed above,

$$P_Y''(f) = | H''(f) |^2 P_w''(f)$$

$$P_w''(f) = \frac{V_w}{2f_{Ny}} = hV_w$$

$$V_Y = 2 \int_0^{f_{Ny}} P_Y''(f) \, df$$

arriving at

$$\frac{V_Y}{V_w} = \frac{1}{f_{Ny}} \int_0^{f_{Ny}} | H''(f) |^2 \, df \tag{5.8}$$

which gives the ratio of the variance of the process output Y to the variance of the process input X when that input is white noise. This equation is effectively the average value of $| H''(f) |^2$ over the Nyquist interval.

If the process is enlarged to consist of the components within the dotted lines in Figure 5.1, then the process input is N (which for the time being is white noise), and the process output is E. The transfer function $H''(f)$ is replaced with $E_T''(f)$, which is one of the noise transmission functions. The power spectral density for E is given by

$$P_E''(f) = | E_T''(f) |^2 hV_w \tag{5.9}$$

which says that the power spectral density for E is directly proportional to the absolute value of the error transmission function squared. A plot of $| E_T''(f) |^2$ versus frequency should therefore give a curve having a shape similar to that of the line spectrum based on empirical data.

The white noise variance ratio for this case gives the amount by which a white noise disturbance is amplified by a controlled process, thereby providing a tool for answering some of the questions posed in Chapter Three.

A graphical interpretation of Equation (5.8) perhaps can give a better feel for what a control algorithm must do in the face of white noise stochastic disturbances. The variance ratio is equal to the area under the $| H''(f) |^2$ or $| E_T''(f) |^2$ curve over the Nyquist interval divided by f_{Ny}. Assume that for some combination of controller and process the $| E_T''(f) |^2$ curve is unity over the entire Nyquist interval. Consequently, the area under the $| E_T''(f) |^2$ curve will be $1 * f_{Ny}$

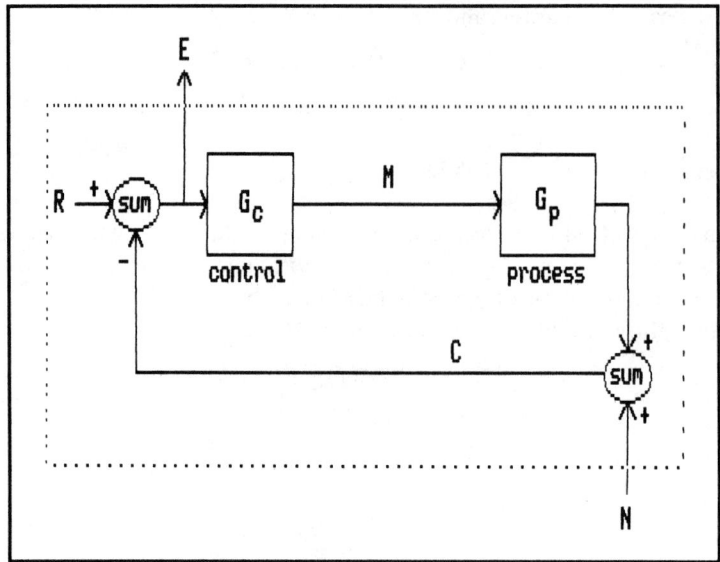

Figure 5.1 Input/output concept of E/N.

and, since the area is divided by f_{Ny}, the variance ratio will be unity. The variance ratio can never be less than unity, so the area under the $|E_T''(f)|^2$ curve can never be less than f_{Ny}. Thus it appears that, whatever controller is causing this type of $|E_T''(f)|^2$ behavior, it must be optimal. However, most control algorithms will contain an integral component (or something equivalent to the integral component) that will remove all power at zero frequency, that is, remove offsets and make the error sequence have a zero mean. The algorithm will probably also attenuate power at frequencies near zero. Therefore, to keep the area under the $|E_T''(f)|^2$ curve just above f_{Ny}, the $|E_T''(f)|^2$ curve will have to be above unity at some other portion of the Nyquist interval.

5.4 NOISE TRANSMISSION CURVES

As was shown in Chapter Four, the noise transmission ratios have two members. The first, E/N or E_T, provides information on how the controlled process translates disturbances represented by N into E, which is the difference between the set point and the controlled variable. The second, M/N or M_T, provides information on how the controlled process translates disturbances into manipulated variable activity. These two transfer functions, when their absolute values are plotted against frequency, will providing a method for determining how a disturbance with a certain spectral distribution of variance will affect the process under control. That is, will the control algorithm be able to attenuate the disturbance, will it amplify it, or will it pass it unattenuated? Likewise, how will this disturbance affect the activity of the manipulated variable? If the disturbance is white noise, these curves will show how the presence of white noise will affect the controlled variable and the manipulated variable. This was a question that was addressed from a different viewpoint in Chapters Two and Three. Also, as mentioned at the end of the last section, a plot of $|E_T''(f)|^2$ versus frequency will give a continuous analog of the line spectrum associated with the controlled variable.

For conventional feedback control loop strategies, the two components of the noise transmission ratios have already been derived in Section 4.1:

$$\frac{E(z)}{N(z)} = -\frac{1}{1 + G_c G_p} = E_T(z)$$

$$\frac{M(z)}{N(z)} = -\frac{G_c}{1 + G_c G_p} = M_T(z)$$

By making the substitution $z = \exp(j\Omega)$ followed by $\Omega = 2\pi f h$, these ratios become

$$\frac{E''(f)}{N''(f)} = E_T''(f) = -\frac{1}{1 + G_c''(f)G_p''(f)}$$

and

$$\frac{M''(f)}{N''(f)} = M_T''(f) = -\frac{G_c''(f)}{1 + G_c''(f)G_p''(f)}$$

where the double prime indicates that the argument f is a result of two steps of variable transformation: from z to Ω to f.

A plot of $|E_T''(f)|$ versus frequency could be used to determine how the control algorithm characterized by G_c would deal with a disturbance having harmonic intensity at any frequency f in the Nyquist interval. A value of $|E_T''(f)|$ less than unity means that disturbances having harmonic intensity at frequency f would be attenuated by the controlled process. Conversely, a value of $|E_T''(f)|$ greater than unity means that the disturbances having harmonic intensity at frequency f will be amplified. Finally, a value of $|E_T''(f)|$ equal to unity means that the disturbance is being passed through the system unattenuated.

In general, $|E_T''(f)|$ would be expected to be significantly smaller than unity near zero frequency simply because a control algorithm with an integral component would remove an offset that is a disturbance, having its harmonic intensity at zero frequency.

At frequencies near the Nyquist frequency, $|E_T''(f)|$ might be expected to be greater than unity for processes with small time constants relative to the control interval, since the process would not be able to act as a low-pass filter. Conversely, for large time constant processes, that is, processes that act as low-pass filters, $|E_T''(f)|$ would be expected to be nearly equal to unity.

If the control algorithm is tuned poorly or if the deadtime-to-time constant ratio is significant, one would expect to see peaks in the plot of $|E_T''(f)|$ versus

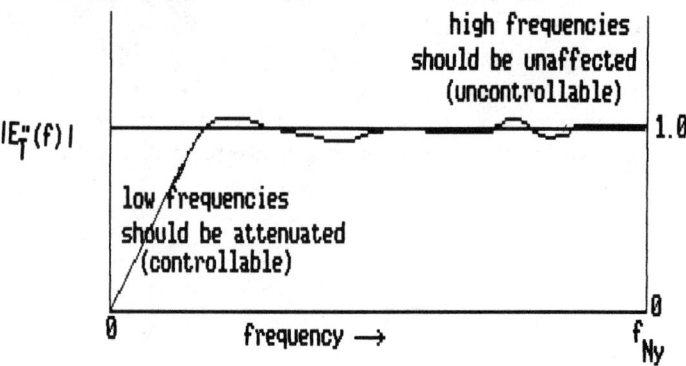

Figure 5.2 Generic $E_T(f)$ curve.

frequency, indicating that disturbances having harmonic intensity in certain frequency bands would be amplified. Instability occurs when the height of the peaks become unbounded. Some of the foregoing comments are illustrated in Figure 5.2.

Finally, as pointed out in the last section, the integral of $\mid E''_T(f)\mid^2/f_{Ny}$ over the Nyquist interval will give the ratio of the variance of the controlled variable to that of the disturbance if it is white noise:

$$\frac{V_E}{V_N} = \frac{1}{f_{Ny}} \int_0^{f_{Ny}} \mid E''_T(f)\mid^2 \, df \qquad (5.10)$$

A similar analysis can be carried out for the case where the process input is N and the process output is M. Here one would be able to analyze the effect that the disturbance has on the activity of the manipulated variable. A plot of $\mid M''_T(f)\mid$ versus frequency should show which frequencies in the manipulated variable will be excited by the disturbance. The integral of $\mid M''_T(f)\mid^2/f_{Ny}$ over the Nyquist interval will give the ratio of the variance of the manipulated variable, V_m, to that of the white noise disturbance. This ratio can be useful in comparing control algorithms and has the following form:

$$\frac{V_M}{V_N} = \frac{1}{f_{Ny}} \int_0^{f_{Ny}} \mid M''_T(f)\mid^2 \, df \qquad (5.11)$$

Equations (5.10)) and (5.11) will be used repeatedly in what follows, so a few comments on computational matters is pertinent. The integrands in these equations, $E_T(z)$ and $M_T(z)$, are usually rather straightforward functions of z. Since the user knows the form of the transfer functions, such as $G_c(z)$ and $G_p(z)$, which frequently occur in $E_T(z)$ and $M_T(z)$, their construction is relatively easy. However, when the transition to the frequency domain is made using

$$z = \exp(j2\pi fh) = \cos(2\pi fh) + j\sin(2\pi fh)$$

the integrands can become complicated algebraic relations containing real and imaginary parts.

To numerically evaluate the integrals in Equations (5.10) and (5.11), one should use a computer language such as FORTRAN that has complex variables. In this case, one can avoid developing the algebra to the point where the real and imaginary parts of the integrand are derived in closed form. Instead, it is simpler to evaluate each component transfer function as a complex quantity and plug the results into the integrand.

If one uses a language such as QUICKBASIC, which has been used exclusively here and which does not have complex variables but has attractively simple graphics ability, then it is usually simpler to write subroutines that carry out complex multiplication and division and use these to calculate the real and imaginary parts of each component transfer function. The results of computations can be inserted into the integrand just as with FORTRAN, except that the real and imaginary parts must be tracked explicitly. Once the integrand is computed, the integral can be evaluated using the trapezoid rule. By far the simplest approach is to use one of the higher-level languages such as MATLAB™ or MATHE-MATICA™ that not only deal in complex variables but provide means for integral evaluation.

5.4.1 Noise Transmission Curves and the White Noise Sigma Ratio for the FOWDT Model with PI Control

Figure 5.3 shows $| E_T''(f) |$ plotted against frequency for the case where the process is represented by

$$G_p(z) = \frac{Bz^{-1-d}}{1 - Az^{-1}}$$

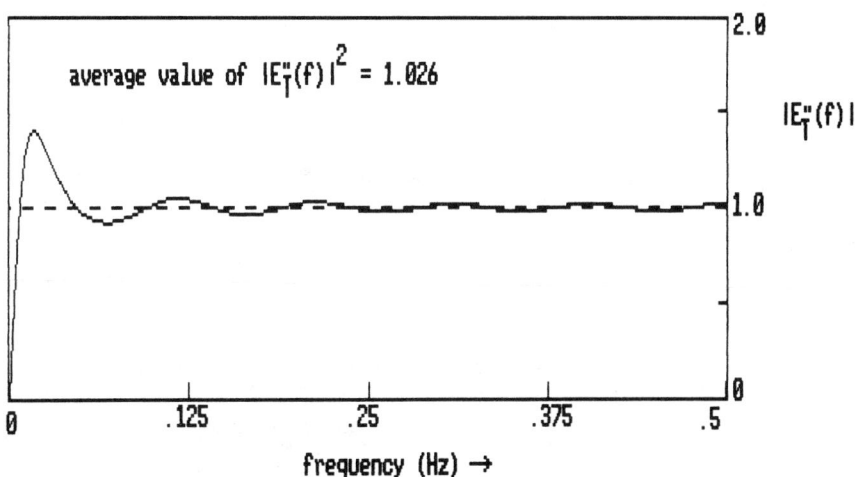

Figure 5.3 $| E_T |$ curve of model 7 ($T = 40$, $D = 10$).

and the control algorithm is represented by

$$G_c(z) = P + \frac{Ih}{1 - z^{-1}}$$

The sampling/control interval is 1 second, the time constant is 40 seconds, the deadtime is 10 seconds, and the process gain is 2 (model 7). The reader will remember this example process was discussed in Section 3.9.3 where it was subjected to white noise. There and here the proportional gain is 0.71 and the integral gain is 0.0175. Evaluation of Equation (5.10) gives a variance ratio of 1.026, which says that the impact of white noise on the controlled variable is to increase its standard deviation over that of white noise by a factor of 1.013. This latter quantity, the square root of the variance ratio, will be called the *white noise sigma ratio*. Note in Figure 5.3, which is plotted on linear axes, how little of the Nyquist interval contains frequencies that are significantly attenuated; that is, the effect of the control algorithm is to attenuate disturbances in only a small portion of the Nyquist interval near zero frequency.

Figure 5.4 shows a plot of $| E_T''(f) |^2$ versus frequency, which can be compared to Figure 3.78, which gave the average line spectrum for 15 simulations. The shape of the average line spectrum in Figure 3.78, although noisy, is similar to that in Figure 5.4, which is not unexpected since, according to Equation (5.9), $| E_T''(f) |^2$ is proportional to the power spectral density of the controlled variable when it is subject to white noise. Note, however, that the noise in Figure 3.78 obscures the ripple, which shows up clearly in Figure 5.4.

Figure 5.5 shows the log–log version of Figure 5.3, where the low-frequency region is expanded. This type of plot will be useful later when the effect of changing a parameter is investigated.

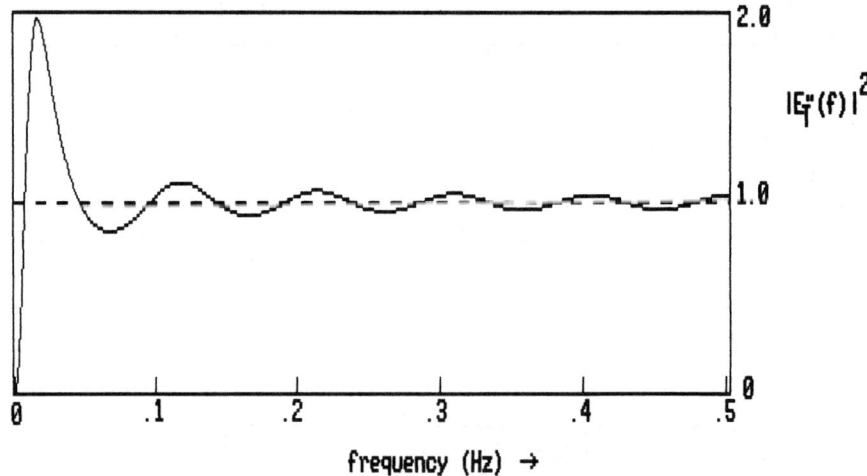

Figure 5.4 $| E_T(f) |^2$ curve for model 7.

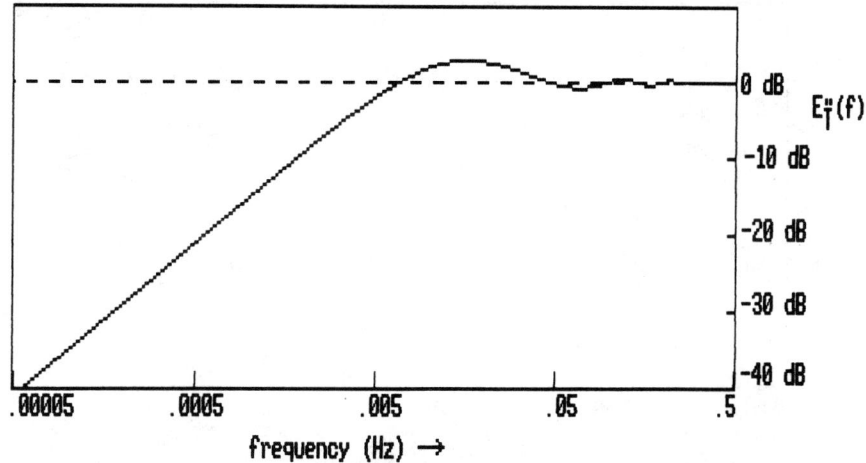

Figure 5.5 $|E_T|$ for model 7 on log–log scale.

The $|M_T''(f)|$ plot in Figure 5.6 indicates that, as the frequency increases, the activity of the manipulated variable settles out. This will not be the case when the use of the derivative component is discussed later.

Figure 5.7 shows the effect of changing the control gains by a factor of 2. When the control gains are doubled, there is significantly more attenuation at lower frequencies at the expense of a higher peak at approximately 0.01 Hz, and the white noise sigma ratio is increased to 1.04. When the control gains are halved, there is much less attenuation at the lower frequencies and a lower peak, and the white noise sigma ratio is decreased to 1.005. Therefore, more aggressive control

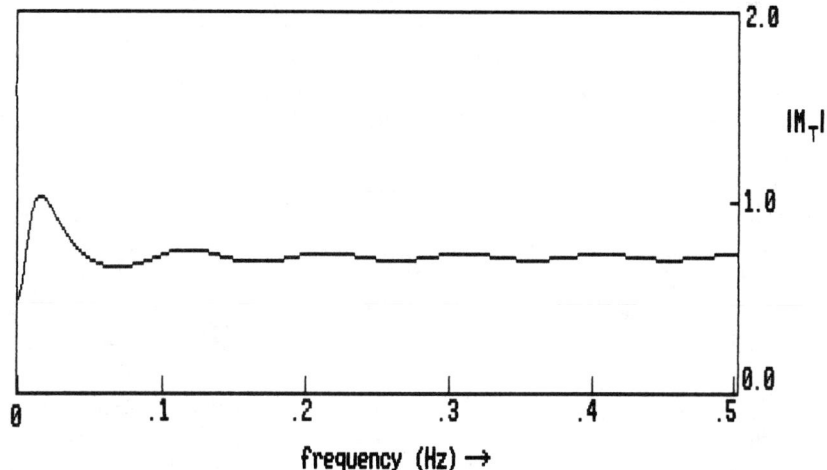

Figure 5.6 $|M_T|$ curve for PI control applied to model 7.

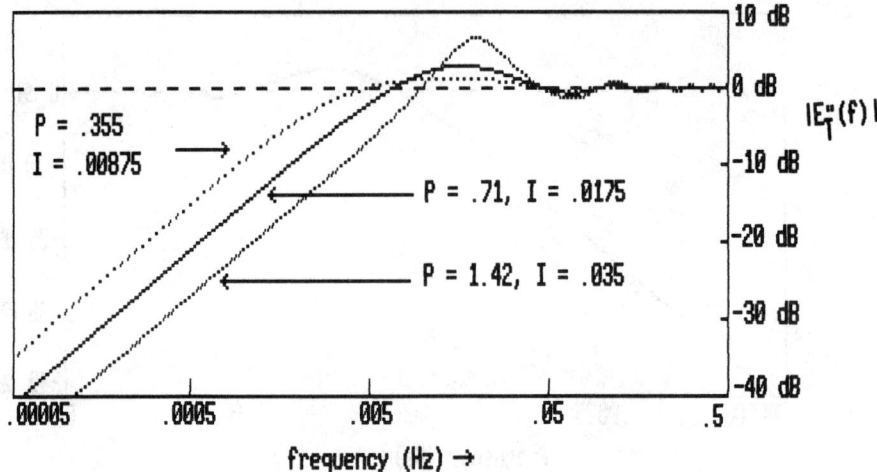

Figure 5.7　Effect of changing control gains by 100%.

gains cause more low-frequency attenuation, higher peaks, and higher amplification of white noise. The converse is true for more conservative control gains. The problem will be revisited when this controlled system is subjected to heavily autocorrelated noise.

5.4.2 Overcontrol

In Section 2.10, an example process (model 6) with a time constant of 0.5 second, a gain of 2.0, and no deadtime was discussed. Integral-only control was applied at 1-second intervals to the model when it was subjected to both white and autocorrelated noise. When the integral gain was chosen to be the reciprocal of the process gain, the autocorrelation showed a strong negative value at lag 1, which was suggested as an indicator of overcontrol. Later, in Section 3.9.3, the same process model was analyzed by using the line spectrum, where it was shown that the effect of overcontrol in the frequency domain was to amplify the higher frequencies. These approaches were empirical in the sense that a simulation was run, samples were gathered, and the autocorrelation or line spectrum was calculated from these samples. In this section, the power spectral density, which can be determined directly from the transfer function of the controlled process, will be dealt with.

　　Figure 5.8 shows how the $|E_T|$ curve changes when the integral control gain is given three different values. For the first case, $I = 0.5$, which is the reciprocal of the process gain. Here the curve has a shape similar to that of Figure 3.76 except that it does not suffer from the noise that results from sampling a simulation. The white noise sigma ratio for this gain is 1.344, so the effect of the control action is to amplify the white noise by about 34%, which corresponds to the observations made in Section 2.10.

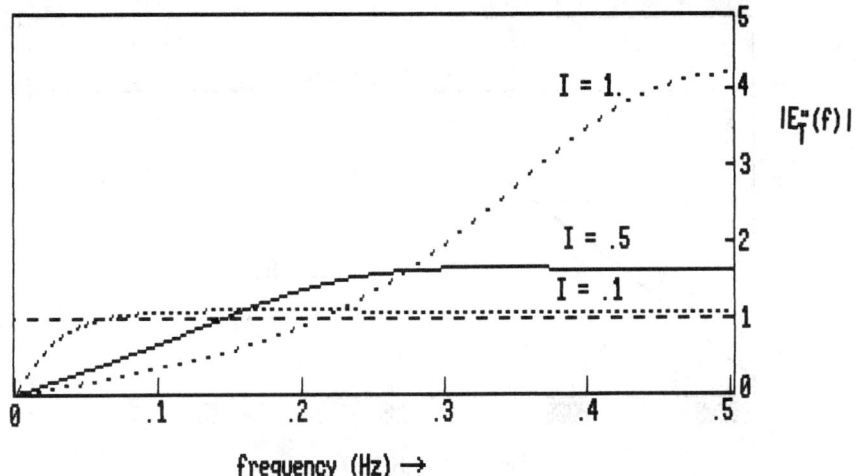

Figure 5.8   Effect of changing integral control gain for model 6.

Figure 5.8 shows the $|E_T|$ curve for the case where the integral gain is lowered to 0.1. Here the white noise sigma ratio is 1.001. Therefore, the white noise is effectively not amplified at all. Also, the curve shows that, except for a narrow band of frequencies near zero, most of the frequencies are passed without attenuation or amplification. This means that the control gain of 0.5 is significantly better in attenuating the low frequencies. However, it also shows that, in the face of white noise, integral-only control with a conservative control gain can be applied without developing an overcontrol situation. When the control gain is set to 1.0, the improvement in the low-frequency attenuation is more than offset by the excessive amplification of the higher frequencies, and the white noise sigma ratio is 2.279.

5.4.3 Effect of Filtered and Unfiltered Derivative Control

As a vehicle for investigating the benefits and detriments of derivative control, the simple third-order model (model 4) that was used in Section 2.6.3 will be used. This model's equation was developed in Section 4.3.1 and has time constants of 25, 26, and 27 seconds and a process gain of 2.0. Figure 5.9 shows three $|E_T''(f)|$ curves plotted on log–log scales in order to accentuate their differences at low frequencies. In the first curve, labeled PI, simple PI control is applied with control gains of $P = 0.54, I = 0.009$. In the second curve, labeled PID, derivative control is added with control gains of $P = 1.1, I = 0.012, D = 15$. An improvement in attenuation in the lower frequencies is seen, but for frequencies greater than 0.05 Hz there is no attenuation for either PI or PID. In the third curve, labeled PIfD, a filtering coefficient of 0.1 is applied to the derivative term and no other changes to the gains are made. At low frequencies this last curve is barely dis-

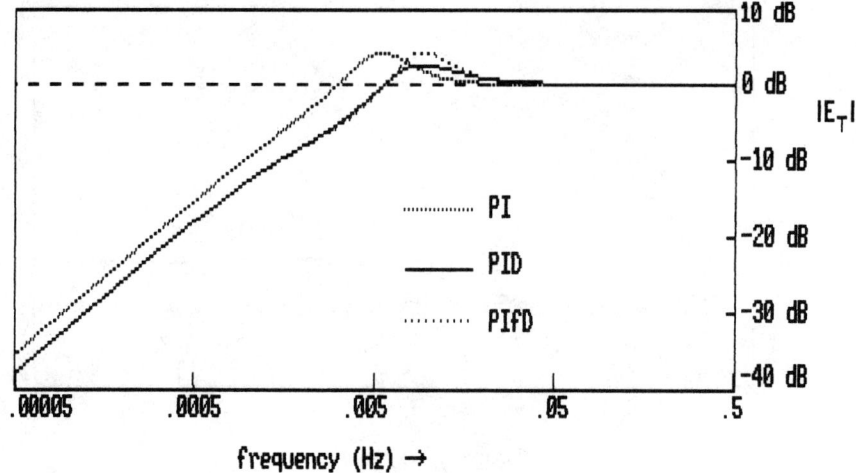

Figure 5.9   PI, PID, and PIfD applied to model 4.

tinguishable from the PID curve; however, near 0.005 Hz, PIfD amplifies disturbances more than PID. At higher frequencies, PID and PIfD are again indistinguishable.

Figure 5.10 shows the three corresponding $| M_T''(f) |$ curves also plotted on
log–log scales. Here the difference between PI and PID is dramatic. The price
one pays for the addition of derivative is a significant increase in the manipulated
variable activity at the higher frequencies. This figure also shows that for disturbances with power near the Nyquist frequency the activity of the manipulated
variable, when derivative is used, is amplified by almost 30 dB relative to that of

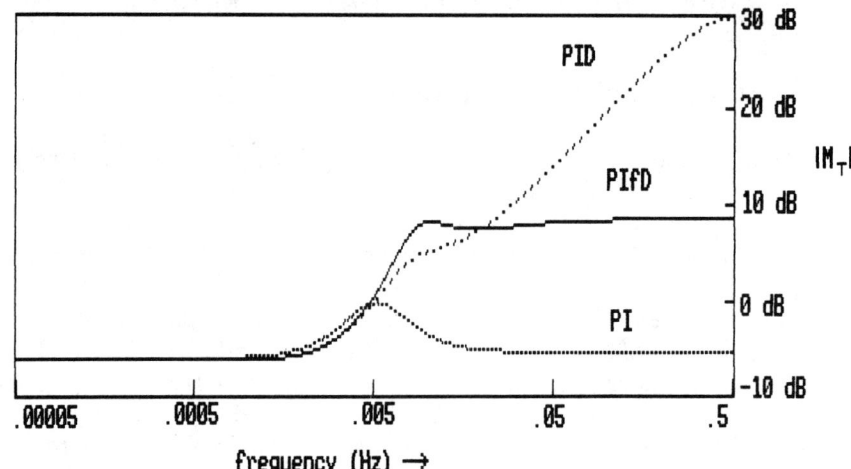

Figure 5.10   PI, PID, and PIfD applied to model 4.

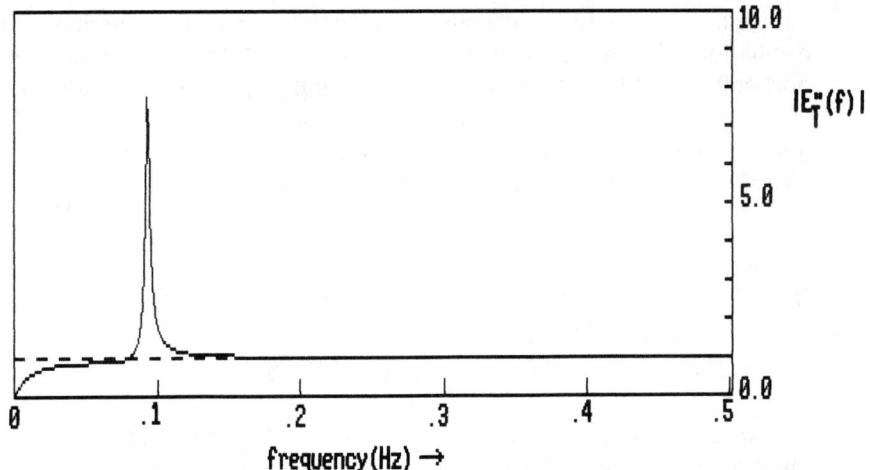

Figure 5.11   PI control applied to an underdamped process.

PI control. The effect of the filtered derivative on the activity of the manipulated variable is significantly less than that of the unfiltered PID, but it is greater than that for PI.

5.4.4 Noise Transmission Curves for Underdamped and Inverse Responding Processes

Using the same control gains as in Section 2.6.6 and the process transfer functions developed in Section 4.3.1, the noise transmission curves are shown in Figures 5.11 and 5.12. For the underdamped process there is a sharp peak at the frequency

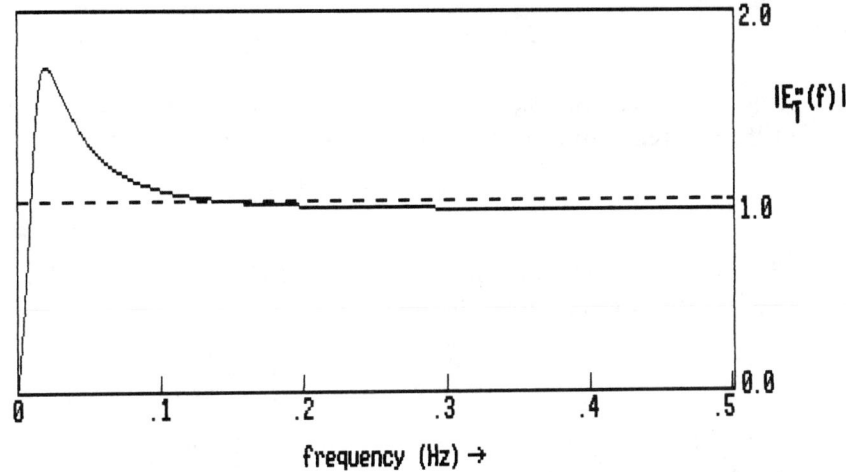

Figure 5.12   PI applied to an inverse responding process.

having a period of 10.89 seconds, which corresponds to the period of the damped oscillation. This means that any disturbance with harmonic power in the neighborhood of that frequency will be significantly amplified. The noise transmission curve for the inverse responding process simply indicates that such a process under PI control will have a difficult time controlling disturbances having a significant amount of low-frequency harmonic power.

5.5 NOISE TRANSMISSION CURVES FOR PROCESSES SUBJECT TO AUTOREGRESSIVE AND NONSTATIONARY DISTURBANCES

The approach in this section is similar to that of Section 5.2, except that now the input to the enlarged process in Figure 5.1 is an autocorrelated disturbance that can be stationary or nonstationary. If the disturbance is stationary and autoregressive, the noise sequence has a power spectral density, derived in Section 5.2 [Equation (5.7)], of

$$P_N''(f) = \frac{[V_w/(2f_{Ny})]}{1 - 2\alpha \cos(2\pi f h) + \alpha^2} \tag{5.12}$$

By combining Equation (5.12) with Equation (5.1), the power spectral density of the controller error can be written as

$$P_E''(f) = \frac{|E_T''(f)|^2[V_w/(2f_{Ny})]}{1 - 2\alpha \cos(2\pi f h) + \alpha^2} \tag{5.13}$$

The variance of E can be obtained from Equation (5.13) by using Equation (5.12):

$$V_E = 2\frac{V_w}{(2f_{Ny})} \int_0^{f_{Ny}} \frac{|E_T''(f)|^2}{1 - 2\alpha \cos(2\pi f h) + \alpha^2} \, df \tag{5.14}$$

If there were no control, the variance of the control error would be equal to that of the autoregressive disturbance, which was shown in Section 1.3.2 to be

$$V_N = \frac{V_w}{1 - \alpha^2} \tag{5.15}$$

Therefore, by combining Equations (5.14) and (5.15) to eliminate V_w, the variance ratio for the control error is found to be

$$\frac{V_E}{V_N} = \frac{1 - \alpha^2}{f_{Ny}} \int_0^{f_{Ny}} \frac{|E_T''(f)|^2}{1 - 2\alpha \cos(2\pi f h) + \alpha^2} \, df \tag{5.16}$$

which says that the variance ratio is effectively equal to the average over the Nyquist interval of an appropriately modified error transmission function multiplied by $1 - \alpha^2$. The square root of this ratio would give the standard deviation or sigma ratio. A similar equation can be derived for V_M/V_N:

$$\frac{V_M}{V_N} = \frac{1 - \alpha^2}{f_{Ny}} \int_0^{f_{Ny}} \frac{|M_T''(f)|^2}{1 - 2\alpha \cos(2\pi f h) + \alpha^2} \, df \tag{5.17}$$

Since the variance of the manipulated variable for a nonstationary disturbance is not defined, an expression for the variance ratios will not be developed. However, expressions for the variance of the controller error can be derived that have a form similar to that of Equation (5.14). Consider the nonstationary sequence presented in Section 1.3.4 as

$$x_k = (1 + \alpha)x_{k-1} + \alpha x_{k-2} + w_k + \beta w_{k-1} \tag{5.18}$$

Formally applying the Z-transform to Equation (5.18) gives

$$\frac{X(z)}{W(z)} = H(z) = \frac{1 + \beta z^{-1}}{1 - (1 + \alpha)z^{-1} - \alpha z^{-2}}$$

and the power spectral density of the sequence can be written as

$$P_x(z) = |H(z)|^2 P_w$$

$$= |H(z)|^2 h V_w$$

or, in the frequency domain,

$$P_x''(f) = |H''(f)|^2 h V_w$$

If the controlled system is subject to a disturbance described by Equation (5.18), the power spectral density of the controller error can be written as

$$P_E''(f) = |E_T''(f)|^2 |H''(f)|^2 h V_w \tag{5.19}$$

and the variance of the controller error can be written as

$$V_E = 2 \int_0^{f_{Ny}} P_E''(f) \, df$$

$$= 2 \int_0^{f_{Ny}} |E_T''(f)|^2 |H''(f)|^2 h V_w \, df \tag{5.20}$$

To compare different control algorithms, one can compare (1) plots of $P_E''(f)$ over the Nyquist interval and (2) values of the variances V_E.

Even though the disturbance is nonstationary, Equations (5.19) and (5.20) are useful because it is reasonable to expect that the control algorithm, with an integral component or something equivalent, will be able to remove power at zero frequency and make the controller error stationary. The same cannot be said about the manipulated variable, because in making the controller error stationary the manipulated variable becomes nonstationary. However, there are at least two possible approaches. First, even though the manipulated variable is nonstationary, an expression similar to Equation (5.19) can be written for the manipulated variable having the form

$$P''_M(f) = |M''_T(f)|^2 |H''(f)|^2 h V_w \qquad (5.21)$$

$P''_M(f)$ is not defined at $f = 0$, but it can be evaluated for

$$0 < f \leq f_{Ny}$$

and can be plotted over most of the Nyquist interval. As will be shown in Section 5.9, a partial variance V'_M can be computed by integrating $P''_M(f)$ over this subportion of the Nyquist interval. Therefore, control algorithms can be compared by graphing $P''_M(f)$ over some selected portion of the Nyquist interval. If $P''_M(f)$ were to be plotted against the logarithm of the frequency, as in Figure 5.10, part of this problem could be bypassed.

Alternatively, even though M is nonstationary, $(1 - z^{-1})M(z)$ may not be, especially if the disturbance has the form given in Equation (5.18). Thus one could compare control algorithms by studying the power spectral density and variance of ΔM.

The balance of this section will contain examples of how Equations (5.16) and (5.17) can be applied for stationary disturbances. Later, in Sections 5.8 and 5.9, the equations for nonstationary disturbances will be used when the Box–Jenkins control algorithms are given another look.

5.5.1 Short Time Constant Process

The first example process (model 6) is the one with the short time constant ($T = 0.5$ second, $I = 0.5$) that was used to illustrate the concept of overcontrol in Sections 2.10.1 and 5.4.2. For the case when the autoregressive coefficient is 0.95, Figure 5.13 shows a plot of $|E''_T(f)|^2$, which, as has been earlier suggested, is proportional to the line spectrum. When the proportional gain is equal to 0.1, the shape of the $|E''_T(f)|^2$ curve is relatively flat, except for the strong attenuation of the frequencies near zero due to the integral component, and the standard deviation ratio associated with it is 0.317. When $P = 0.0$, Equation (5.16) gives a standard deviation ratio of 0.321, while it yields 0.333 when $P = 0.2$. This example supports the contention that the minimum standard deviation is associated with the conditions that make the error curve the "whitest."

When the disturbance is white noise, the $|E_T|$ plot, shown in Figure 5.14, indicates that the smallest standard deviation ratio is obtained when $P = 0.0$. This example illustrates the point that the best gains depend on the nature of the disturbance. Proportional control is beneficial when the disturbance is heavily autocorrelated because the controlled variable spends much of its time drifting toward or away from the target, and there is a need for the braking effect that the proportional component can provide. On the other hand, when the disturbance is white noise, the deviation of the controlled variable from its target is, by definition, never dependent on past deviations, and so there is no drifting and therefore no need for a braking effect.

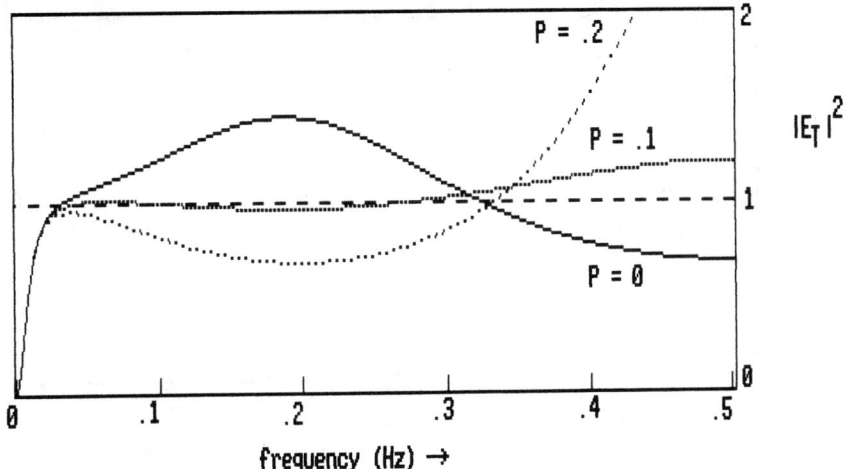

Figure 5.13 PI control of model 6 subject to autoregressive disturbances.

5.5.2 Long Time Constant Process

This section can be compared with Section 5.4.1 where the effect of white noise on model 7 while under PI control was studied. Here, the same system will be subjected to heavily autocorrelated noise (autoregressive coefficient equal to 0.999, nearly a random walk). Figure 5.15 shows the $\mid E_T''(f) \mid$ curve plotted for the first fifth of the Nyquist interval using linear axes for three sets of control gains. The linear axes were chosen in order to provide a comparison with the Fourier line spectrum that one might compute from experimental data.

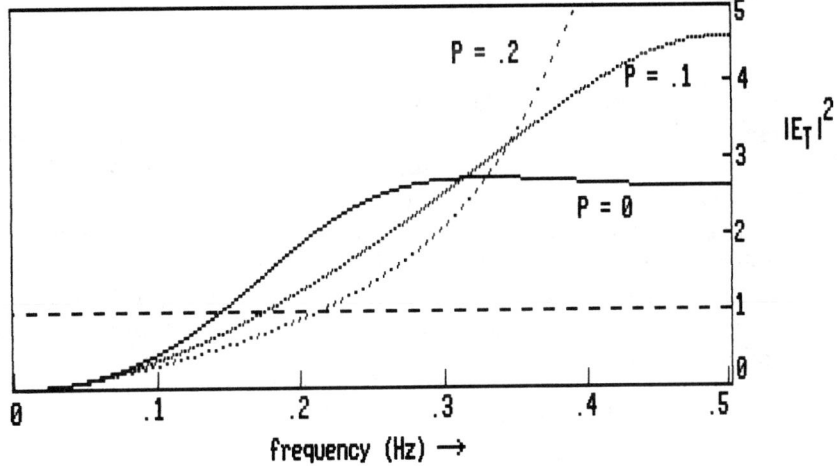

Figure 5.14 Effect of proportional gain on model 6 subject to white noise.

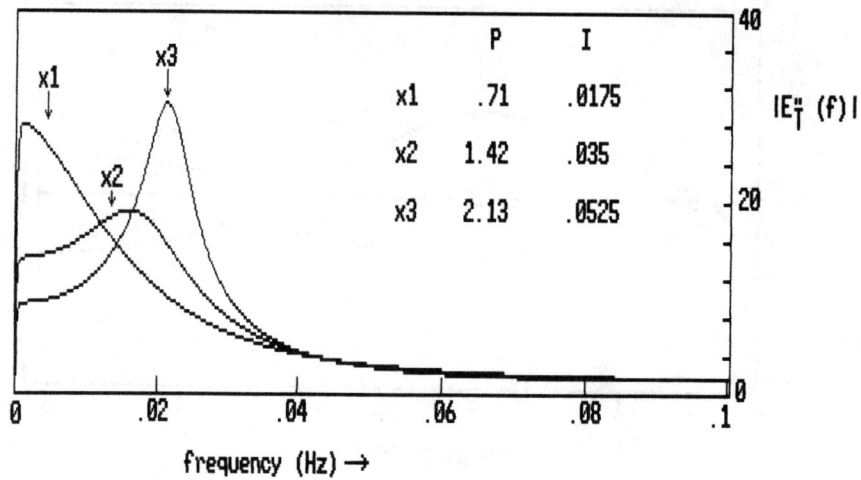

Figure 5.15   Effect of tuning on an autoregressive disturbance.

Note how the peak in the spectrum moves toward higher frequencies as the gains increase and how the previous best set of control gains ($P = 0.71$, $I = 0.0175$) does not appear to do as well as the aggressive set, which are twice as great. The following compares the sigma ratios for four sets of gains:

		Sigma ratios	
P	I	E/N	M/N
0.355	0.00875	0.256	0.364
0.71	0.0175	0.200	0.732
1.42	0.035	0.178	1.502
2.13	0.0525	0.203	2.411

This table suggests that the (1.42, 0.035) PI pair may be the best set of control gains for this type of disturbance. The E/N sigma ratio is minimized for this combination, while the M/N sigma ratio, although high, may not be unreasonable.

Figure 5.16 shows how this best set of control gains handles a set point change in the absence of noise. The point to be made here is that the best set of control gains depends on whether the set point is going to be changed and on the nature of the stochastic disturbances, although it might be argued that for this case the "times two" control gains are preferable for both the set point change and the disturbance rejection.

5.5.3 Effect of Derivative on a FOWDT Model

This section is similar to Section 5.4.3, where the effect of PI, PID, and PIfD on a third-order model subject to a white noise disturbance was examined. Here, PI,

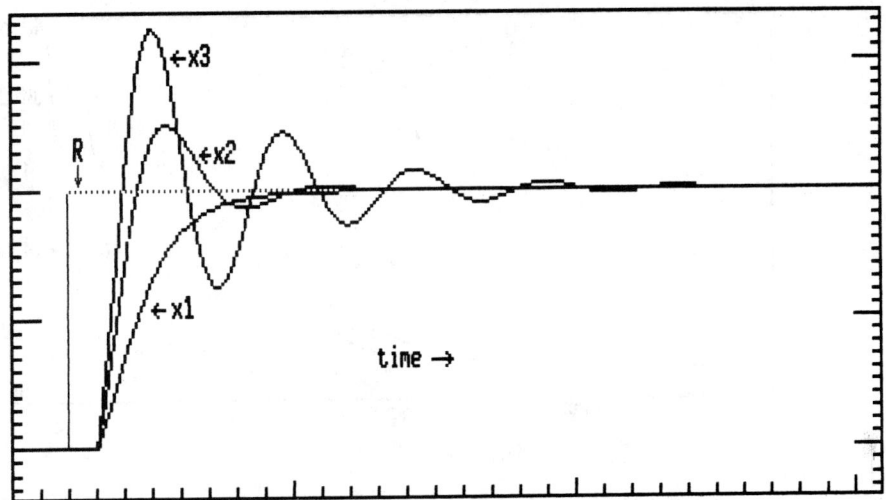

Figure 5.16   Effect of tuning on noiseless set point response.

PID, and PIfD will be applied to a FOWDT model (model 7, time constant = 40, deadtime = 10) subject to an autocorrelated stochastic disturbance with an autoregressive coefficient of 0.95. When PI is applied, the control gains are P = 0.71 and I = 0.0175; when PID is applied, the control gains are P = 1.1, I = .02, and D = 10; finally, when PIfD is applied, the derivative filter coefficient is 0.1. The $|E_T|$ curve in Figure 5.17, compared to Figure 5.9, shows a hump at intermediate frequencies due to the effect of the autocorrelated disturbance. Furthermore, at high frequencies, the $|E_T|$ curve for PID appears to become cyclical.

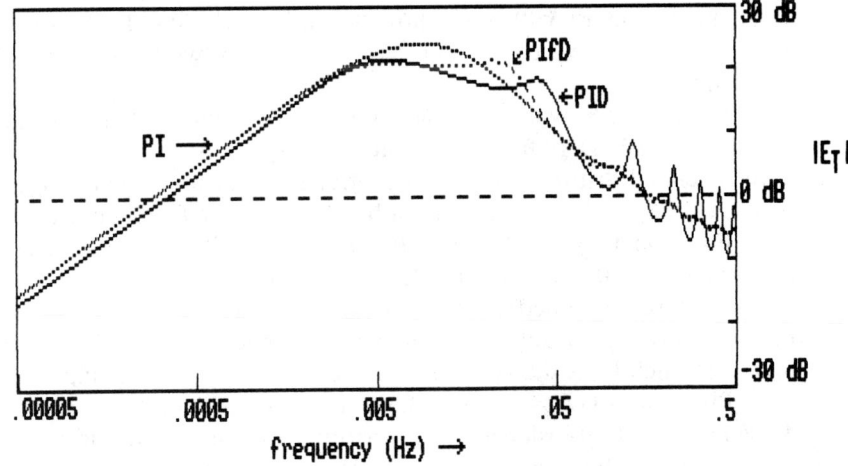

Figure 5.17   PI, PID, and PIfD applied to model 7 subject to an autoregressive disturbance.

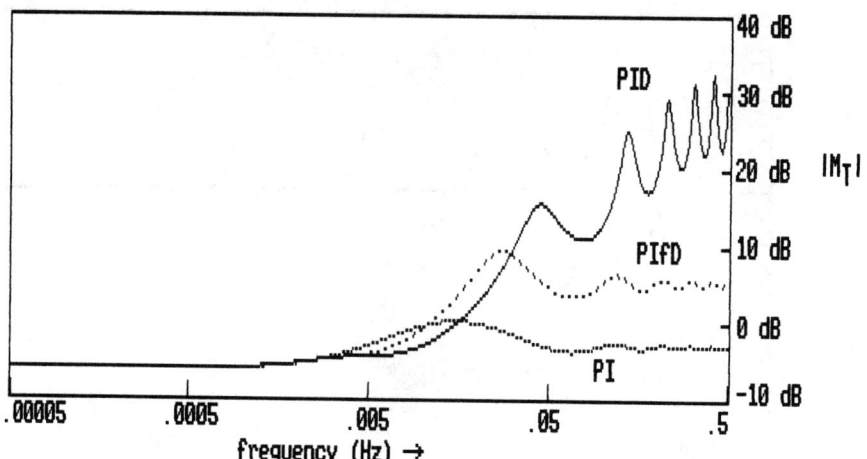

Figure 5.18 PI, PID, and PIfD applied to model 7 subject to autoregressive disturbances.

The $|M_T|$ curve shown in Figure 5.18 supports the idea that PID causes increased activity in the higher frequencies.

5.5.4 Dependence of Controllability on Degree of Autocorrelation

Equation (5.16) can be used to investigate the ability of control algorithms to attenuate autocorrelated disturbances as a function of the degree of autocorrelation for process models 6 and 7, which have been dealt with extensively earlier in the book. Figure 5.19 shows how the ratio of the standard deviations changes as the autoregressive coefficient varies from zero (white noise) to 0.99 (nearly a random walk).

For model 6 (which has the short time constant relative to the control interval: $T = 0.5$, $D = 0$, $h = 1$), integral-only control ($I = 0.5$) amplifies lightly autocorrelated disturbances significantly. It is not until the autoregressive coefficient exceeds 0.5 that attenuation by means of feedback control is possible.

For model 7 ($T = 40$, $D = 10$, $G = 2$) the situation is different. PI control ($P = 0.71$, $I = 0.0175$) is not able to attenuate autocorrelated disturbances until the autoregressive coefficient exceeds 0.9. However, the lightly autocorrelated noise is only lightly amplified compared to model 6. Note how the sigma ratio increases slightly as the autoregressive coefficient increases until it exceeds 0.9.

PIfD control ($P = 1.1$, $I = 0.012$, $D_g = 15$, $\alpha = 0.1$), when applied to model 4 (which is a third-order model), performs similarly to the last case and is not plotted since it falls right on top of the trace for model 7, except when the autoregressive coefficient exceeds 0.9, and then it is insignificantly different.

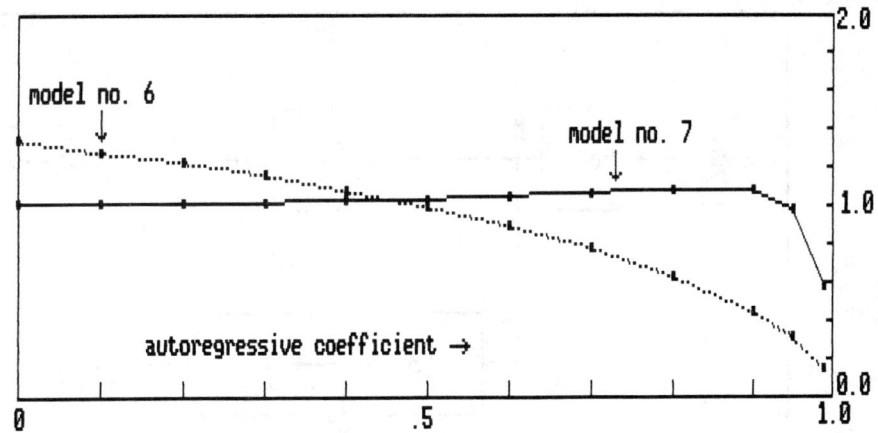

Figure 5.19 Effect of autocorrelation on degree of controllability (standard deviation ratio vs. autoregressive coefficient).

These data suggest that, when the mean is on target and the disturbance is lightly autocorrelated, no control at all is better. However, in industrial situations it is rare when such a situation exists. Most often the disturbance is at best heavily autocorrelated and usually nonstationary.

5.6 ANOTHER LOOK AT THE SMITH PREDICTOR

In Section 2.7 the PI control algorithm was augmented by the Smith predictor, and the simulation results suggested, both for set point changes and for controlling in the face of autocorrelated stochastic disturbances, that this modification was an improvement if one could accurately determine the model parameters. In this section, the Smith predictor will be given another look using some of the tools developed in this chapter. The results of this analysis, using the error transmission curves, will be compared to the results of Chapter Two, which used time-domain simulation. The main goal of this section is to demonstrate how the error transmission curves and the white noise sigma ratios can be used to analyze a control algorithm. Unlike the methods of Chapters Two and Three, where the empirical standard deviations and line spectra calculated from simulations were subject to sampling variation, the methods used in this chapter yield theoretical quantities.

5.6.1 Transfer Functions

Before determining the error transmission curve, the transfer function relating the control error E to the disturbance N will be derived. Referring to Figure 2.33, which is repeated here with slight modifications as Figure 5.20, the transfer functions can be combined as follows:

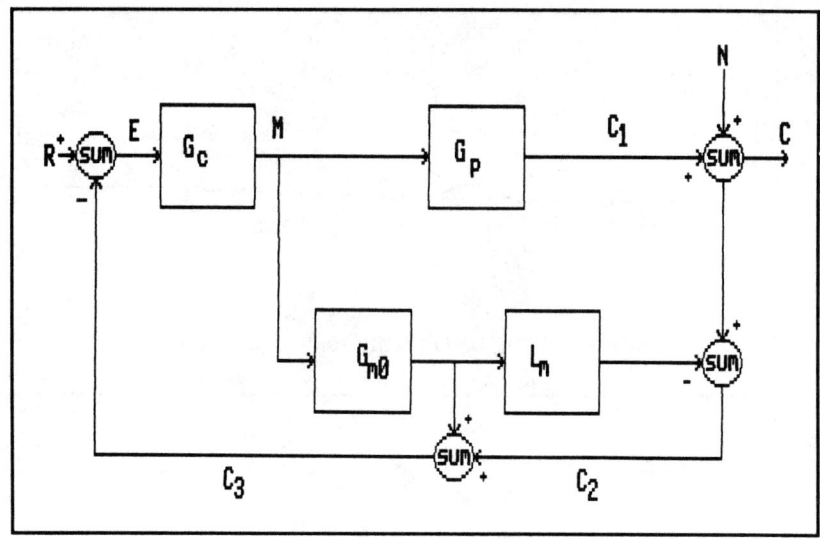

Figure 5.20 Block diagram of Smith predictor structure.

$$E = R - C_3$$

$$= R - G_{m0}M - (-G_{m0}L_mM + C)$$

$$= R - G_{m0}M + G_{m0}L_mM - (N + G_pM)$$

$$= R + (-G_{m0} + G_{m0}L_m - G_p)G_cE - N$$

where G_{m0} is the transfer function for the first-order model without the deadtime, L_m is the deadtime factor, G_c is the control algorithm (PI) transfer function, and G_p is the process transfer function. The components of G_{m0}, G_c, and L_m have been developed in preceding sections and will be summarized here as follows:

$$G_{m0} = \frac{g_m(1 - A_m)z^{-1}}{1 - A_mz^{-1}}$$

$$A_m = \exp(-h/T_m)$$

$$L_m = z^{-d_m}$$

$$G_c = P + \frac{Ih}{1 - z^{-1}}$$

A few comments about the notation are appropriate. The reference model consists of two parts: the first-order component without deadtime, G_{m0}, and the first-order component *with* deadtime, $G_{m0}L_m$. The model parameters T_m, g_m, and d_m are chosen to match the actual process, represented by G_p, as accurately as possible. In the previous chapters, the symbol G represented the process gain; here it is used with subscripts to represent transfer functions in the Z-domain, and g_m represents the reference model gain.

The above expression for E can be solved for the relation between E and N and the relation between E and R at the same time:

$$\frac{E}{N} = -\frac{1}{1 + G_{m0}G_c + G_c(G_p - G_{m0}L_m)} = -\frac{E}{R}$$

If the model matches the process exactly, that is, if $G_{m0}L_m = G_p$, then the two transfer functions become

$$\frac{E}{N} = -\frac{1}{1 + G_{m0}G_c} = -\frac{E}{R}$$

which shows that, in effect, the process to be controlled is now no longer represented by G_p, which contains a deadtime, but by G_{m0}, which is a first-order process without deadtime and therefore much easier to control.

Note that *if* regular feedback control were applied to a process without deadtime having a process transfer function of G_{m0}, then, in the absence of set point changes,

$$\frac{C}{N} = \frac{1}{1 + G_{m0}G_c} \tag{5.22}$$

which will be referred to below.

The variable E occurring in the foregoing equations is the error input to the PI control algorithm, not the error between C and its target R. Since it is this latter quantity that is of interest, the error transmission function E_T needs to be redefined as

$$\frac{R - C}{N}$$

or simply as

$$-\frac{C}{N}$$

Removing R (since for this analysis it is a constant) makes sense because an offset can always be added to both the controlled variable and the target so that the target is effectively zero. Therefore, the transfer function showing how C depends on N for the case where the PI control algorithm is augmented by the Smith predictor will be developed

Referring to Figure 5.20,

$$C = C_1 + N = G_pM + N = G_pG_cE + N = N - G_pG_cC_3$$

where $E = -C_3$ has been assumed since the target R does not figure into this analysis. Continuing the Z-domain algebraic analysis gives

$$C = N - G_pG_c(C_2 + G_{m0}M)$$

$$= N - G_pG_cC + G_pG_cG_{m0}(L_m - 1)M$$

$$C(1 + G_pG_c) = N - G_pG_cG_{m0}(1 - L_m)M$$

The above expression gives C as a function of M. To complete the exercise, return to Figure 5.20 and develop a relationship between M and C. Then combine the two expressions to eliminate M.

$$M = -G_c C_3 = -G_c(C_2 + G_{m0}M) = -G_c C_2 - G_c G_{m0}M$$

$$= -G_c(C - G_{m0}L_m M) - G_c G_{m0}M \tag{5.23}$$

$$= -\frac{G_c C}{1 + G_c G_{m0}(1 - L_m)}$$

Having found C as a function of M and M as a function of C, the variable M can be eliminated.

$$\frac{C}{N} = \frac{1 + G_c G_{m0}(1 - L_m)}{1 + G_p G_c + G_c G_{m0}(1 - L_m)} \tag{5.24}$$

This rather complicated expression gives the dependence of the process output (which is also the variable that is to be kept on target) on the disturbance N. Based on the earlier argument, it is also the error transmission function, E_T.

If the model deadtime is equal to zero, then $L_m = 1$ and the error transmission function simplifies to

$$\frac{C}{N} = \frac{1}{1 + G_p G_c}$$

Referring to Figure 5.20, this makes sense because the right and left legs on the second tier of the diagram cancel out, leaving a conventional feedback control structure.

If one assumes exact model–process matching and sets $G_{m0}L_m = G_p$ in the error transmission function, Equation (5.24) becomes

$$\frac{C}{N} = \frac{1 - G_p G_c + G_{m0}G_c}{1 + G_{m0}G_c}$$

This shows that even with exact reference model matching the error transmission function still depends on the process deadtime and is not the same as the error transmission function for the case when regular feedback control was applied to a first-order process without deadtime [Equation (5.22)].

The other member of the error transmission curves, M_T, can be derived as follows starting with the identity

$$M_T = \frac{M}{N} = \frac{C}{N}\frac{M}{C}$$

The first factor is given by Equation (5.24), and the second member can be obtained from Equation (5.23). When these factors are replaced, the result is

$$M_T = \frac{M}{N} = -\frac{G_c}{1 + G_p G_c + G_c G_{m0}(1 - L_m)} \tag{5.25}$$

Note that with perfect model–process matching, Equation (5.25) becomes

$$M_T = \frac{M}{N} = -\frac{G_c}{1 + G_{m0}G_c}$$

which says that the M_T curve is independent of the process deadtime.

5.6.2 Smith Predictor Applied to a FOWDT Process

Figure 5.21 shows the error transmission curves plotted on log–log axes for two approaches (PI and Smith predictor) to controlling, at 1-second intervals, a FOWDT process having a time constant of 40 seconds, a deadtime of 40 seconds, and a gain of 2.0 (model 10). A third curve, showing the most attenuation and no peaks, is the reference curve of PI applied to a first-order process without deadtime having the same time constant and using $P = 1.7$, $I = 0.05$. These two control gains were determined using the formulas of Section 2.5 with $T_d = 11$, $T = 40$, and $D = 0$ seconds.

The curve with the least attenuation at low frequencies but with only one broad peak is for the application of PI to the FOWDT process using $P = 0.18$, $I = 0.0045$. These two gains were determined using the formulas of Section 2.5 with $T_d = 40$, $T = 40$, and $D = 40$ seconds.

In between these two curves at low frequencies is the PI + Smith predictor with exact matching using $P = 1.7$, $I = 0.05$. These gains are the same as were used for the case where there was no deadtime. In other words, it is assumed that the reference model matches the process exactly so that it is a first-order process without deadtime that is under control. Note that this last curve has a peak higher than that for PI and also has ripples at higher frequencies. In com-

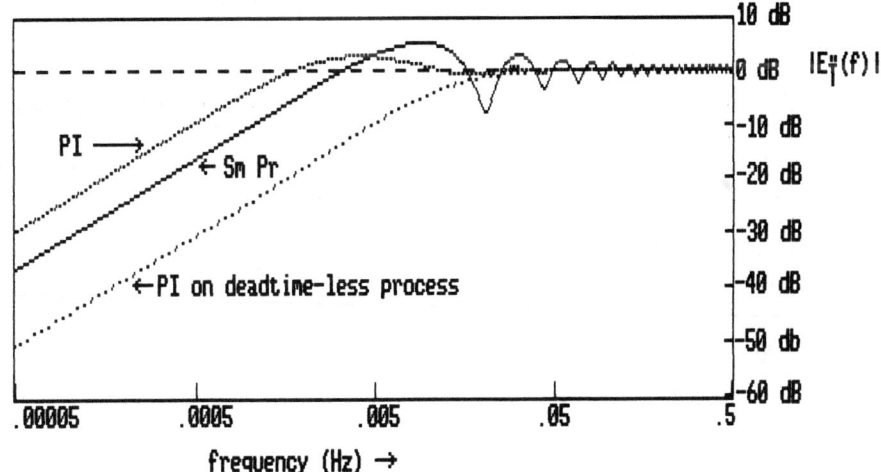

Figure 5.21 Comparison of PI and Smith predictor applied to model 10.

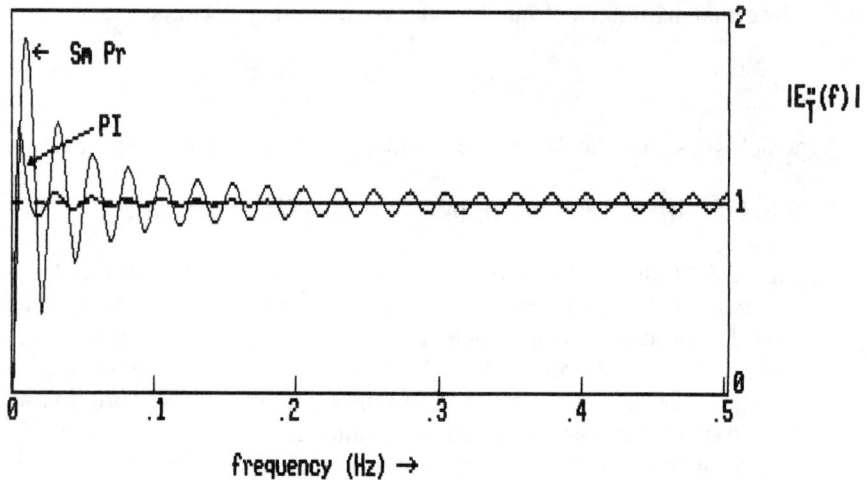

Figure 5.22   Comparison of PI and Smith predictor using linear axes.

paring these three curves, one sees the dramatic difference that having a deadtime makes on the ability to attenuate disturbances having power in the lower frequencies.

Using linear axes, Figure 5.22 shows the same E_T curves for the Smith predictor and the PI algorithms. Here the Smith predictor ripples are equally spaced, having a separation of $\frac{1}{40}$ Hz, that is, a frequency equal to the reciprocal of the deadtime. Note that the E_T curve for the PI algorithm also exhibits ripples at the lower frequencies. The choice between Smith predictor and regular PI would depend on where in the Nyquist interval the disturbances appeared. Since most disturbances have their power in the lower frequencies, the Smith predictor would probably be preferable, assuming that the model parameters can be estimated.

Before moving on, the reader should be warned that interpreting these error transmission curves should be done with care. For example, one can plot the error transmission curve for PI applied to this process (model 10) using the aggressive gains of $P = 1.7$ and $I = 0.05$ and be completely misled because, although the error transmission curve will indicate superior attenuation at all frequencies, the choice of the gains makes the system unstable. Therefore, in questionable cases, one should either apply the stability techniques of Section 4.5 or the time-domain simulation techniques of Chapters Two and Three in order to verify that the system is stable before jumping to conclusions based on an error transmission curve.

For this example process, the sensitivity of the Smith predictor to changes in the reference model deadtime is shown in Figure 5.23, where the error transmission curves are plotted for exact matching in the time constant and gain but with model deadtimes of 36, 40 (exact matching), and 44. The curves are essentially indistinguishable from each other, indicating a low sensitivity.

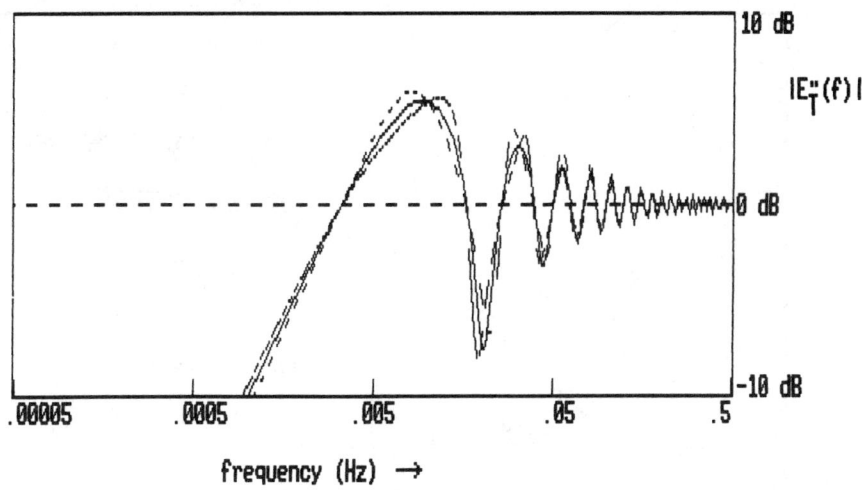

Figure 5.23 Effect of varying Smith predictor model deadtimes.

5.6.3 Smith Predictor Applied to a Third-order Process

The previous examples in this section have dealt with FOWDT processes, and their error transmission curves have exhibited a ripple related to the deadtime. Now try a third-order process without an explicit deadtime but with time constants of 25, 26, and 27 (model 4). Figure 5.24 shows how one might approximate this third-order process with a FOWDT reference model having a deadtime of 29 and a time constant of 52. This approximation was developed by finding a nonlinear least-squares fit over the entire range of data shown in Figure 5.24, and as a result the estimate of the deadtime is probably a little larger than it should be for control purposes. This will cause the control gains to be more conservative than those one might obtain from a visual fit.

The error transmission curves E_T will be examined for three cases. First, the third-order process is controlled by PI with the control gains of $P = 0.33$, $I = 0.006$. These gains were determined using the formulas of Section 2.5 with $T_d = 29$, $T = 52$, and $D = 29$. This curve (in Figure 5.25) shows the least attenuation at the lower frequencies and the smallest resonant peak at the middle frequencies.

Second, the PI control algorithm is augmented by the Smith predictor using the FOWDT reference model shown in Figure 5.24 and is applied to the third-order process. Using PI gains of $P = 2.5$ and $I = 0.05$, this curve shows more attenuation at the lower frequencies and a resonant hump similar to that for PI. These control gains were determined using the formulas of Section 2.5 with $T_d = 29$, $T = 52$, and $D = 0$; that is, it was assumed that the reference model matched the third-order process model, leaving a first-order deadtimeless process with a time constant of 52 seconds.

Finally, PIfD is applied to the third-order process, and it is seen that the

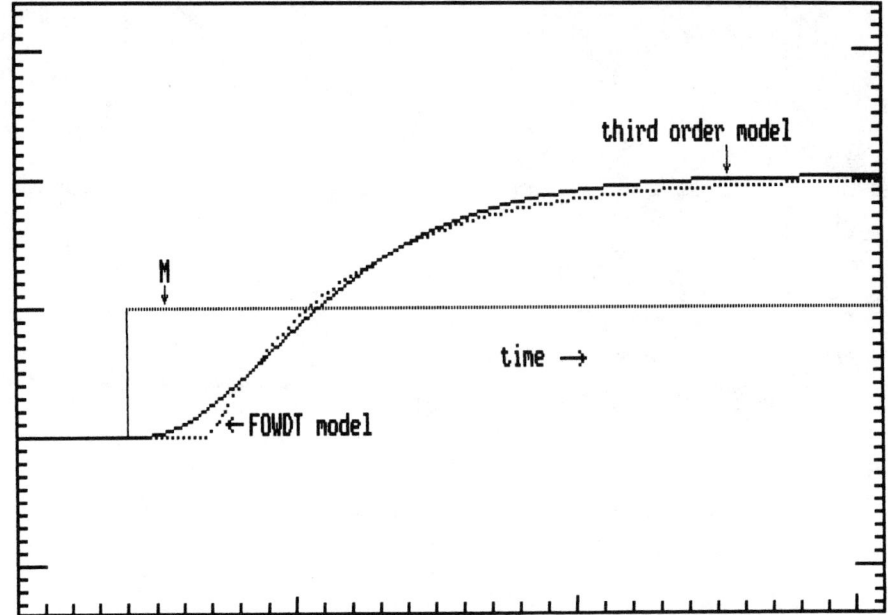

Figure 5.24 Approximation of a FOWDT model to a third-order model.

error transmission curve shows the greatest attenuation at the lower frequencies and the highest resonant hump. The control gains, $P = 1.1, I = 0.012, D_g = 15$, and $a = 0.1$, were the same as those used in Section 2.6.3, where this model was treated earlier.

This example suggests that PIfD, which requires the estimation of four parameters (P, I, α, and D_g), may be preferable to the PI + Smith predictor, which requires the estimation of five parameters (P, I, g_m, T_m, and d_m).

As mentioned earlier, the square root of the integral of $|E_T|^2/f_{Ny}$ over the Nyquist interval gives the ratio of the standard deviation of the controlled variable to that of the white noise disturbance [Equation (5.10)], so the three cases can be compared as follows:

PI	1.003
PIfD	1.010
Sm Pr	1.007

Therefore, given the accuracy of the numerical evaluation of the above-mentioned integral, all three methods perform about the same from the point of view of amplifying a white noise disturbance.

To complete the comparison of these approaches to controlling a third-order process, the M_T curves should also be studied. Using Equation (5.11), the square root of the integral of $|M_T|^2/f_{Ny}$ over the Nyquist interval, that is, the ratio of

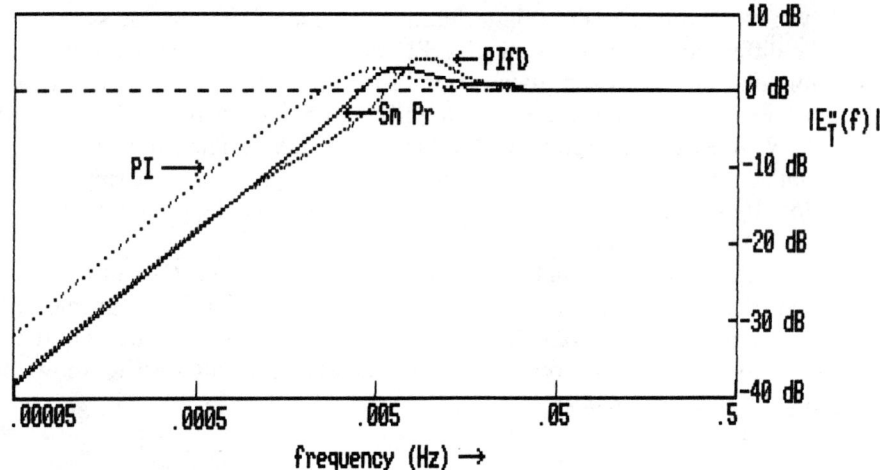

Figure 5.25   Comparison of PI, PIfD, and Smith predictor applied to model 4.

the standard deviation of the manipulated variable to that of the white noise disturbance, for the three cases gives

PIfD	2.64
PI	0.337
Sm Pr	2.64

Note that the Smith predictor is significantly more active than PI but effectively the same as PIfD.

Figure 5.26 shows the M_T curves for these three cases. Note that the Smith predictor and PIfD give about the same overall level of activity, but the distribution

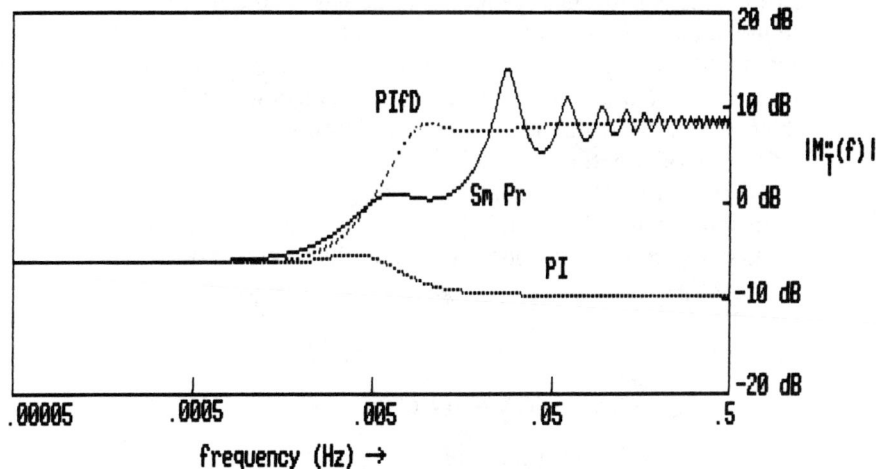

Figure 5.26   Comparison of PI, PIfD, and Smith predictor applied to model 4.

over the Nyquist interval is different. The use of the log–log scales emphasizes the differences. The conservative PI control algorithm gives the minimum variation and the most even distribution of manipulated variable activity, while PIfD, because of the high-frequency amplifying characteristic of the derivative term, shows most of its strength at the high frequencies. The Smith predictor shows a similar amplification of high frequencies but also shows a ripple, which when plotted on linear axes has its peaks separated by $\frac{1}{29}$ Hz, which corresponds to the deadtime of the model.

Therefore, one might conclude that, based on the E_T and M_T curves, aside from the trouble of estimating more parameters, the Smith predictor and PIfD appear to perform about the same. Keep in mind that as the deadtime/effective time constant ratio increases, the Smith predictor will start to show more of an edge in performance.

5.7 INTERNAL MODEL CONTROL (IMC) ALGORITHM

The IMC algorithm (see Garcia and Morari, 1982, and Morari and Zaferiou, 1989) appears to have been inspired by the Smith predictor and in recent years has developed a significant following in the chemical process industries. Because the development of its structure requires a knowledge of the Z-transform, it has not been discussed until this chapter. For one-dimensional systems, it performs about the same as the Smith predictor.

5.7.1 Derivation of the Control Algorithm

Figure 5.27 shows the structure of the internal model control scheme. The fundamental idea is similar to that of the Smith predictor in that a reference model of the process is used to cancel out the signal coming from the actual process. If the model, represented by the transfer function G_m, exactly matches the process, represented by G_p, the signal fed to the summing point will consist only of the disturbance. (Note that the IMC model represented by G_m is the same as $G_{m0}L_m$, which was used in the section on the Smith predictor, where G_{m0} represented a FOWDT model without deadtime.)

The ability of this control strategy to attenuate disturbance power in the Nyquist interval will be studied. To do this, one must derive the error transmission function E_T. Start with following block diagram algebra:

$$M = G_c(R - C_2) = G_cR - G_cC_2 = G_cR - G_c(C - G_mM)$$

$$= G_cR - G_c(N + C_1) + G_cG_mM$$

$$= G_cR - G_cN - G_cG_pM + G_cG_mM$$

which can be solved for M to give

$$M = \frac{G_cR - G_cN}{1 + G_c(G_p - G_m)} \tag{5.26}$$

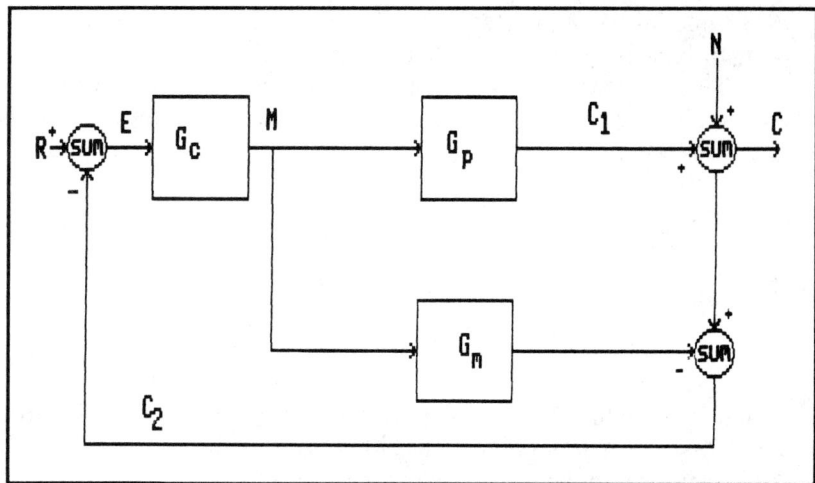

Figure 5.27   Block diagram of the IMC structure.

If $R = 0$, the difference between the set point and the controlled variable is now $-C$, so the error transmission function can be written as

$$E_T = -\frac{C}{N}$$

Thus an expression for C/N will be equivalent to the error transmission function. (Note that, as in the case of the Smith predictor, the error fed to the controller is not the difference between the set point and the controlled variable.)

The derivation proceeds as follows:

$$C = N + G_p M$$

$$= N + \frac{G_p(G_c R - G_c N)}{1 + G_c(G_p - G_m)}$$

Since the interest is in C/N, the set point R will be set to zero, giving

$$\frac{C}{N} = \frac{1 - G_c G_m}{1 + G_c(G_p - G_m)} \tag{5.27}$$

First, note that if $G_m = 0$ the above expression collapses to the transfer function for conventional feedback control. Second, if the model exactly matches the process, Equation (5.27) gives

$$\frac{C}{N} = 1 - G_p G_c \tag{5.28}$$

If it is specified that the control algorithm, represented by G_c, satisfy the condition

$$G_c = \frac{1}{G_p} = \frac{1}{G_m} \tag{5.29}$$

then Equation (5.28) indicates that the control algorithm would remove all effects of disturbances on the controlled variable.

There are two problems with this idea. First, in specifying Equation (5.29), an unrealizable control algorithm would result. For example, if the process model is a FOWDT model

$$G_m(z) = \frac{g_m(1 - A)z^{-1-d_m}}{1 - Az^{-1}}, \qquad A = \exp(-h/T_m)$$

then G_c would be

$$G_c = \frac{z^{1+d}(1 - Az^{-1})}{G(1 - A)}$$

which is unrealizable since it requires future values of the controlled variable. Second, even if this control algorithm were realizable, it must be checked for unrealistic manipulated variable activity.

To make the control algorithm realizable, the process model is factored as follows:

$$G_m(z) = G_-(z)G_+(z)$$

where $G_+(z)$ contains all the time delays and all the roots that lie outside the unit circle in the Z-domain. For example, if the model is FOWDT, then one can factor $G_m(z)$ as follows:

$$G_m(z) = \frac{g_m(1 - A)z^{-1-d_m}}{1 - Az^{-1}} = G_+(z)G_-(z)$$

$$G_+(z) = z^{-1-d_m}, \qquad G_-(z) = \frac{g_m(1 - A)}{1 - Az^{-1}}$$

With this factoring in mind, the control algorithm now is specified as

$$G_c(z) = \frac{1}{G_-(z)}$$

For the case where the model is FOWDT,

$$\frac{M}{E} = G_c = \frac{1}{G_-} = \frac{1 - Az^{-1}}{g_m(1 - A)}$$

or

$$M = \frac{E - AEz^{-1}}{g_m(1 - A)}$$

Remembering that z^{-1} is a backshift operator, this expression can be realized in the time domain as

$$M_i = \frac{E_i - AE_{i-1}}{g_m(1 - A)}$$

where $E_i = R_i - C_{2i}$ and where C_2 is the difference between the model and the actual process output, which in general is contaminated by noise. Note that in the case of perfect model–process matching E will equal $R - N$ and, in the absence of set point changes, the control algorithm will be acting on only the disturbance. Also note that this control algorithm does not appear to contain clearly identifiable proportional and integral components.

As a consequence of the factoring, Equation (5.28) becomes

$$\frac{C}{N} = 1 - G_p G_c$$

$$= 1 - \frac{G_-(z)G_+(z)}{G_-(z)} \tag{5.30}$$

$$= 1 - G_+(z)$$

To complete the specification of IMC, it is required to have no offset in the face of constant disturbances. This requirement will be satisfied if

$$G_+(1) = 1$$

To show the basis for this condition, consider the case where the disturbance N is a step change, as in

$$N(z) = \frac{z}{z - 1}$$

For such a disturbance, Equation (5.30) becomes

$$C(z) = [1 - G_+(z)]\frac{z}{z - 1}$$

Zero offset means that C_∞, the asymptotic value of C, must be zero. To find this asymptotic value, the final value theorem (discussed in Chapter Four) can be applied as follows:

$$C_\infty = \lim_{z \to 1} (1 - z^{-1})C(z)$$

$$= \lim_{z \to 1} (1 - z^{-1})[1 - G_+(z)]\frac{z}{z - 1} = 1 - G_+(1)$$

Therefore, if $G_+(1) = 1$, there will be no offset.

In summary, the IMC control algorithm is described by the following equations:

$$G_m(z) = G_-(z)G_+(z)$$

$$G_c(z) = \frac{1}{G_-(z)}$$

$$G_+(1) = 1$$

5.7.2 Analysis of the Control Algorithm in the Frequency Domain

Consider the case where the process can be exactly matched with a FOWDT model. Equation (5.30) gives the error transmission function, which is seen to be independent of the time constant and the gain of the process model. The analysis can be performed in the frequency domain by making the usual substitution of

$$z = \exp(j2\pi fh)$$

Unlike the expressions for the Smith predictor, the mathematics here is extremely simple:

$$E_T(z) = 1 - G_+(z) = 1 - z^{-1-d}$$

$$E_T''(f) = 1 - \cos[(d + 1)2\pi fh] + j \sin[(d + 1)2\pi fh] \qquad (5.31)$$

$$|E_T''(f)|^2 = 2 - 2 \cos[(d + 1)2\pi fh]$$

$$= 4 \sin^2[(d + 1)\pi fh]$$

Several observations can be made. First, the error transmission curve depends only on the deadtime. Second, independent of the value of the deadtime, Equation (5.31) satisfies

$$|E_T''(0)| = 0$$

Thus the zero offset condition is satisfied. Third, unlike any of the previous control algorithms, a closed form for the variance ratio can be obtained. If the control interval is assumed to be unity so that the Nyquist frequency is 0.5, it follows that

$$\frac{V_E}{V_N} = \frac{1}{0.5} \int_0^{0.5} \{4 \sin^2[(d + 1)\pi f]\} \, df$$

$$= 2 \frac{1 - \sin[(d + 1)\pi]}{(d + 1)\pi}$$

which equals 2.0 no matter what the deadtime index, d, is. Therefore, this algorithm amplifies the white noise standard deviation by a factor of 1.414.

For the case of zero deadtime, the E_T curve in Figure 5.28 shows how the low frequencies are significantly attenuated but how the high frequencies are amplified. The E_T curve for a deadtime of ten control intervals is shown in Figure 5.29. As was pointed out in the previous paragraph, the variance ratio is the same for both cases. Comparing Figure 5.29 with Figure 5.3, which dealt with PI applied to model 7 ($T = 40$, $G = 2$, $D = 10$), one sees that the internal model control algorithm is slightly better at low frequencies but much worse at high frequencies. It is interesting to note that, under the conditions of exact model–process matching, the IMC algorithm will give the same performance independent of the process time constant and the process gain.

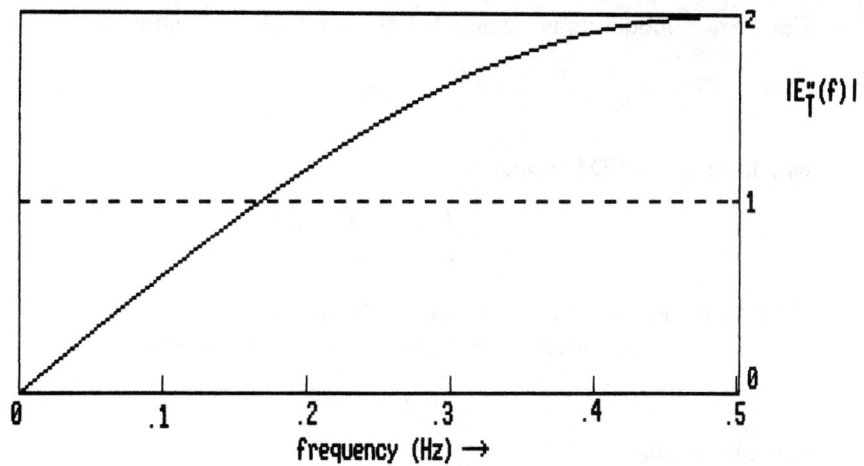

Figure 5.28   IMC $|E_T|$ curve for exact matching with $D = 0$.

The companion to E_T is $M_T = M/N$, which can be derived as follows. Start with Equation (5.26):

$$M = \frac{G_c R - G_c N}{1 + G_c(G_p - G_m)}$$

which, after setting $R = 0$, yields

$$\frac{M}{N} = -\frac{G_c}{1 + G_c(G_p - G_m)} \tag{5.32}$$

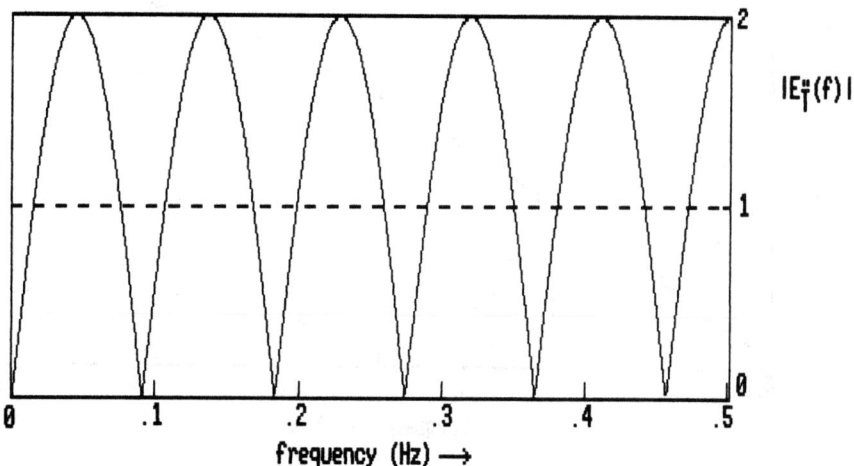

Figure 5.29   IMC $|E_T|$ curve for exact matching with $D = 10$.

Since exact modeling is assumed, Equation (5.32) simplifies to

$$\frac{M}{N} = -G_c = -\frac{1}{G_-}$$

which for a FOWDT model is

$$\frac{M}{N} = -\frac{(1 - Az^{-1})}{g_m(1 - A)}$$

Note that there is no dependence on the deadtime.

To move to the frequency domain, the usual substitution of

$$z = \exp(j2\pi fh)$$

is made, giving

$$|M_T''|^2 = \frac{1 - 2A\cos(2\pi fh) + A^2}{[g_m(1 - A)]^2}$$

For the case where the process model's time constant is 40.0 and its gain is 2.0, the M_T'' curve is shown in Figure 5.30. The higher frequencies are magnified as though the control algorithm had a derivative component. Using Equation (5.11) shows that the ratio of the standard deviations is 28.3. The reader should compare this curve to Figure 5.6, which shows the M_T'' curve for the same process model when PI is applied. Note how much less activity there is in this latter case at the higher frequencies. In fact, for this latter example Equation (5.11) shows that the ratio of the standard deviations is 0.73.

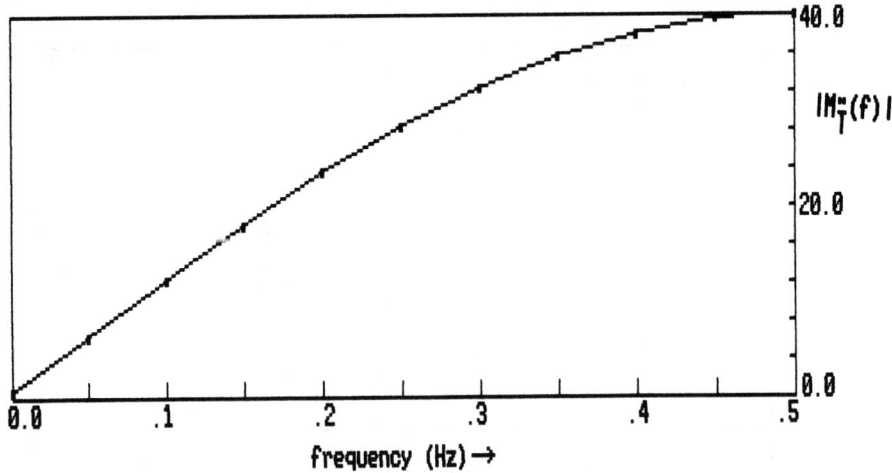

Figure 5.30 $|M_T|$ curve for IMC applied to model 7 (exact model matching).

5.7.3 Effect of Putting a Low-pass Filter in the Feedback Loop

The previous section showed that the internal model control algorithm, as it presently stands, will not be successful because of the excessive activity at high frequencies. The problem is solved by inserting a low-pass filter in the feedback path, as shown in Figure 5.31. (Note that for set point changes the filter is best placed after the summing point so that it operates on E, not C_2.) With this filter, whose transfer function is F, Equation (5.26) becomes

$$M = \frac{G_c R - F G_c N}{1 + F G_c (G_p - G_m)} \tag{5.33}$$

and Equation (5.27) becomes

$$\frac{C}{N} = \frac{1 - F G_c G_m}{1 + F G_c (G_p - G_m)} \tag{5.34}$$

Under exact model–process matching and zero set point, that is,

$$R = 0$$

$$G_p = G_m$$

$$G_m = G_+ G_-$$

$$G_c = \frac{1}{G_-}$$

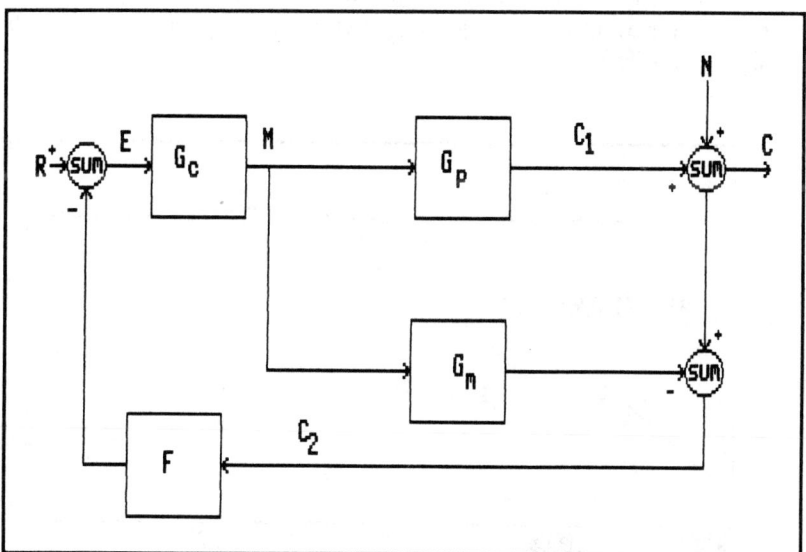

Figure 5.31   IMC with filter.

Equation (5.34) becomes

$$\frac{C}{N} = 1 - G_+ F \tag{5.35}$$

and

$$\frac{M}{N} = -\frac{F}{G_-} \tag{5.36}$$

The low-pass filter F is frequently chosen to be the one that has been dealt with throughout this book; that is,

$$F = \frac{\alpha}{1 - (1 - \alpha)z^{-1}}$$

Note that for the case of exact model–process matching, with a FOWDT model, if the value of α is chosen so that the time constant of the filter matches that of the process model, that is, if

$$\alpha = 1 - A, \qquad A = \exp\left(-h/T_m\right)$$

then F will cancel G_- and M_T will be constant, having a value of $1/g_m$.

The effect of the filter is shown in Figures 5.32 and 5.33, where the E_T and M_T curves are shown for IMC applied to model 7 with exact model–process matching and with filter coefficients of 0.1 and 0.025. The latter coefficient value provides a filter time constant that matches that of the process. Note that with more filtering the low-frequency attenuation is poorer but that the white noise sigma ratio is better.

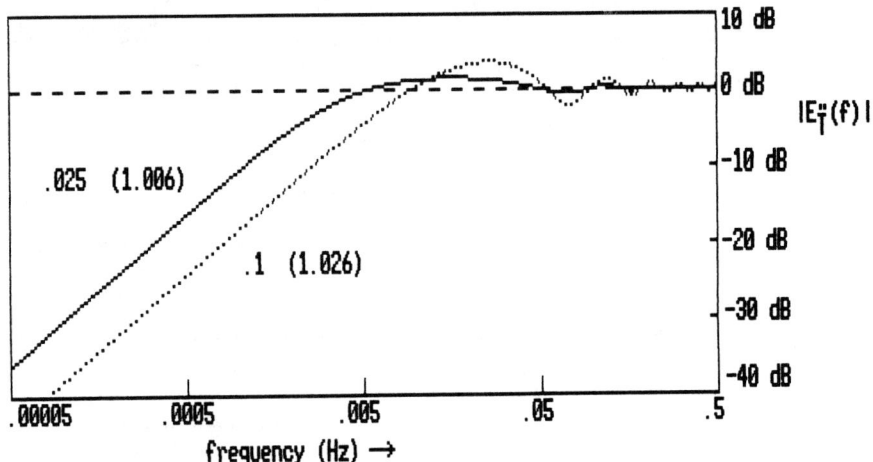

Figure 5.32 Effect of filtering on IMC (white noise sigma ratio in parentheses).

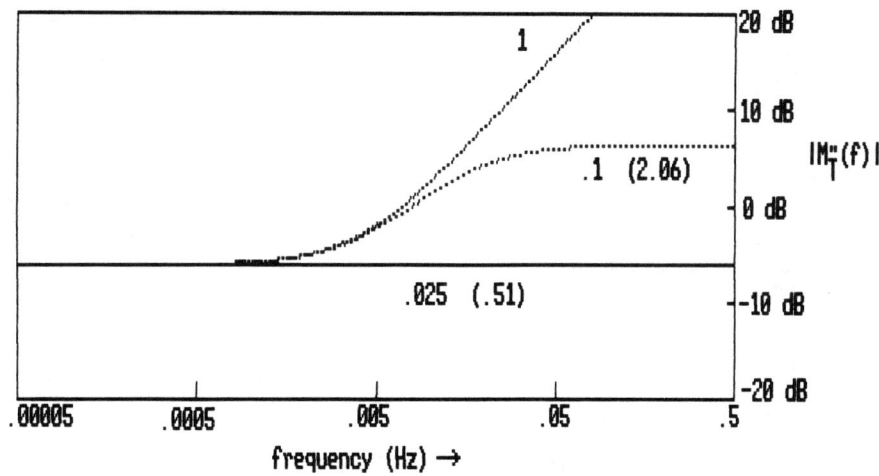

Figure 5.33 Effect of filtering on IMC (white noise sigma ratio in parentheses).

The M_T curve shows that for less filtering there is more activity, with the white noise sigma ratio being four times greater. In Figure 5.33, the M_T curve with no filtering (coefficient equal to unity) is shown for reference purposes.

Figures 5.34 and 5.35 compare the IMC approach with the Smith predictor when both techniques are applied to model 10 ($G = 2$, $T = 40$, $D = 40$) using exact model–process matching. For IMC, the filtering coefficient is 0.1, and for the Smith predictor the control gains are $P = 1.7$ and $I = 0.05$. The E_T and M_T curves are similar, and in fact in Figure 5.34 the curves are indistinguishable. This example suggests that, in the case where a good model has been developed, the

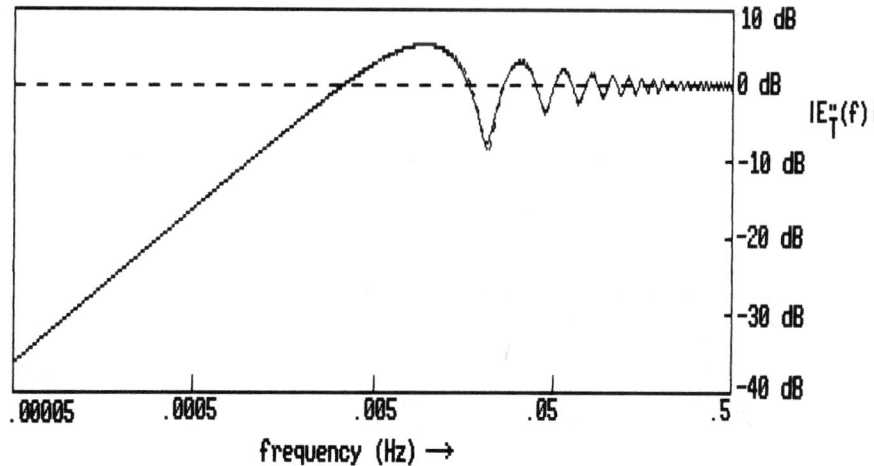

Figure 5.34 Comparison of IMC and Smith predictor applied to model 10.

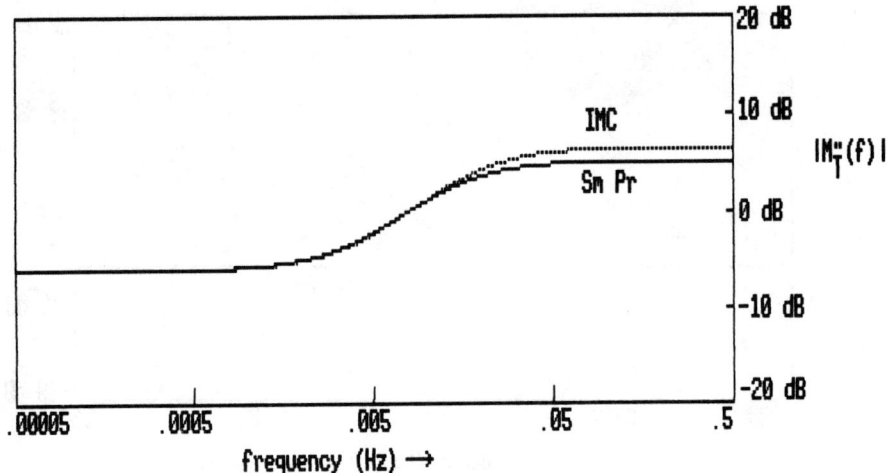

Figure 5.35 Comparison of IMC and Smith predictor applied to model 10.

IMC algorithm may have an advantage in that only one additional parameter, the filtering coefficient, must be estimated as compared to the Smith predictor, where two PI control gains must be estimated.

5.7.4 Effect of Model–Process Mismatch

The sensitivity of the IMC approach to model–process mismatch can be illustrated by calculating the E_T curve using Equation (5.34) for the case where the process is model 7 ($G = 2$, $T = 40$, $D = 10$) and the reference FOWDT model has errors

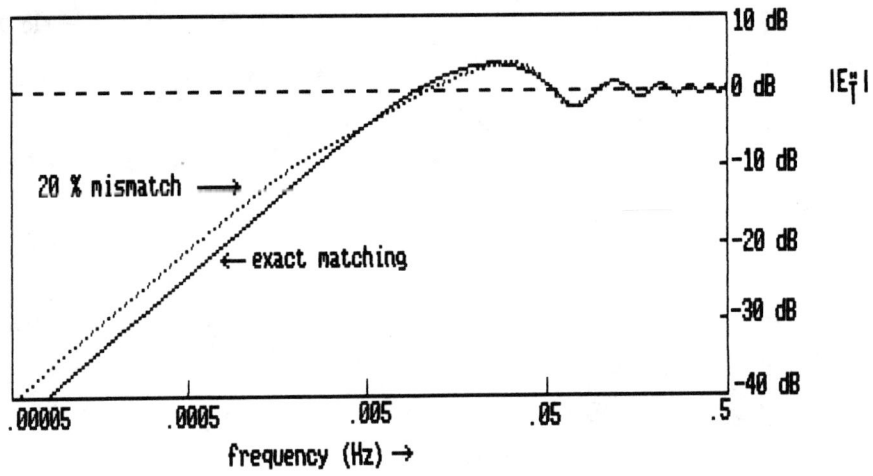

Figure 5.36 Model–process mismatch for IMC.

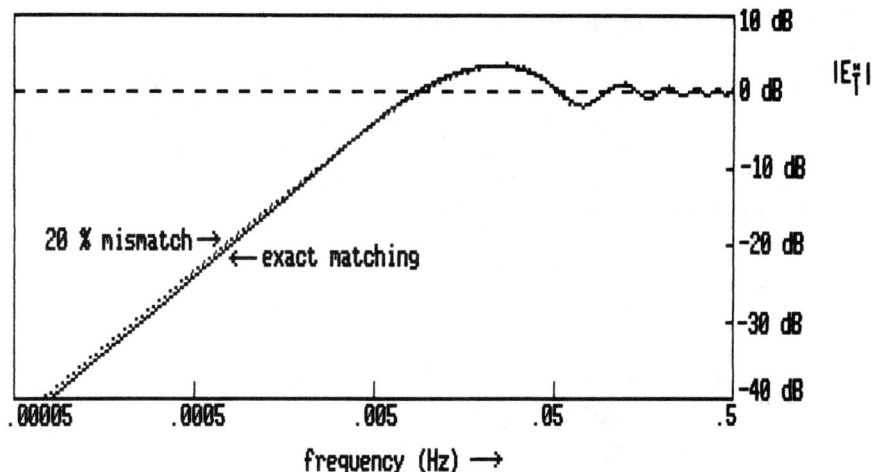

Figure 5.37   Model–process mismatch for the Smith predictor.

in its T, D, or G. Consider Figure 5.36, which shows a 20% error in both the time constant and the deadtime; that is, the model time constant is 48, and the model deadtime is 12. At low frequencies the mismatched version attenuates disturbances less than the exactly matched version of the IMC.

Figure 5.37 shows the results of carrying out the same exercise for the Smith predictor. The same control gains that were used in Section 5.6.2 are used here, and the model mismatch is the same as in the previous example with IMC. It appears that the Smith predictor, at least for this case, is less sensitive to modeling errors than the IMC.

5.8 ANOTHER LOOK AT THE BOX–JENKINS CONTROL ALGORITHMS

In Chapter Two, the Box–Jenkins approach to designing control algorithms was presented, and it was pointed out that, first, their approach was the only one in this book that took into account the characteristics of the disturbances, and, second, their algorithms sometimes cause unacceptably high activity in the manipulated variable. To derive the Box–Jenkins control algorithms, a slightly modified form of the Z-transform methodology will be used. To the reader who has struggled with Chapter Four, this modification may be confusing. However, this notation is widely accepted in the literature (see Chapter 12 in the text by Box and Jenkins, 1970) and it does make the derivation easier. A FOWDT process model and an integrated moving average disturbance will be dealt with first. As the derivation proceeds, generalizations will be made so that extensions to other process and noise models can easily be made.

5.8.1 Development of a Hybrid Notation

In Chapter Four it was shown how handy the Z-transform is when dealing with equations defined on the discrete time domain. The Z-transform was also used extensively in this chapter. To develop the Box–Jenkins equations, it is convenient to mix time and Z-domain quantities. For example, the FOWDT model in the time domain can be written as

$$C_k = AC_{k-1} + BM_{k-1-d} \tag{5.37}$$

where

$$
\begin{aligned}
A &= \exp(-h/T) \\
h &= \text{control interval} \\
T &= \text{time constant} \\
B &= G(1 - A) \\
G &= \text{gain} \\
d &= \text{deadtime index} \\
D &= \text{deadtime} = dh
\end{aligned}
$$

In the Z-domain the FOWDT model can be written as

$$\frac{C}{M} = \frac{Bz^{-1-d}}{1 - Az^{-1}} = \frac{Q_2(z^{-1})z^{-1-d}}{Q_1(z^{-1})} \tag{5.38}$$

where C is the Z-transform of C and M is the Z-transform of M. The transfer function C/M has also been written as a ratio of generalized polynomials in z^{-1}, where

$$Q_2(z^{-1}) = B \quad \text{and} \quad Q_1(z^{-1}) = 1 - Az^{-1}$$

which will occur in the general form of the control algorithm. Later, should process model changes be desired, the polynomials Q_1 and Q_2 can be changed without changing the overall form of the algorithm.

In the hybrid notation, time-domain and Z-domain quantities will be mixed so that the FOWDT model can be written as

$$C_k = Az^{-1}C_k + Bz^{-1-d}M_k \tag{5.39}$$

or

$$C_k = \frac{Bz^{-1-d}}{1 - Az^{-1}}M_k \tag{5.40}$$

or

$$C_k = \frac{B}{1 - Az^{-1}}M_{k-1-d} \tag{5.41}$$

where the time-domain quantities C_k and M_k occur along with z^{-1}, which acts as a backshift operator. Equations (5.37) through (5.41) are all equivalent representations of the FOWDT model.

5.8.2 Derivation of the Box–Jenkins Equations for a FOWDT Process Model and an Integrated Moving Average Noise Model

The derivation in this section is by far the most convoluted in the book and the reader is advised to proceed with caution. To soften the blow, the following six step outline of the derivation precedes the actual derivation:

1. Subtract a constant from the controlled variable such that the set point or target is zero. Augment the FOWDT process model given in the preceding section with a noise or disturbance term. Since the target is zero, the controlled variable is equal to the negative of the controller error.

2. Set the controlled variable equal to the desired set point (zero) and solve for the manipulated variable as a function of the disturbance.

3. Break up the disturbance term into the known component and the unknown component. Replace the unknown component with its expected value. At this point the manipulated variable depends on the estimate of the disturbance which is written in terms of the white noise that drives the disturbance.

4. Combine equations derived in the previous steps to relate the controlled variable (which is also the controller error) to the white noise that drives the disturbance term.

5. Using the results of the last two steps, relate the controller error to the estimate of the disturbance.

6. Using the results of steps 2 and 5, derive the Box–Jenkins control algorithm relating the manipulated variable to the controller error.

At this point the detailed derivation begins.

Start with Equation (5.41) after modifying it to include noise:

$$C_k = \frac{B}{1 - Az^{-1}}M_{k-1-d} + N_k$$

$$= \frac{Q_2(z^{-1})}{Q_1(z^{-1})}M_{k-1-d} + N_k \tag{5.42}$$

where the noise is assumed to be an integrated moving average nonstationary sequence

$$N_k = \frac{1 + \beta z^{-1}}{1 - z^{-1}}w_k \tag{5.43}$$

and w_k is a zero-mean white noise sequence. Since, for this analysis, the target

R is not going to change, a constant has been subtracted from the controlled variable C_k such that the target is zero.

Based on the information at time $k - 1 - d$, what must M_{k-1-d} be to drive C_k to the zero (the target)? The naive answer would be to set C_k equal to zero in Equation (5.42) and solve for M_{k-1-d}:

$$M_{k-1-d} = - \frac{(1 - Az^{-1})}{B} N_k \tag{5.44}$$

or

$$M_{k-1-d} = - \frac{Q_1(z^{-1})}{Q_2(z^1)} N_k$$

This is not feasible because at time $k - 1 - d$ the value of the noise or disturbance N_k is unknown, as are $N_{k-1}, N_{k-2}, \ldots, N_{k-d}$. Instead, N_k in Equation (5.44) is replaced with $\hat{N}_k$, which is the estimated value of the disturbance at time k, and Equation (5.44) becomes

$$M_{k-1-d} = - \frac{(1 - Az^{-1})}{B} \hat{N}_k$$

or

$$M_{k-1-d} = - \frac{Q_1(z^{-1})}{Q_2(z^{-1})} \hat{N}_k \tag{5.45}$$

An expression for the estimated noise at time k can be derived from Equation (5.43) as follows. First, the $1/(1 - z^{-1})$ term is expanded as an infinite series so that Equation (5.43) becomes

$$N_k = (1 + \beta z^{-1})(1 + z^{-1} + z^{-2} + \cdots)w_k \tag{5.46}$$

Then Equation (5.46) is separated into two parts: one that contains known noise components and one that contains the $d + 1$ unknown noise components (as of time $k - 1 - d$)

$$N_k = w_k + (1 + \beta)w_{k-1} + \cdots + (1 + \beta)w_{k-d}$$
$$+ (1 + \beta)z^{-1-d}(1 + z^{-1} + z^{-2} + \cdots)w_k \tag{5.47}$$

Since the second component in Equation (5.47) is an infinite series, it can be written as

$$N_k = w_k + (1 + \beta)w_{k-1} + \cdots + (1 + \beta)w_{k-d}$$
$$+ \frac{(1 + \beta)z^{-d-1}}{(1 - z^{-1})} w_k$$

or

$$N_k = Q_4(z^{-1})w_k + Q_3(z^{-1})z^{-d-1}w_k \tag{5.48}$$

where

$$Q_4(z^{-1}) = 1 + (1 + \beta)z^{-1} + \cdots + (1 + \beta)z^{-d}$$

and

$$Q_3(z^{-1}) = \frac{1 + \beta}{1 - z^{-1}}$$

As with Q_1 and Q_2, these two new generalized polynomials, $Q_3(z^{-1})$ and $Q_4(z^{-1})$, have been introduced for later use when noise models are changed. The first part of Equation (5.48), $Q_4 w_k$, contains white noise components w_k, w_{k-1}, $\ldots$, w_{k-d}, which are unknown at time $k - 1 - d$. The best estimate that can be made for these components is that they will be equal to their means, which are zero. The second part of Equation (5.48), $Q_3 z^{-d-1} w_k$, contains the known white noise components, so the best estimate of N_k is

$$\hat{N}_k = \frac{(1 + \beta)z^{-d-1}}{1 - z^{-1}} w_k$$
$$= Q_3(z^{-1})z^{-d-1} w_k \tag{5.49}$$

Before proceeding, a relationship between C_k and the white noise components w_k must be developed. The control algorithm should remove much of the correlation in the disturbance, so C_k should be as close to a white noise sequence as possible. Combining Equations (5.42) and (5.45) to eliminate M_{k-1-d} gives

$$C_k = -\hat{N}_k + N_k \tag{5.50}$$

Using Equations (5.48) and (5.49), Equation (5.50) becomes

$$C_k = [1 + (1 + \beta)(z^{-1} + z^{-2} + \cdots + z^{-d})]w_k$$
$$= Q_4(z^{-1})w_k \tag{5.51}$$

Thus, if the process has no deadtime, that is, if $d = 0$, then the control algorithm will be able to turn the nonstationary integrated moving average disturbance into a white noise sequence. Otherwise, the controlled variable will be a stationary moving average sequence.

Since the target is zero,

$$C_k = -E_k$$

and Equation (5.51) can be inverted to give

$$w_k = -\frac{E_k}{1 + (1 + \beta)(z^{-1} + z^{-2} + \cdots + z^{-d})}$$
$$= -\frac{E_k}{Q_4(z^{-1})} \tag{5.52}$$

To obtain the algorithm equation for M_k in terms of M_{k-1}, $\ldots$, and E_k, $\ldots$, the variable w_k can be eliminated between Equations (5.49) and (5.52), giving

$$\hat{N}_k = -\frac{(1 + \beta)z^{-1-d}}{(1 - z^{-1})[1 + (1 + \beta)(z^{-1} + \cdots + z^{-d})]}E_k$$

$$= -\frac{Q_3(z^{-1})z^{-1-d}E_k}{Q_4(z^{-1})}$$

(5.53)

Next, Equations (5.45) and (5.53) are combined to eliminate $\hat{N}_k$ to give

$$M_{k-1-d} = \frac{[(1 - Az^{-1})/B](1 + \beta)z^{-1-d}}{(1 - z^{-1})[1 + (1 + \beta)(z^{-1} + \cdots + z^{-d})]}E_k$$

or

$$M_k = \frac{[(1 - Az^{-1})/B](1 + \beta)}{(1 - z^{-1})[1 + (1 + \beta)(z^{-1} + \cdots + z^{-d})]}E_k$$

(5.54)

or

$$M_k = \frac{Q_1(z^{-1})Q_3(z^{-1})}{Q_2(z^{-1})Q_4(z^{-1})}E_k$$

(5.55)

Equation (5.55) is the general Box–Jenkins control algorithm equation.

For the time being, the derivations will proceed with Equation (5.54), which is peculiar to FOWDT process models and to integrated moving average noise models. Making use of the identity

$$(1 - z^{-1})(z^{-1} + z^{-2} + \cdots + z^{-d}) = z^{-1} - z^{-1-d}$$

allows Equation (5.54) to be written as

$$(1 - z^{-1})M_k = -(1 + \beta)(M_{k-1} - M_{k-1-d})$$

$$+ \frac{(1 + \beta)}{B}[A(1 - z^{-1})E_k + (1 - A)E_k]$$

or

$$(1 - z^{-1})M_k = -(1 + \beta)(M_{k-1} - M_{k-1-d})$$

$$+ P(1 - z^{-1})E_k + IhE_k$$

(5.56)

where

$$E_k = R - C_k$$

$$I = \frac{(1 + \beta)(1 - A)}{hB}$$

$$P = \frac{(1 + \beta)A}{B}$$

The form of Equation (5.56) occurs also in the Smith predictor and in the Dahlin algorithm, which was mentioned in Chapter Four. The form of the Smith predictor, augmented by the tuning rules of Section 2.5, is

$$(1 - z^{-1})M = (1 - A_d)(z^{-1-d} - z^{-1})M + P(1 - z^{-1})E + IhE \quad (5.57)$$

where

$$P = [1 - \exp(-h/T_d)]/[G(1 - \exp(-h/T))]$$

$$I = [1 - \exp(-h/T_d)]/(Gh)$$

$$A_d = \exp(-h/T_d)$$

T_d = time constant of desired response of the controlled variable to a step in the target

Thus, for the case of a FOWDT process model and an integrated moving average noise model, the Smith predictor and Box–Jenkins algorithms have identical form.

In several places in this book, integral-only control has been applied to model 6, which has no deadtime, a process gain of 2, and a time constant of 0.5. In the face of white noise disturbances, the integral control gain has been set equal to the reciprocal of the process gain; that is, $I = 0.5$. In Section 5.5.1, the disturbance was autoregressive and the power spectral density graphs suggested that performance was improved when a proportional component with a control gain of 0.1 was added. For this process model, where there is no deadtime, the Box–Jenkins algorithm given in Equation (5.56) simplifies to PI. If the disturbance is a random walk so that the parameter β is zero, the Box–Jenkins control gains can be computed as follows:

$$G = 2 \quad \text{(process gain)}$$

$$T = 0.5 \quad \text{(process time constant)}$$

$$h = 1 \quad \text{(control interval)}$$

$$A = \exp(-h/T) = 0.1353$$

$$B = G(1 - A) = 1.7293$$

$$P = \frac{(1 + \beta)A}{B} = 0.078 \quad \text{(B–J proportional control gain)}$$

$$I = \frac{(1 + \beta)(1 - A)}{B} = 0.5 \quad \text{(B–J integral control gain)}$$

Thus both PI and the Box–Jenkins approach yield essentially the same control gains.

For model 10, which has a time constant of 40, a deadtime of 40, a process gain of 2, the situation is slightly different. In Section 5.6.2, where regular PI control was applied, the following parameter values were used:

$$G = 2 \qquad \text{(process gain)}$$

$$T = 40 \qquad \text{(process time constant)}$$

$$D = 40 \qquad \text{(process deadtime)}$$

$$P = 0.18 \qquad \text{(PI proportional control gain)}$$

$$I = 0.0045 \qquad \text{(PI integral control gain)}$$

When the Smith predictor was applied in Section 5.6.2, the following control gains of

$$P = 1.7 \qquad \text{(SP proportional control gain)}$$

$$I = .05 \qquad \text{(SP integral control gain)}$$

were used. In applying the Box–Jenkins control algorithm, with $\beta = 0$, Equation (5.56) yields

$$A = \exp(-h/T) = 0.97531$$

$$B = G(1 - A) = 1.0247$$

$$P = \frac{(1 + \beta)A}{B} = 0.952 \qquad \text{(B–J proportional control gain)}$$

$$I = \frac{(1 + \beta)(1 - A)}{B} = 0.0458 \qquad \text{(B–J integral control gain)}$$

Thus, for the case of a random-walk disturbance, both the Smith predictor and the Box–Jenkins approach give similar control gains and both are significantly more aggressive than the PI.

5.8.3 Summary with Applications for Different Process Models and Different Noise Models

In the previous section the Box–Jenkins control algorithm was shown to have the general form of

$$M_k = \frac{Q_1 Q_3}{Q_2 Q_4} E_k \qquad (5.58)$$

The polynomials Q_1 and Q_2 in z^{-1} define the process model

$$C_k = \left[\frac{Q_2}{Q_1} \right] M_{k-1-d} \qquad (5.59)$$

The polynomials Q_3 and Q_4 in z^{-1} appear in the definition of the noise model

$$N_k = Q_4 w_k + Q_3 z^{-d-1} w_k \tag{5.60}$$

where $Q_4 w_k$ is the unknown component of the disturbance at time $k - 1 - d$ and $Q_3 z^{-d-1} w_k$ is the known component. As a consequence of the control algorithm, the controlled variable becomes a moving average stochastic sequence

$$C_k = Q_4(z^{-1}) w_k \tag{5.61}$$

For the case of the FOWDT model

$$Q_2 = B \quad \text{and} \quad Q_1 = 1 - A z^{-1}$$

For the case of the integrated moving average disturbance

$$Q_3 = \frac{1 + \beta}{1 - z^{-1}}$$

and

$$Q_4 = 1 + (1 + \beta) z^{-1} + \cdots + (1 + \beta) z^{-d}$$

To change noise models, only the polynomials Q_3 and Q_4 have to be changed. Consider the case where the noise model is given by

$$N_k = (1 + \alpha) N_{k-1} - \alpha N_{k-2} + w_k$$

or

$$N_k = \frac{1}{(1 - z^{-1})(1 - \alpha z^{-1})} w_k \tag{5.62}$$

A generalization of Equation (5.62) as yet another ratio of polynomials in z^{-1} would be

$$N_k = \frac{U_2(z^{-1})}{(1 - z^{-1}) U_1(z^{-1})} w_k$$

where for Equation (5.62)

$$U_1(z^{-1}) = 1 \quad \text{and} \quad U_2(z^{-1}) = 1 - \alpha z^{-1}$$

To break up Equation (5.62) into a known and unknown component to fit the format of Equation (5.60), some tedious algebra has to be carried out. In general, $Q_4(z^{-1})$ will have the form of a moving average,

$$Q_4(z^{-1}) = 1 + q_1 z^{-1} + \cdots + q_d z^{-d} \tag{5.63}$$

while $Q_3(z^{-1})$ will look like

$$Q_3(z^{-1}) = \frac{U_3(z^{-1}) z^{-d}}{(1 - z^{-1}) U_2(z^{-1})}$$

where both the order and the coefficients of $U_3(z^{-1})$ can be determined by comparing coefficients of different powers of z^{-1} in

$$\frac{U_2(z^{-1})}{(1 - z^{-1})U_1(z^{-1})} = Q_4(Z^{-1}) + \frac{U_3(z^{-1})z^{-1-d}}{(1 - z^{-1})U_2(z^{-1})} \qquad (5.64)$$

For the case of the noise model being given by Equation (5.62) and for the case of no process deadtime, that is, $d = 0$, Equation (5.64) gives

$$\frac{1}{(1 - z^{-1})(1 - \alpha z^{-1})} = 1 + \frac{(u_1 + u_2 z^{-1})z^{-1}}{(1 - z^{-1})(1 - \alpha z^{-1})}$$

or

$$1 = (1 - z^{-1})(1 - \alpha z^{-1}) + (u_1 + u_2 z^{-1})z^{-1} \qquad (5.65)$$

Equating coefficients of z^{-1} and z^{-2} gives

$$u_1 = 1 + \alpha$$

$$u_2 = -\alpha$$

Therefore,

$$Q_4(z^{-1}) = 1$$

and

$$Q_3(z^{-1}) = \frac{(1 + \alpha) - \alpha z^{-1}}{(1 - z^{-1})(1 - \alpha z^{-1})}$$

Having found expressions for Q_3 and Q_4, Equation (5.58) can be used to write the control algorithm:

$$M_k = \frac{Q_1 Q_3}{Q_2 Q_4} E_k$$

$$Q_2 = B$$

$$Q_1 = 1 - A z^{-1} \qquad (5.66)$$

$$Q_4 = 1$$

$$Q_3 = \frac{(1 + \alpha) - \alpha z^{-1}}{(1 - z^{-1})(1 - \alpha z^{-1})}$$

$$M_k = \frac{(1 - A z^{-1})[(1 + \alpha) - \alpha z^{-1}]}{B[(1 - z^{-1})(1 - \alpha z^{-1})]}$$

Note that, since Q_4 is unity, Equation (5.61) shows that the control algorithm will be capable of turning a nonstationary disturbance into a white noise sequence.

To be specific, consider an example presented by MacGregor, Harris, and Wright (1984), where

$$A = 0.7 = \exp(-1/T)$$

$$T = 2.804 \quad \text{(process time constant)}$$

$$B = 0.3$$

$$G = 1 \quad \text{(process gain)}$$

$$\alpha = 0.2$$

$$D = 0 \quad \text{(process deadtime)}$$

Therefore, Equation (5.66) becomes

$$\Delta M_k = \frac{4(1 - 0.7z^{-1})(1 - 0.1667z^{-1})}{1 - 0.2z^{-1}} E_k \qquad (5.67)$$

or

$$M_k = 1.2M_{k-1} - 0.2M_{k-2} + 4(E_k - 0.867E_{k-1} + 0.1169E_{k-2}) \quad (5.68)$$

Unlike the algorithm obtained in the previous section, Equation (5.67) cannot be easily manipulated to fit into a modified PI format.

For purposes of comparison, PI control will be applied to this example. The tuning rules of Section 2.5 will be used, so a value for the time constant, T_d, that describes the desired response of the controlled variable to a set point change must be chosen. Since this example process has no deadtime, a desired time constant can be chosen that is significantly less than the process time constant. The PI control gains for two choices of the desired time constant are

$$
\begin{array}{ll}
T_d = 0.5 & T_d = 0.1 \\
P = 2.88 & P = 3.33 \\
I = 0.864 & I = 1
\end{array}
$$

At this point it would be instructive to compare the performance of Equation (5.68) with PI control in the frequency domain. First, the controlled variable or the controller error can be studied using Equations (5.19) and (5.20):

$$P_E''(f) = |E_T''(f)|^2 |H''(f)|^2 h V_w \qquad (5.19)$$

$$V_E = 2 \int_0^{f_{Ny}} |E_T''(f)|^2 |H''(f)|^2 h V_w \, df \qquad (5.20)$$

where

$$E_T(z) = \frac{1}{1 + G_c G_p}$$

$$G_p(z) = \frac{Bz^{-1}}{1 - Az^{-1}} = \text{process transfer function}$$

$$G_c(z) = \text{control algorithm transfer function}$$

$$= \frac{P(z - 1) + Ihz}{z - 1}, \quad \text{for PI}$$

$$= \frac{4(z^2 - 0.867z + 0.1169)}{z^2 - 1.2z + 0.2}, \quad \text{for B--J}$$

$$H(z) = \frac{1}{(1 - z^{-1})(1 - 0.2z^{-1})} = \text{noise transfer function}$$

V_w = variance of white noise that drives the noise

V_E = variance of the controller error

$P_E''(f)$ = power spectral density of controller error

Equation (5.61) shows that the Box–Jenkins controller error is white noise; therefore, the power spectral density of the Box–Jenkins controller error is constant. Figure 5.38 shows the power spectral density for the Box–Jenkins controller and PI controller for the above two choices of control gains for the case of unit white noise variance V_w. Note that PI control passes more low-frequency power, and the more aggressive set of PI control gains tends to amplify power at higher frequencies. Using Equation (5.20), a ratio of the variance for PI to that for Box–

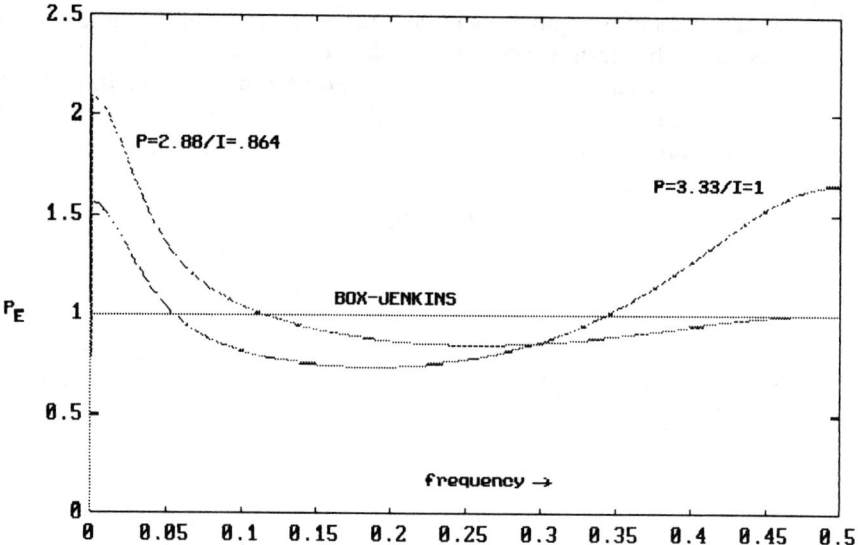

Figure 5.38 Comparison of controller error using power spectral density.

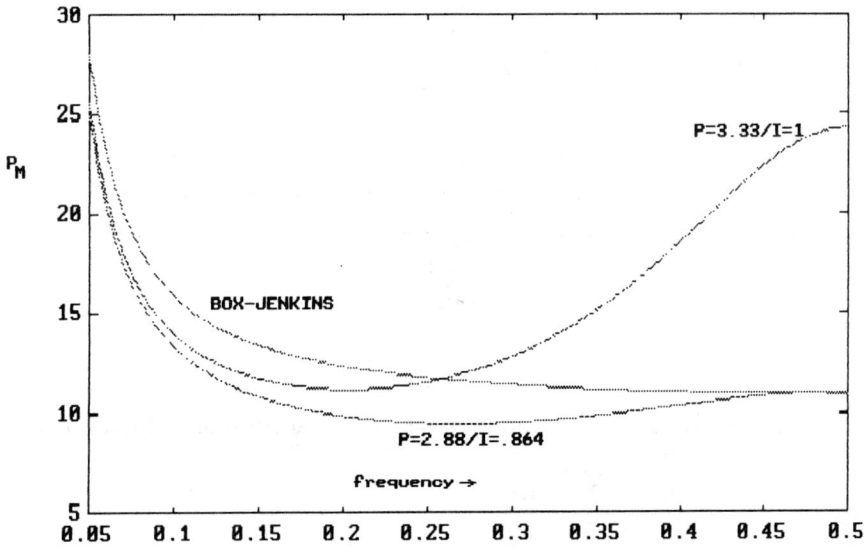

Figure 5.39   Comparison of manipulated variable activity using power spectral density.

Jenkins can be calculated. With the more conservative set of PI control gains, the ratio is 1.026, and with the more aggressive set of PI control gains the ratio is 1.04.

A similar analysis can be made, as mentioned in Section 5.5, for the manipulated variable. Figure 5.39 shows the power spectral densities of the manipulated variable over a portion of the Nyquist interval that excludes zero frequency for the case of unit white noise variance V_w. Note that at lower frequencies the Box–Jenkins algorithm has significantly higher power than the PI algorithms. However, at the higher frequencies only the more conservative set of PI control gains continues to show less activity than the Box–Jenkins algorithm.

The examples discussed so far indicate that the Box–Jenkins algorithm performance is similar to PI for first-order processes without deadtime and similar to PI + Smith predictor for FOWDT processes. A final example will show how the Box–Jenkins approach performs when the process has an order greater than 1. Consider model 4, which is third order, having three time constants of 25, 26, and 27 seconds, a process gain of 2.0, and no deadtime. This process model has been used repeatedly in this book because, although it has no explicit deadtime, it does appear to have an effective deadtime due to the inflection point that shows up in the step change response (see Figure 1.31).

Using the Equations (5.58) through (5.61), the Box–Jenkins algorithm can be set up as follows:

$$\frac{C}{M} = \frac{Bz^{-1}}{1 - A_1z^{-1} - A_2A^{-2} - A_3z^{-3}} = \frac{Q_2z^{-1}}{Q_1}$$

where

$$Q_1 = 1 - A_1z^{-1} - A_2z^{-2} - A_3z^{-3}$$
$$Q_2 = B$$
$$B = G(1 - A_1 - A_2 - A_3)$$
$$G = \text{process gain} = 2$$

A_1, A_2, A_3 defined in terms of the time constants (see Section 4.3.1)

The disturbance model will be the same as in the previous example, so

$$Q_4 = 1$$

$$Q_3 = \frac{(1 + \alpha) - \alpha z^{-1}}{(1 - z^{-1})(1 - \alpha z^{-1})}$$

The control algorithm can then be written according to Equation (5.58):

$$M_k = \frac{Q_1 Q_3}{Q_2 Q_4} E_k$$

$$= \frac{(1 - A_1z^{-1} - A_2z^{-2} - A_3z^{-3})[(1 + \alpha) - \alpha z^{-1}]E_k}{B(1 - z^{-1})(1 - \alpha z^{-1})}$$

or

$$M_k = (1 + \alpha)M_{k-1} - \alpha M_{k-2} + \frac{1}{B}\{(1 + \alpha)E_k$$

$$- [A_1(1 + \alpha) + \alpha]E_{k-1} + [\alpha A_1 - A_2(1 + \alpha)]E_{k-2}$$

$$+ [\alpha A_2 - A_3(1 + \alpha)]E_{k-3} + \alpha A_3 E_{k-4}\}$$

This Box–Jenkins algorithm will be compared to three algorithms: two versions of PI and one of PID. The first PI algorithm has the conservative control gains of

$$P = 0.54, \qquad I = 0.009$$

(see Sections 2.6.2 and 2.6.3 for time-domain graphs showing how these control gains perform). For the second PI algorithm, values of P and I were found that made the controller error variance, as derived from Equation (5.19), minimum. This brute force minimization can be done by variety of ways. For this example, the MATLAB™ minimization algorithm was used and the variance minimizing values were found to be

$$P = 1.43, \qquad I = 0.0098$$

The PID algorithm had the gains used in Chapter Two of

$$P = 1.1, \qquad I = 0.012, \qquad D = 15$$

with no filter. Figure 5.40 shows the performance of these three algorithms for

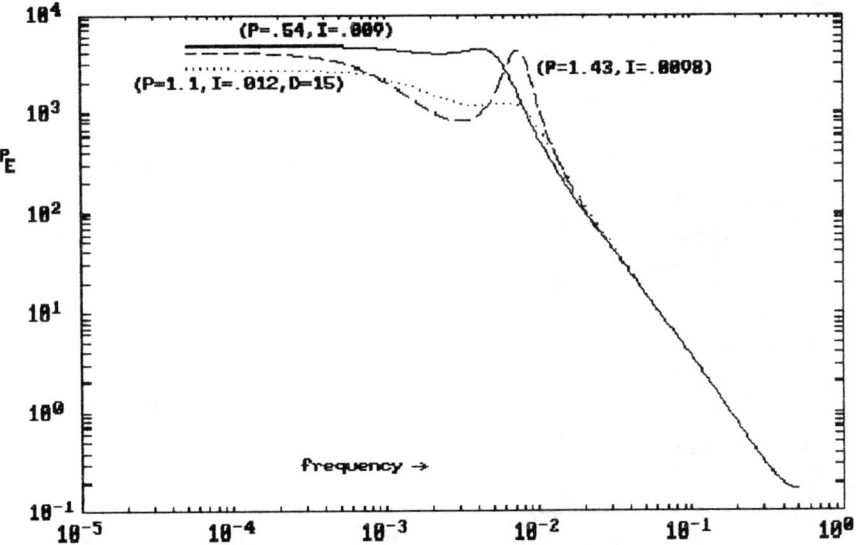

Figure 5.40   Controller error for PI applied to model 4 subject to nonstationary noise.

the case of unity variance white noise. Note that the PID algorithm has the best performance in terms of noise rejection.

One might wonder why the minimization procedure was not applied to the PID algorithm. This is an interesting idea, but the results are a bit discouraging. The presence of the derivative component causes all three control gains to be unbounded. In other words, with the anticipation feature of the derivative, the proportional and integral control gains can be extremely high without causing instability; the only penalty is the unsatisfactorily high activity level of the manipulated variable. (A modification of this approach will be discussed in Section 5.9.) Figure 5.41 shows the spectral density of the manipulated variable for these three cases. Note the power at the higher frequencies for the PID control algorithm. This power increases dramatically as the derivative control gain increases.

For the Box–Jenkins algorithm, the controller error power spectral density need not be plotted since Equation (5.61) indicates that it would be flat with a value of unity. However, there is a severe price to be paid for making the controller error white, and Figure 5.42 reveals the enormous (and unacceptable) activity of the manipulated variable. This is a consequence of asking a control algorithm to make the controlled variable of a sluggish process, subject to a highly nonstationary disturbance, behave as a white noise sequence. In effect, the Box–Jenkins algorithm does what the PID algorithm would do if it could be allowed to minimize the variance. The only difference is that with the PID algorithm one sees trouble immediately in the form of the unreasonably high values of the control gains.

To overcome the significantly higher manipulated variable activity of the

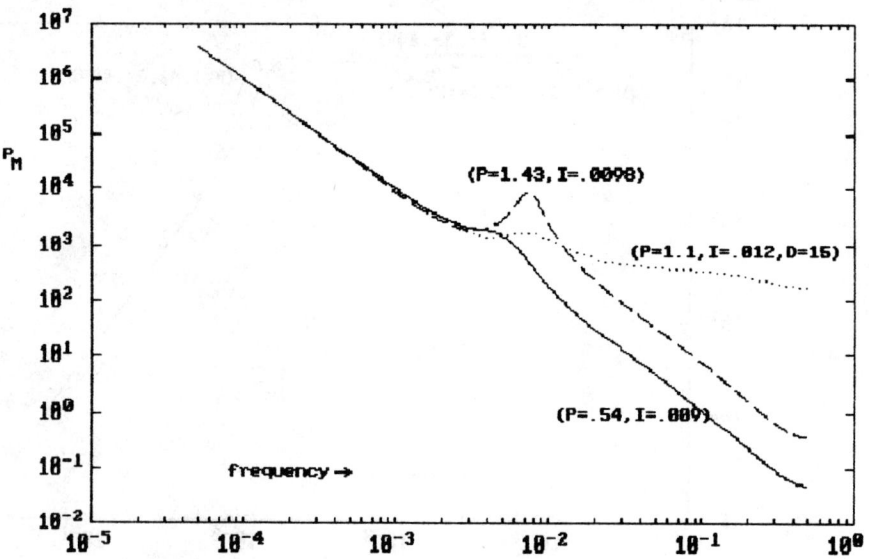

Figure 5.41 Manipulated variable activity for PI applied to model 4 subject to nonstationary noise.

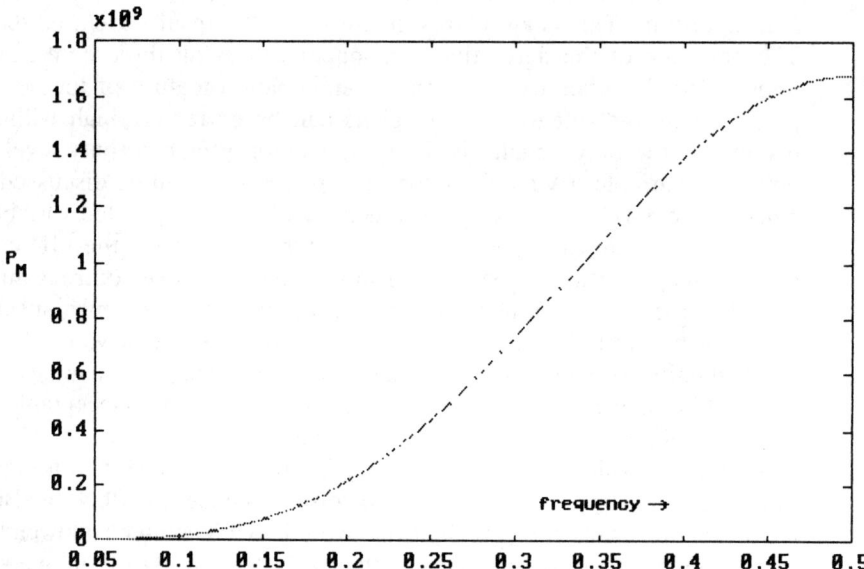

Figure 5.42 Manipulated variable activity for Box–Jenkins applied to model 4 subject to nonstationary noise.

Box–Jenkins algorithms, MacGregor and Harris (1987) have modified the approach by minimizing a weighted sum of the variance of the controlled variable and that of the manipulated variable. The minimization method uses spectral factorization and is beyond the scope of this text.

These examples suggest that, in spite of the greater sophistication of the Box–Jenkins approach, a PI, PID, or PI plus Smith predictor control algorithm can usually be found that has equivalent or sometimes superior performance, particularly if a model noise is known. In addition, the simplicity of the PID approach is often found to be attractive in an industrial atmosphere where the ability to tune the algorithm on line and the ease of installation and maintenance are highly valued, especially in those cases where a priori knowledge of the process and noise models is unknown.

5.9 DETERMINATION OF PID CONTROL GAINS FROM PROCESS AND NOISE MODELS

The idea of minimizing a weighted sum of the controller error variance and the manipulated variable variance is not unique to the spectral factorization extension of the Box–Jenkins algorithms. In fact, it is a widely used criterion in the design of linear optimal control systems (see Kwakernaak and Sivan, 1972). Although the criterion is frequently part of methods that usually yield analytical expressions for control algorithms, it will be used here strictly as a criterion whose minimum is to be determined numerically. This is a realistic goal now that there are so many reasonably priced software packages available for personal computers that perform higher-level functions such as minimization.

If models are known for the process, the control algorithm and the noise then Equations (5.19), (5.20), and (5.21) can be used to calculate variances for the controller error and the manipulated variable. If the noise is nonstationary so that the variance of the manipulated variable is unbounded, then either the integral of $P''_M(f)$ can be taken over a subportion of the Nyquist interval or M can be replaced with ΔM, whose variance may be bounded.

Consider the example discussed at the close of the last section. The process model (model 4) is

$$\frac{C}{M} = \frac{Bz^{-1}}{1 - A_1 z^{-1} - A_2 z^{-2} - A_3 z^{-3}}$$

the noise model is

$$N_k = \frac{1}{(1 - z^{-1})(1 - 0.2z^{-1})} w_k$$

and the control algorithm is PID. The variance of the manipulated variable will be unbounded because the disturbance is nonstationary. To determine a partial

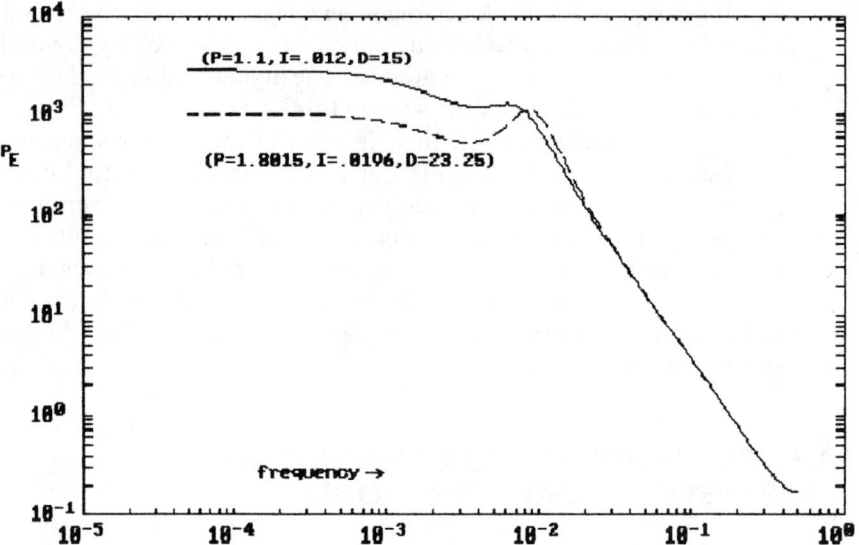

Figure 5.43   Controller error power spectral density for PID algorithms applied to model 4 subject to nonstationary noise.

variance of the manipulated variable, V'_M, the integration of $P''_M(f)$ was taken from 0.01 to 0.5 Hz instead of 0 to 0.5 Hz. Values for the PID control gains were determined by minimizing the following sum:

$$V_E + 0.01 V'_M$$

where V_E is the variance of the controller error and where the weighting factor of 0.01 was chosen by trial and error. The results are

$$P = 1.8015, \qquad I = 0.0196, \qquad D = 23.25$$

Figure 5.43 shows the power spectral density of the controller error for this set of control gains along with the set used earlier. The difference in the standard deviation of the controller error for these two sets of control gains is 14%.

The method of this section does not guarantee that the resulting control gains will make the controlled system stable. Therefore, one should always check the performance using time-domain simulation and the stability analysis methods of Section 4.5.

5.10 SUMMARY

The idea of the spectral density as the distribution of the variance over the Nyquist interval was introduced and used to develop expressions for variance ratios and the noise transmission curves of controlled systems. With this tool in hand, the

PID, the Smith predictor, the internal model control approach (which was also introduced in this chapter), and the Box–Jenkins algorithms were reviewed. Finally, in conjunction with minimization procedures, the power spectral density was used to find control gains for PID controllers in those situations where models for the process and noise are available. The main idea of this chapter was not to show which algorithm was best but to present a method that the reader should be able to use to study various algorithms and pick the appropriate one for his or her application.

5.11 DISCUSSION OF THE REFERENCES

A rigorous derivation of the power spectral density using the Fourier transform of the autocorrelation function is given in

ASTROM, KARL J., *Introduction to Stochastic Control Theory*, Academic Press, New York, 1970,

and in

FRANKLIN, G. F., and POWELL, J. DAVID, *Digital Control of Dynamic Systems*, Addison-Wesley, Reading, Mass., 1980.

The basis for the alternative definition of the power spectral density is derived from the approach presented in

MARPLE, S. L., *Digital Spectral Analysis with Applications*, Prentice Hall, Englewood Cliffs, N.J., 1987.

A text that admirably complements the above book is

KAY, S. M., *Modern Spectral Estimation, Theory and Application*, Prentice Hall, Englewood Cliffs, N.J., 1988.

The derivation of the spectral density of stochastic processes driven by white noise appears more rigorously in

JENKINS, G. M., and WATTS, D. G., *Spectral Analysis and Its Applications*, Holden-Day, San Francisco, 1968.

Many of the comments in Section 5.1 about the relation between the line spectrum and the power spectral density are presented in much greater detail in

OPPENHEIM, A. V., and SCHAFER, R. W., *Digital Signal Processing*, Prentice Hall, Englewood Cliffs, N.J., 1975.

The more recent text by these two authors on the same subject matter is

OPPENHEIM, A. V., and SCHAFER, R. W., *Discrete Time Signal Processing*, Prentice Hall, Englewood Cliffs, N.J., 1989.

The idea of designing controllers by studying the variance as derived from an integral in the frequency domain first appeared in

NEWTON, G. C., GOULD, L. A., and KAISER, J. F., *Analytical Design of Linear Feedback Controls*, Wiley, New York, 1957.

However, all their work was done for analog controllers in the continuous time domain; therefore, the path to the frequency domain used the Laplace transform instead of the Z-transform as in this book.

A review of the internal model control strategy (where the multidimensional case is dealt with extensively) is given in

GARCIA, C. E., and MORARI, M., "Internal Model Control. 1. A Unifying Review and Some New Results," *Ind. Eng. Chem. Process Des. Dev.*, 21, 1982, pp. 308–323.

and in

MORARI, M., and ZAFERIOU, E., *Robust Process Control*, Prentice Hall, Englewood Cliffs, N.J., 1989.

The starting point for the Box–Jenkins control algorithm is Chapter 12 of

BOX, G. E. P., and JENKINS, G. M., *Time Series Analysis, Forecasting and Control*, Holden-Day, San Francisco, 1970.

The example of Section 5.8.3 was taken from

MACGREGOR, J. F., HARRIS, T. J., and WRIGHT, J. D., "Duality between the Control of Processes Subject to Randomly Occurring Deterministic Disturbances and ARIMA Stochastic Disturbances," *Technometrics*, vol. 26, no. 4, November 1984.

Spectral factorization and other advanced concepts for multivariable control are discussed in

MACGREGOR, J. F., and HARRIS, T. J., "Design of Multivariable Linear Quadratic Controllers Using Transfer Functions," *A.I.Ch.E.J.*, vol. 33, no. 9, 1987, pp. 1481–1495.

An application of these concepts is given in

MACGREGOR, J. F., and KOZUB, D. J., "Applications of LQ and IMC Controllers to a Packed-Bed Reactor," *A.I.Ch.E.J.*, vol. 33, no. 9, 1987, pp. 1496–1506.

More detail on the derivation of the Box-Jenkins control equations in Section 5.8 can be found in

MACGREGOR, J. F., "Optimal Stochastic Control: Theory and Application" (Lecture notes prepared for *An Intensive Short Course on Digital Computer Techniques for Process Identification and Control*, McMaster University, Hamilton, Ont., 1988.)

The minimization of a weighted sum of the controller error variance and manipulated variable variance as a basis for developing control algorithms is discussed in

KWAKERNAAK, H., and SIVAN, R., *Linear Optimal Control Systems*, Wiley-Interscience, New York, 1972.

APPENDIX

MATHEMATICAL STRUCTURE OF THE PROCESS MODELS USED FOR SIMULATION STUDIES

The tools for mathematically describing the process models that have been used to test control algorithms were not developed until Section 4.3, but, since the process models were used throughout the book, their mathematical structure is summarized in this appendix. Eleven of the thirteen process models used can be treated as special cases of the third-order model developed in Section 4.3.1, which in the time domain has the form

$$C_i = A_1 C_{i-1} + A_2 C_{i-2} + A_2 C_{i-3} + B M_{i-1-d}$$

As was shown in Section 4.3.1, the Z-domain form is

$$C(z) = A_1 z^{-1} C(z) + A_2 z^{-2} C(z) + A_3 z^{-3} C(z) + B z^{-1-d} M(z)$$

which can be written in factored form as

$$C(z)[(z - r_1)(z - r_2)(z - r_3)] = B z^2 z^{-d} M(z)$$

where the roots r_i, $i = 1, 2, 3$, can be related to time constants as $r_i = \exp(-h/T_i)$ and to the first three coefficients as

$$A_1 = r_1 + r_2 + r_3$$

$$A_2 = -(r_1 r_2 + r_2 r_3 + r_1 r_3)$$

$$A_3 = r_1 r_2 r_3$$

The last coefficient can be determined from

$$B = G(1 - A_1 - A_2 - A_3)$$

Thus, if values of the six parameters h, T_1, T_2, T_3, G, and d can be specified, all the coefficients in the model can be determined. Note that the approach is the same even when, instead of relating the roots to time constants, they are placed in the z-plane according to the desires of the modeler. The only constraint is that two of the roots must be conjugates so that the A_i's have no imaginary parts.

In all the models the sampling/control interval h is unity and the process gain G is 2.0.

The following table gives the structure of the models:

Model no.	T1	T2	T3	d
1	5	0.6	0.7	0
2	5	0.6	0.7	4
3	40	0	0	20
4	25	26	27	0
5	15	0.6	0.7	15
6	0.5	0	0	0
7	40	0	0	10
8	1.44	0	0	0
9	25	26	27	0
10	40	0	0	40
11	0.5			10

Model 12 has roots that are complex conjugates and the roots are $r_1 = 0.9$, $r_2 = 0.8 + j.5$, $r_3 = 0.8 - j.5$, with no deadtime.

Model 13 is the inverse responding model composed of the major component with a time constant of 19.5 seconds, a major gain of 3.0, a minor time constant of 2.8 seconds, and a minor gain of 1.0.

INDEX